Industrial Microbiology: An Introduction

Industrial Microbiology: An Introduction

Michael J. Waites
BSc, PhD, CBiol, MIBiol

Neil L. Morgan
BSc, PhD, MIFST

John S. Rockey
BSc, MSc, PhD

Gary Higton
BSc, PhD

All:
School of Applied Science, South Bank University, London, UK

Blackwell
Science

BLACKWELL PUBLISHING
350 Main Street, Malden, MA 02148-5020, USA
9600 Garsington Road, Oxford OX4 2DQ, UK
550 Swanston Street, Carlton, Victoria 3053, Australia

First published 2001

9 2009

Library of Congress Cataloging-in-Publication Data

Introduction to Industrial Microbiology / Michael J. Waites . . . [et al.].
p. cm.
ISBN 978-0-632-05307-0 (pbk.)
1. Introduction to Industrial Microbiology.
I. Waites, Michael J.
QR53 .I522 2001
660.6′2—dc21

2001025874

ISBN-13: 978-0-632-05307-0 (pbk.)

A catalogue record for this title is available from the British Library

Set by Best-set Typesetter Ltd, Hong Kong

For further information on
Blackwell Publishing, visit our website:
www.blackwellpublishing.com

Dedication

This book is dedicated to the memory of our great friend and co-author Gary Higton who, at the age of only 40 years, died unexpectedly during the final stages of its preparation. Gary was a knowledgeable microbiologist, a fine teacher, supportive colleague and a loyal friend. He is greatly missed by us all.

Contents

Preface

Industrial microbiology is primarily associated with the commercial exploitation of microorganisms, and involves processes and products that are of major economic, environmental and social importance throughout the world. There are two key aspects of industrial microbiology, the first relating to production of valuable microbial products via fermentation processes. These include traditional fermented foods and beverages, such as bread, beer, cheese and wine, which have been produced for thousands of years. In addition, over the last hundred years or so, microorganisms have been further employed in the production of numerous chemical feedstocks, energy sources, enzymes, food ingredients and pharmaceuticals. The second aspect is the role of microorganisms in providing services, particularly for waste treatment and pollution control, which utilizes their abilities to degrade virtually all natural and man-made products. However, such activities must be controlled while these materials are in use, otherwise consequent biodeterioration leads to major economic losses.

This textbook is intended to be an introduction to industrial microbiology. In writing it, the authors have drawn on their experience teaching industrial microbiology and other aspects of applied microbiology to undergraduates and masters students on a range of applied biology, microbiology, biotechnology, food science and chemical engineering courses. It is assumed that the reader will have an elementary knowledge of microbiology and biochemistry. Nevertheless, even those students with only a basic knowledge of chemistry and cell biology, and those who are not specialist microbiologists, should find the material accessible.

The book is divided into three sections. Part 1 is designed to underpin the main content. Its aim is to provide a sufficient, albeit brief, overview of microbial structure, physiology and biochemistry to enable the student to fully appreciate the diversity of microorganisms and their metabolic capabilities. The reader should soon come to recognize the versatility of microorganisms, their ability to grow on a wide range of substrates and to produce an extensive array of products, many of which are commercially available.

Part 2 is devoted to bioprocessing. It examines the commercial fermentation operations and requirements for large-scale cultivation of microorganisms. This involves the acquisition and development of suitable production strains that must then be provided with nutrients, especially appropriate carbon and energy sources. The object of any industrial fermentation is then to optimize either growth of the organism or the production of a target microbial product. This is normally achieved by performing fermentations under rigorously controlled conditions in large fermenters with culture capacities often in excess of several thousand litres. The design and operation of these fermentation processes is discussed, along with the downstream processing operations necessary for the recovery and purification of the target products. Aspects of safety and good manufacturing practices are also examined.

Over the last twenty years, many traditional and established industrial fermentation processes have been advanced through the contribution of genetic engineering (*in vitro* recombinant DNA technology). This technology has also facilitated the development of many novel processes and products. It not only accelerates strain development, leading to improvements in the production microorganisms, but can aid downstream processing and other elements of the process. Initially, it involved the manipulation of bacteria, but has moved to cloning in yeasts, filamentous fungi, and plant and animal cells. Developments in this field continue to grow rapidly.

There are several good texts that cover the fundamental aspects of genetic manipulation of microorganisms. Consequently, we have not attempted to give a detailed account of such techniques here, they are merely introduced and further reading is suggested. Nevertheless, many examples of the application and potential of genetically engineered microorganisms within industrial processes are discussed.

Part 3 explores the wide range of industrial microbial processes and products, including human food and animal feed production, the provision of chemical feedstocks, alternative energy sources, enzymes and products for application in human and animal health. We consider that the economic and scientific importance of traditional and long-established fermentation processes can be somewhat overlooked, as attention is often dominated by more recent developments. Therefore we have aimed to give a balanced and comprehensive coverage of current industrial processes and their products. The production of valuable commodities from animal and plant cell culture is also included, primarily because the culture and handling of such cells involves techniques similar to those used for the propagation of microorganisms.

An additional aspect of industrial microbiology, examined here, is the application of extensive degradative abilities of microorganisms, particularly the harnessing of these properties in waste treatment and pollution control. The necessity to limit these activities in inappropriate situations, i.e. the prevention of biodeterioration of materials while still in use, is also discussed.

Acknowledgements

We would like to thank the following authors and the publishers for allowing us to use figures and tables from their publications:
Figs 1.1 and 3.12 are from Dawes, I. W. & Sutherland, I. W. (1992) *Microbial Physiology*, 2nd Edition. Blackwell Scientific Publications.
Fig. 1.2 is from Poxton, I. R. (1993) *Journal of Applied Bacteriology* Supplement **74**, 1S–11S.
Figs 4.3, 7.8 and 7.9 are from Brown, C. M., Campbell, I. & Priest, F. G. (1987) *Introduction to Biotechnology*. Blackwell Scientific Publications.
Fig. 12.7 is from Lewis, M. & Young, T. W. (1995) *Brewing*. Chapman & Hall.
Tables 12.3 and 12.4 are based on tables in Bamforth, C. W. (1985) The use of enzymes in brewing. *Brewers' Guardian* **114**, 21–26.
In addition, we wish to thank our colleague Dr Tom Coultate for his advice and enthusiastic support.

Introduction to industrial microbiology

Traditional fermentation processes, such as those involved in the production of fermented dairy products and alcoholic beverages, have been performed for thousands of years. However, it is less than 150 years ago that the scientific basis of these processes was first examined. The birth of industrial microbiology largely began with the studies of **Pasteur**. In 1857 he finally demonstrated beyond doubt that alcoholic fermentation in beer and wine production was the result of microbial activity, rather than being a chemical process. Prior to this, **Cagniard-Latour**, **Schwann** and several other notable scientists had connected yeast activities with fermentation processes, but they had largely been ignored. Pasteur also noted that certain organisms could spoil beer and wine, and that some fermentations were aerobic, whereas others were anaerobic. He went on to devise the process of pasteurization, a major contribution to food and beverage preservation, which was originally developed to preserve wine. In fact, many of the early advances of both pure and applied microbiology were through studies on beer brewing and wine making. Pasteur's publications, *Études sur le Vin* (1866), *Études sur la Bière* (1876) and others, were important catalysts for the progress of industrial fermentation processes. Of the further advances that followed, none were more important than the development of pure culture techniques by **Hansen** at the Carlsberg Brewery in Denmark. Pure strain brewing was carried out here for the first time in 1883, using a yeast isolated by Hansen, referred to as Carlsberg Yeast No. 1 (*Saccharomyces carlsbergensis*, now classified as a strain of *Saccharomyces cerevisiae*).

During the early part of the 20th century, further progress in this field was relatively slow. Around the turn of the century there had been major advancements in the large-scale treatment of sewage, enabling significant improvement of public health in urban communities. However, the first novel industrial-scale fermentation process to be introduced was the acetone–butanol fermentation, developed by **Weizmann** (1913–15) using the bacterium *Clostridium acetobutylicum*. In the early 1920s an industrial fermentation process was also introduced for the manufacture of citric acid, employing a filamentous fungus (mould), *Aspergillus niger*. Further innovations in fermentation technology were greatly accelerated in the 1940s through efforts to produce the antibiotic penicillin, stimulated by the vital need for this drug during World War II. Not only did production rapidly move from small-scale surface culture to large-scale submerged fermentations, but it led to great advances in both media and microbial strain development. The knowledge acquired had a great impact on the successful development of many other fermentation industries.

More recent progress includes the ability to produce **monoclonal antibodies** for analytical, diagnostic, therapeutic and purification purposes, pioneered by **Milstein** and **Kohler** in the early 1970s. However, many of the greatest advances have followed the massive developments in **genetic engineering** (recombinant DNA technology) over the last 20 years. This technology has had, and will continue to have, a tremendous influence on traditional, established and novel fermentation processes and products. It allows genes to be transferred from one organism to another and allows new approaches to strain improvement. The basis of gene transfer is the insertion of a specific gene sequence from a donor organism, via an expression vector, into a suitable host. Hosts for expression vectors can be prokaryotes such as the bacterium *Escherichia coli*; alternatively, where post-translational processing is required, as with some human proteins, a eukaryotic host is usually required, e.g. a yeast.

A vast range of important products, many of which were formerly manufactured by chemical processes, are now most economically produced by microbial fermentation and biotransformation processes. Microorganisms also provide valuable services. They have proved to be particularly useful because of the ease of their mass cultivation, speed of growth, use of cheap substrates that in many cases are wastes, and the diversity of potential products. In addition, their ability to readily undergo genetic manipulation has opened up almost limitless possibilities for new products and services from the fermentation industries.

Successful development of a fermentation process requires major contributions from a wide range of other disciplines, particularly biochemistry, genetics and molecular biology, chemistry, chemical and process engineering, and mathematics and computer technology. A typical operation involves both **upstream processing** (USP) and **downstream processing** (DSP) stages (Fig. i). The USP is associated with all factors and processes leading to and including the fermentation, and consists of three main areas.

1 **The producer microorganism.** Key factors relating to this aspect are: the strategy for initially obtaining a suitable industrial microorganism, strain improvement to enhance productivity and yield, maintenance of strain purity, preparation of a reliable inoculum and the continuing development of selected strains to improve the economic efficiency of the process. For example, the production of stable mutant strains that vastly overproduce the target compound is often essential.

Some microbial products are primary metabolites, produced during active growth (the **trophophase**), which include amino acids, organic acids, vitamins and industrial solvents such as alcohols and acetone. However, many of the most important industrial products are secondary metabolites, which are not essential for growth, e.g. alkaloids and antibiotics. These compounds are produced in the stationary phase of a batch culture, after microbial biomass production has peaked (the **idiophase**).

2 **The fermentation medium.** The selection of suitable cost-effective carbon and energy sources, and other essential nutrients, along with overall media optimization are vital aspects of process development to ensure maximization of yield and profit. In many instances, the basis of industrial media are waste products from other industrial processes, notably sugar processing wastes, lignocellulosic wastes, cheese whey and corn steep liquor.

3 **The fermentation.** Industrial microorganisms are normally cultivated under rigorously controlled conditions developed to optimize the growth of the organism or production of a target microbial product. The synthesis of microbial metabolites is usually tightly regulated by the microbial cell. Consequently, in order to obtain high yields, the environmental conditions that trigger regulatory mechanisms, particularly repression and feedback inhibition, must be avoided.

Fermentations are performed in large fermenters often with capacities of several thousand litres. These range from simple tanks, which may be stirred or unstirred, to complex integrated systems involving varying levels of computer control. The fermenter and associated pipework, etc., must be constructed of materials, usually stainless steel, that can be repeatedly sterilized and that will not react adversely with the microorganisms or with the target products. The mode of fermenter operation (batch, fed-batch or continuous systems), the method of its aeration and agitation, where necessary, and the approach taken to process scale-up have major influences on fermentation performance.

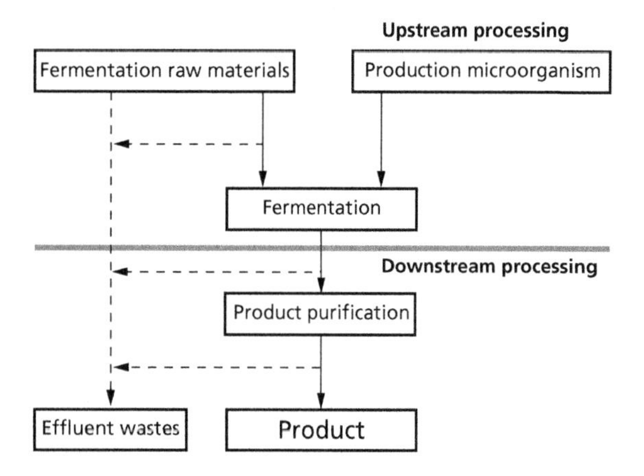

Fig. i Outline of a fermentation process.

Conventional DSP includes all unit processes that follow fermentation. They involve cell harvesting, cell disruption, product purification from cell extracts or the growth medium, and finishing steps. However, attempts are now being made to integrate fermentation with DSP operations, which often increases process productivity. Overall, DSP must employ rapid and efficient methods for the purification of the product, while maintaining it in a stable form. This is especially important where products are unstable in the impure form or subject to undesirable modifications if not purified rapidly. For some products, especially enzymes, retention of their biological activity is vital. Finally, there must be safe and inexpensive disposal of all waste products generated during the process.

Fermentation products

The overall economics of fermentation processes are influenced by the costs of raw materials and consumables, utilities, labour and maintenance, along with fixed charges, working capital charges, factory overheads and operating outlay. Fermentation products can be broadly divided into two categories: high volume, low value products or low volume, high value products. Examples of the first category include most food and beverage fermentation products, whereas many fine chemicals and pharmaceuticals are in the latter category.

Food, beverages, food additives and supplements

A wide range of fermented foods and beverages have been produced throughout recorded history. They continue to be major fermentation products worldwide and are of vast economic importance. Fermented dairy products, for example, result from the activities of lactic acid bacteria in milk, which modify flavour and texture, and increase long-term product stability. Yeasts are exploited in the production of alcoholic beverages, notably beer and wine, due to their ability to ferment sugars, derived from various plant sources, to ethanol. Most processes use strains of one species, *S. cerevisiae*, and other strains of this yeast are used as baker's yeast for bread dough production.

Several organic acids derived from microbial action are employed in food manufacture and for a wide range of other purposes. The first human use was for acetic acid, as vinegar, produced as a result of the oxidation of alcoholic beverages by acetic acid bacteria. A further aerobic fermentation involves **citric acid** production by the filamentous fungus, *A. niger*, which has become a major industrial fermentation product, as it has numerous food and non-food applications. Also, most of the **amino acids** and **vitamins** used as supplements in human food and animal feed are produced most economically by microorganisms, particularly if high-yielding overproducing strains are developed. In addition, some microorganisms contain high levels of protein with good nutritional characteristics suitable for both human and animal consumption. This so-called '**single-cell protein**' (**SCP**) can be produced from a wide range of microorganisms cultivated on low-cost carbon sources.

Health-care products

In terms of providing human benefit, antibiotics are probably the most important compounds produced by industrial microorganisms. Most are secondary metabolites synthesized by filamentous fungi and bacteria, particularly the actinomycetes. Well over 4000 antibiotics have now been isolated, but only about 50 are used regularly in antimicrobial chemotherapy. The best known and probably the most medically useful antibiotics are the β-lactams, penicillins and cephalosporins, along with aminoglycosides (e.g. streptomycin) and the tetracyclines. New antibiotics are still being sought as resistance to established antibiotics has become a major problem in recent years, through the misuse and overuse of these drugs.

Other important pharmaceutical products derived from microbial fermentation and/or biotransformations are alkaloids, steroids and vaccines. More recently, therapeutic recombinant human proteins such as insulin, interferons and human growth hormone have been produced by a range of microorganisms. This is a rapidly expanding field and many more recombinant therapeutic products are likely to come on to the market over the coming years.

Microbial enzymes

Microbial enzymes, particularly extracellular hydrolytic enzymes, have numerous roles as process aids or in the production of a wide range of specific food and non-food products. Proteases, for instance, are extensively used as additives to washing powders, in the removal of protein hazes from beer and as microbial rennets for the production of cheese. Several carbohydrases are employed in the production of a diversity of sugar syrups from starch. For example, high-fructose corn syrup is produced by hydrolysing corn starch to glucose using α-amylase and amyloglucosidase, and the resulting glucose is then isomerized to a sweeter molecule, fructose, by a glucose isomerase. All of these examples involve the use of 'bulk' enzymes. Smaller quantities of highly purified 'fine' enzymes are used for numerous specialized purposes.

Immobilization of enzymes or whole cells, by their attachment to inert polymeric supports, allows easier recovery and reuse of the biocatalyst, and some enzymes are much more stable in this form. Also, the product does not become contaminated with the enzyme. Applications of immobilized biological catalysts include the production of amino acids, organic acids and sugar syrups.

Industrial chemicals and fuels

Industrial feedstock chemicals supplied through fermentation include various alcohols, solvents such as acetone, organic acids, polysaccharides, lipids and raw materials for the production of plastics. Some of these fermentation products also have applications in food manufacture.

Fossil fuels, especially oil, are likely to become exhausted within the next 50–100 years, resulting in the need to develop alternative sources of energy. Biological fuel generation may make an increasing contribution, particularly in the conversion of renewable plant biomass to liquid and gaseous fuels. This plant biomass can be in the form of cultivated energy crops, natural vegetation, and agricultural, industrial and domestic organic wastes. Currently, methane and ethanol are the main products, although other potential fuels can be generated using microorganisms, including hydrogen, ethane, propane and butanol.

Environmental roles of microorganisms

Microorganisms are particularly important in **waste-water treatment**, which utilizes the metabolic activities of diverse mixed microbial populations capable of degrading any compound that may be presented to them. The two main objectives are to destroy all pathogenic microbes present in the sewage, particularly the causal organisms of the water-borne diseases cholera, dysentery and typhoid. The second objective is to break down the organic matter in waste-water to mostly methane and carbon dioxide, thereby producing a final effluent (outflow) that can be safely discharged into the environment. Microbial activities can also be employed in the degradation of man-made xenobiotic compounds within waste streams and in the bioremediation of environments contaminated by these materials.

Microbial-based 'clean technology' is also being increasingly used in the desulphurization of fuels and the leaching of metals (e.g. copper, iron, uranium and zinc) from low-grade mineral ores and wastes using species of *Thiobacillus* and *Sulfolobus*. Environmental biological control is a further area where microorganisms are employed in an effort to reduce our reliance on synthetic chemical pesticides. Bacteria, fungi, protozoa and viruses are cultivated to produce biomass or cell products for the control of fungal, insect and nematode pests of agricultural crops, along with some vectors of human and animal diseases.

Conclusion

As can be seen from this brief introduction, microorganisms have a major role in providing food, raw materials and essential services. This role is likely to expand due to our increasing requirements for resources and the ability to manipulate microorganisms to improve their yields and the range of their products and activities.

Part 1 Microbial physiology

1 Microbial cell structure and function

The cell is the basic unit of all living organisms, many of which are unicellular, whereas others are multicellular forms, enabling cell specialization. All cells are filled with a fluid matrix and surrounded by a cytoplasmic membrane, primarily composed of lipids and proteins. They also contain nucleic acids, the physical carriers of genetic information, along with ribosomes that take part in protein synthesis. Cells are divided into two categories: those of archaeans and eubacteria are **prokaryotic**, whereas the cells of fungi, protozoa, algae and other plants, and animals are **eukaryotic**.

Prokaryotic cells are normally less than 5 μm in diameter, with a very few notable exceptions such as the marine bacterium, *Thiomargarita namibiensis*, at over 100 times larger. Prokaryotes rarely possess membrane-bound organelles and have little recognizable internal ultrastructure, apart from inclusion bodies (granules of organic or inorganic compounds), various vacuoles and some specialized invaginations of the cell membrane, e.g. the lamellae of photosynthetic cells (Fig. 1.1a). Most prokaryotic cells contain a single chromosome composed of deoxyribonucleic acid (DNA), which is located in a region of the cell referred to as the nucleoid. The chromosome is usually circular, although in some prokaryotes it is linear. Prokaryotic ribosomes are 70 S (Svedberg units), which refers to their rate of sedimentation on centrifugation and is a measure of their size, although density and shape of the particle can also influence this value. Almost all prokaryotes have cell walls or cell envelopes located outside the cytoplasmic membrane, which usually contain some peptidoglycan. Outside this wall they may have capsules or slime coats and propelling flagellae that are less complex than those of eukaryotic cells. Cell division in prokaryotes is normally by simple binary fission.

Eukaryotic cells are generally larger than those of prokaryotes and contain a range of membrane-bound organelles, including mitochondria, lysosomes, Golgi bodies and an extensive endoplasmic reticulum (Fig. 1.1b). Photosynthetic cells also contain chloroplasts. The DNA of eukaryotic cells, in the form of several linear chromosomes, is characteristically complexed with histone proteins and is housed in a double membrane-bound nucleus. Eukaryotic ribosomes, apart from those located within certain organelles, are 80 S, somewhat larger than those of prokaryotes. If cell walls are present, they are composed of materials other than peptidoglycan, such as cellulose and related β-glucans, chitin or silica. Eukaryotic cells divide by a complex process of mitosis and usually have a sexual life-cycle, involving meiosis (reduction division). This process halves the number of chromosomes from diploid (2n) chromosome pairs to produce haploid (n) cells containing a single set of chromosomes, facilitates genetic recombination and results in the formation of gametes.

Prokaryotes

Prokaryotes have been separated into two distinct groups on the basis of the study of phylogenetic (evolutionary) relationships. They are the **archaebacteria** or **archaea** ('ancient' bacteria) and the **eubacteria** ('true' bacteria), the group that contains almost all established industrial prokaryotes (Table 1.1).

Archaea

These prokaryotes are quite different from eubacteria and have some features, especially aspects of the transcription and translation machinery associated with protein synthesis, that are similar to eukaryotic cells. Most archaeans live in extreme environments similar to those that early life forms are thought to have endured. Three basic physiological types are found, namely halophiles (adapted to high salt concentrations), methanogens (methane producers) and thermophiles (adapted to high temperatures), and some of these are also barophiles (adapted to high pressure) (see Chapter 2). Archaeans have relatively small genomes containing

Table 1.1 Microbial genera with established industrial roles (for specific examples of their roles see Table 4.1)

	Eubacteria		Fungi	
Archaeans	Gram-negative	Gram-positive	Filamentous	Yeasts
Methanobacterium	*Acetobacter*	*Actinomyces*	*Acremonium*	*Blakeslea*
Methanococcus	*Acinetobacter*	*Actinoplanes*	*Agaricus*	*Candida*
Pyrococcus	*Agrobacterium*	*Arthrobacter*	*Aureobasidium*	*Hansenula*
Sulfolobus	*Alcaligenes*	*Bacillus*	*Aspergillus*	*Kluyveromyces*
	Azotobacter	*Brevibacterium*	*Claviceps*	*Pachysolen*
	Erwinia	*Clostridium*	*Coniothyrium*	*Phaffia*
	Escherichia	*Corynebacterium*	*Curvularia*	*Pichia*
	Klebsiella	*Lactobacillus*	*Cylindrocarpon*	*Rhodotorula*
	Methylocococcus	*Lactococcus*	*Fusarium*	*Saccharomyces*
	Methylophilus	*Leuconostoc*	*Lentinus*	*Xanthophyllomyces*
	Pseudomonas	*Micrococcus*	*Mortierella*	*Yarrowia*
	Ralstonia	*Mycobacterium*	*Mucor*	*Zygosaccharomyces*
	Salmonella	*Nocardia*	*Paecilomyces*	
	Sphingomonas	*Propionibacterium*	*Penicillium*	
	Spirulina	*Streptococcus*	*Rhizomucor*	
	Thermus	*Streptomyces*	*Rhizopus*	
	Thiobacillus		*Sclerotium*	
	Xanthomonas		*Trametes*	
	Zoogloea		*Trichoderma*	
	Zymomonas		*Trichosporon*	

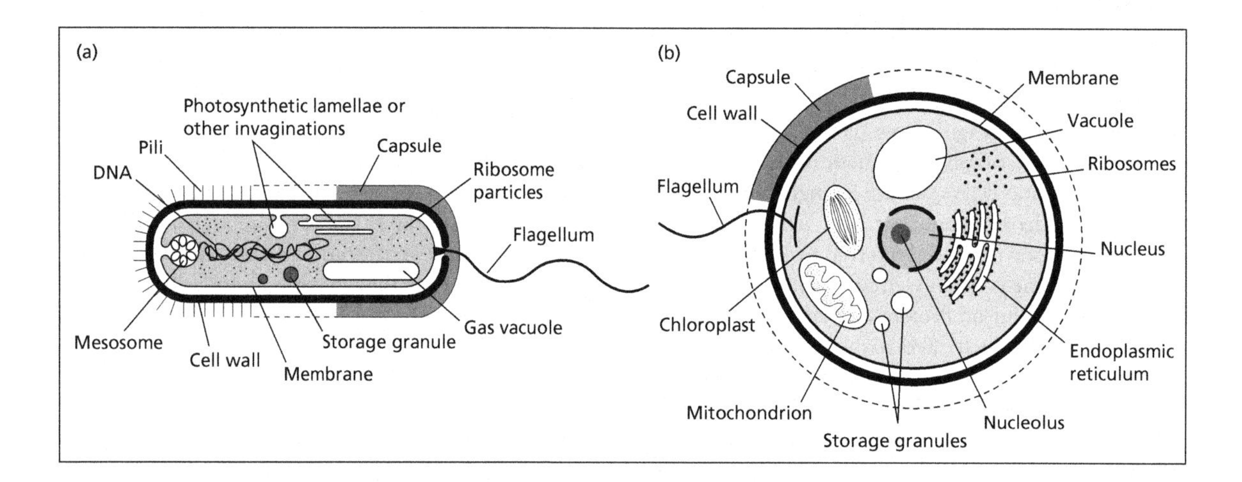

Fig. 1.1 Diagrammatic representation of the main structures of (a) prokaryotic and (b) eukaryotic microbial cells. Not all structures are always present, including capsules, chloroplasts, flagellae, pili, storage granules and vacuoles (from Dawes & Sutherland (1992)).

less than half the DNA of eubacteria. For example, the genome of *Methanococcus jannaschii* has been sequenced and found to contain 1760 genes composed of 1700 kilobase pairs (kbp). Although few archaeans are currently used for industrial purposes, they possess many interesting properties that could be exploited for biotechnological uses in the future. Their enzymes are of particular interest.

All prokaryotes may be initially subdivided according to their characteristics in the Gram staining procedure, which is determined by cell wall/envelope structure. Archaeal cell wall composition varies considerably, some appearing Gram-positive whereas others are Gram-negative. Their cell wall constituents are quite different from those of the eubacteria, in that they contain the unique polymers, methanochondroitin and pseudomurein. They also possess distinctive membrane lipids. The archaea may be divided into three kingdoms.

1 **Euryarchaeota** are primarily methanogens, such as *Methanobacterium* and *Methanosarcina*, and the extreme halophiles *Halobacterium* and *Halococcus*.

2 **Crenarchaeota** are mostly extreme barophiles and thermophiles that include *Pyrodictium*, *Pyrolobus*, *Sulfolobus* and *Thermoproteus*.

3 **Korarchaeota** are hyperthermophiles which, as yet, have not been isolated as pure cultures.

Eubacteria

The eubacteria are a very diverse group that may be divided into 12 subgroups. However, almost all industrial bacteria are contained within just two of them, the proteobacteria and the Gram-positive eubacteria.

1 The **Proteobacteria** is a major kingdom of Gram-negative bacteria that is divided into five groups, α, β, γ, δ and ε. They include purple photosynthetic bacteria and non-photosynthetic relatives, notably the Enterobacteriaceae (e.g. *Escherichia coli*), along with *Hyphomicrobium*, *Nitrobacter*, *Pseudomonas*, *Thiobacillus* and *Vibrio*.

2 The **Gram-positive eubacteria** are composed of two major subdivisions:

(a) the low guanine (G) + cytosine (C) group, which refers to the proportion of this base pair within the organism's DNA and includes *Bacillus*, *Clostridium*, *Lactobacillus*, *Leuconostoc*, *Staphylococcus*, *Streptococcus* and *Mycoplasma*; and

(b) the high G + C group, which contains the actinomycetes (filamentous bacteria, e.g. *Streptomyces*), *Corynebacterium*, *Mycobacterium* and *Micrococcus*.

The other subgroups, which contain few examples of industrial bacteria, are as follows.

3 *Cyanobacteria* and relatives, which are oxygenic phototrophs, e.g. *Anabaena*, *Nostoc* and *Spirulina*.

4 *Chlamydia*, a group of obligate intracellular parasites.

5 *Planctomyces* and *Pirella*, bacteria lacking peptidoglycan; some with membrane-bound nucleoid.

6 *Bacteroides* and *Flavobacteria*, a subgroup that contains a mixture of physiological types.

7 Green sulphur bacteria, such as *Chlorobium*, an anaerobic phototroph.

8 Spirochetes and relatives which are Gram-negative coiled bacteria.

9 Deinococci, radioresistant micrococci and relatives, e.g. *Deinococcus radiodurans* and *Thermus aquaticus*.

10 Green non-sulphur bacteria and anaerobic phototrophs.

11 *Thermotoga* and *Thermosulfobacteria*, thermophiles from hot springs and marine sediments.

12 *Aquiflex*, a group of obligate chemolithotrophic and autotrophic hyperthermophiles.

There is probably no such thing as a typical prokaryotic cell. There is a vast amount of diversity, including: morphological diversity (size and shape; rods, cocci, spirals, filaments, etc.), structural diversity (Gram-positive or Gram-negative cell walls/envelopes, external structures such as flagellae and pili, and the ability to form spores), along with metabolic, ecological and behavioural diversity. However, it is worthwhile considering, in some detail, the key cellular features of examples of industrially valuable Gram-negative and Gram-positive bacteria.

ESCHERICHIA COLI, A GRAM-NEGATIVE BACTERIUM

E. coli was discovered in 1885 by the German bacteriologist Theodor Escherich. It is a major inhabitant of the colon of humans and the lower gut of other warm-blooded animals. Some strains can cause food- and water-borne diseases that produce diarrhoea and can be especially problematical for human infants and young animals. A particularly virulent strain is *E. coli* 0157:H7, the cause of haemorrhagic colitis, which has recently emerged and has been associated with the ingestion of undercooked beef and raw milk.

Extensive information has been accumulated about the biochemistry, physiology and genetics of *E. coli*. This knowledge and the organism's rapid growth rate, with a doubling time as low as 20 min (see Chapter 2), has led to it becoming an important industrial microorganism. *E. coli* has been used extensively as a model for the study of molecular biology and is often considered to be the ideal host in gene-cloning experiments. Consequently, it has proved to be extremely useful for the production of heterologous proteins, derived from other organisms. However, protein secretion by *E. coli* is rather more problematical, due to the nature of its cell envelope (see below), whereas secretion from Gram-positive bacteria is often more readily achieved.

E. coli is a Gram-negative facultative anaerobe, belonging to the family Enterobacteriaceae, whose members are often referred to as enterobacteria or enteric bacteria. The cells are short straight rods, approximately 0.3–1.0 μm wide and 1.0–3.0 μm long, that divide by binary fission after elongating to approximately twice their normal length. Members of the genus *Escherichia* are oxidase-negative (lacking cytochrome c oxidase) and carry out mixed acid fermentation, producing mainly lactate, acetate, succinate and formate. The formate may be further degraded to form H_2 and CO_2.

Outer membrane

The outer coverings of Gram-negative bacterial cells are often referred to as **envelopes**, rather than walls, as they are more complex than the cell walls of Gram-positive bacteria (Fig. 1.2a). They are essentially composed of two layers that protect the cell and provide rigidity. The outermost layer is called the outer membrane, which is approximately 7–8 nm thick, containing lipopolysaccharide and mucopeptide. This structure does not impede the movement of small molecules, charged or uncharged, and is more permeable than the cell/cytoplasmic membrane (see below), but is a barrier to hydrophobic molecules and proteins. It contains porin proteins, composed of three subunits, that form narrow channels of about 1–2 nm diameter through which small molecules can pass. Non-specific porins allow the passage of molecules up to 600–700 Da, whereas specific porins have binding sites for one or more substances of up to 5000 Da. Lipopolysaccharide components of the envelope are effective in protecting the cell from detergents and other antimicrobial agents. The most common outer membrane protein is Braun's lipoprotein, which extends through the outer membrane and links to the underlying peptidoglycan.

Flagellae of motile strains propel the cell through aqueous media. In *E. coli*, flagellae are arranged around the entire cell, referred to as a peritrichous arrangement, whereas other bacteria have polar flagellae or alternative specific arrangements. Each flagellum is several micrometres long, composed of the protein flagellin, and is attached to the outer membrane layer through a basal body, composed of four rings. In addition, fibrils (fimbriae or pili), may be attached to the outer membrane, which are short hair-like projections, 5–7 nm in diameter and 400 nm long. They enable *E. coli* to attach to surfaces, such as intestinal epithelium.

Some strains also possess capsules located outside the outer membrane, which are composed of polysaccharides. Their production is influenced by the chemical and physical conditions within the local environment. These exopolysaccharides may provide a barrier to certain molecules, help protect against desiccation, or aid attachment of pathogenic strains to host cell surfaces.

Peptidoglycan and the periplasmic space

Within the outer membrane of Gram-negative bacteria, and covalently attached to it through lipoprotein, is a thin layer of peptidoglycan some 2–3 nm thick. It constitutes only 5–10% of the cell envelope and is composed of one to three layers, compared with the 20–25 layers of peptidoglycan in the walls of many Gram-positive bacteria. Nevertheless, it is a very important structural component. When the peptiglycan layer is incomplete, bacterial cells may swell and ultimately burst (see Chapter 3, Peptidoglycan biosynthesis).

The peptidoglycan extends down into the underlying periplasmic space, which is approximately 12–15 nm wide. This region is not empty, it contains a range of proteins, binding proteins, chemoreceptors and various enzymes. Binding proteins initiate transport of specific substances into the cell by taking them to their membrane-bound carriers. The chemoreceptors are involved in chemotaxis, which is the movement of a cell towards attractant and away from repellant chemicals. Hydrolytic enzymes, notably alkaline phosphatase, nucleases and proteases, are secreted into the periplasm from the cytoplasm and are retained close to the cell as they cannot normally pass through the outer membrane. They are associated with the breakdown of large impermeable nutrients into smaller molecules that can be transported across the cell membrane into the cell (see Chapter 2, Nutrient uptake). Some detoxifying and defence enzymes are also located here, e.g. penicillinase.

Cell (cytoplasmic) membrane

Below the periplasmic space lies the inner cell (cytoplasmic) membrane that encloses the cytoplasmic matrix (Fig. 1.2a). This structure is highly selective, controlling the entry of nutrients and the secretion of ions and larger compounds. The membrane is in the form of a lipid bilayer, primarily composed of phosphatidyl ethanolamine. It is interspersed with both transport proteins, such as lactose permease, and pores made up of porins that selectively control the entry of molecules and charged ions into cells. Respiratory proteins, including cytochromes and other electron transfer proteins, are also located within this membrane (see Chapter 3).

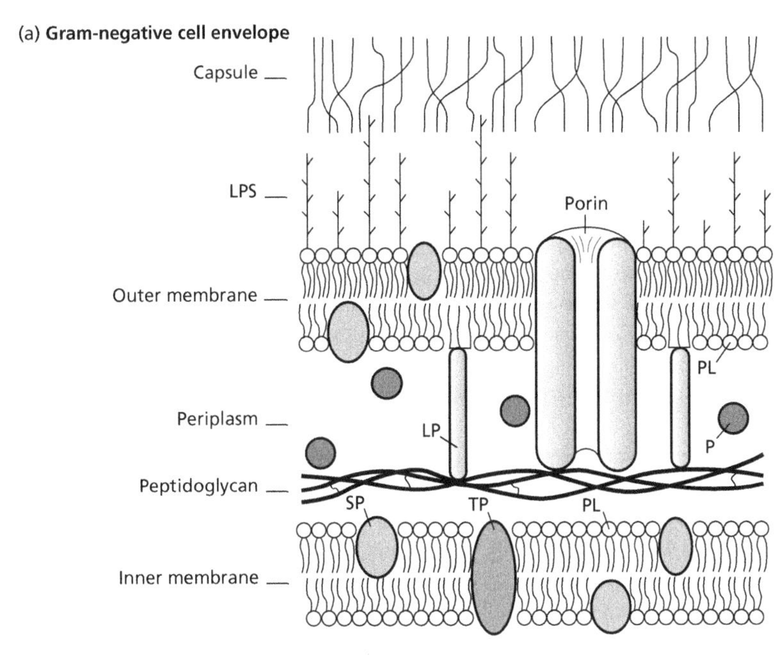

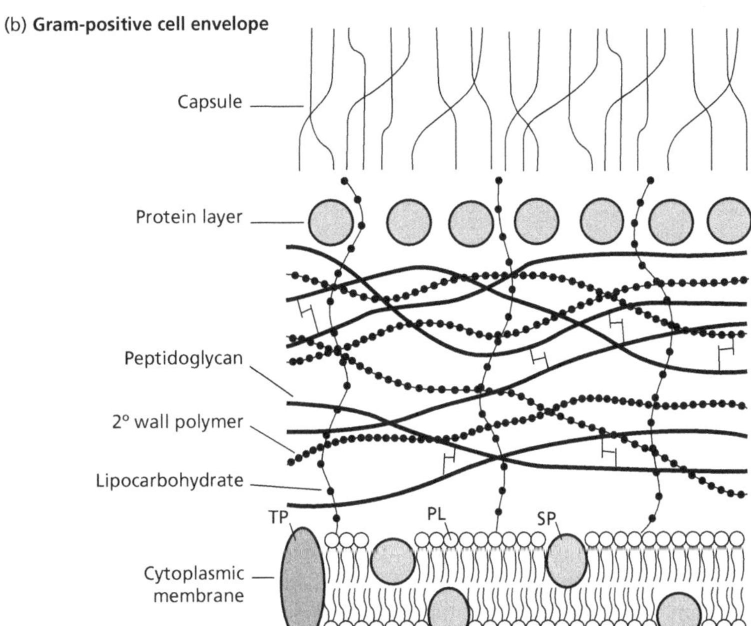

Fig. 1.2 Diagrammatic representation of the structure of typical (a) Gram-negative and (b) Gram-positive cell envelopes. LPS = lipopolysaccharide; LP = lipoprotein; P = protein; PL = phospholipid; SP = surface protein; TP = transmembrane proteins (from Poxton (1993)).

Cytoplasmic matrix and cell contents

The cytoplasmic matrix is maintained at pH 7.6–7.8, with differences between the intracellular and extracellular pH being controlled by the primary proton pumps associated with electron transport and respiration. It contains metabolic intermediates, and the enzymes and coenzymes required for catabolic and anabolic metabolism (see Chapter 3). Machinery for protein synthesis, both transcription and translation, is also located here.

This includes RNA polymerases for transcribing the genetic code of DNA into messenger RNA (mRNA), and around 18 000 ribosomes and transfer RNAs for translating the mRNA message sequences into proteins.

The chromosome resides in the nucleoid region that occupies approximately 10% of the cell's volume. It is a single circular molecule of double-stranded DNA made up of about 4600 kbp, constituting over 4000 genes, and is attached to the inside of the cell membrane at, or close to, the origin of replication. This DNA molecule is approximately 1 mm long and 1 nm thick, but is extensively folded and coiled, and stabilized by associated proteins.

Plasmids, relatively small circular extrachromosomal DNA molecules, may also be present. They include the fertility (F) factor, resistance (R) plasmids and Col plasmids that code for colicins, specific antibacterial bacteriocins. In enteropathogenic strains of *E. coli* at least two toxins are known to be plasmid encoded.

The polysaccharide glycogen is a main store of carbon and energy, and may be seen as inclusion bodies within the cytoplasmic matrix. Under certain circumstances osmoprotective betaines (*N*,*N*,*N*-trimethyl glycine) are also accumulated.

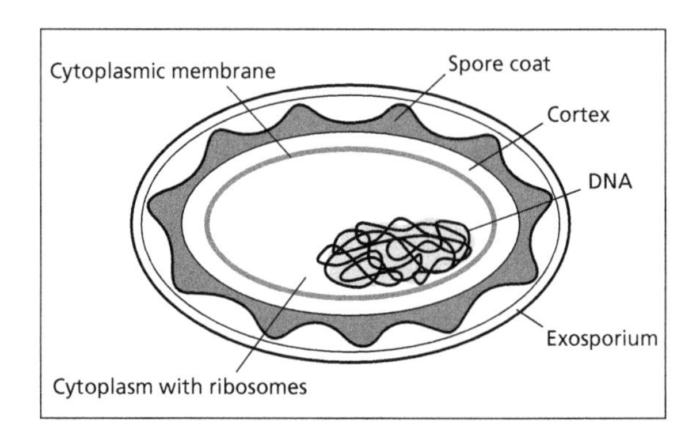

Fig. 1.3 The structure of a typical bacterial endospore.

BACILLUS SUBTILIS, A GRAM-POSITIVE BACTERIUM

The genus *Bacillus* consists of a large number of diverse, rod-shaped, chemoheterotrophic, Gram-positive bacteria. These cells are usually somewhat larger than *E. coli*, at 0.5–2.5 μm wide and 1.2–10 μm long. Some species are strictly aerobic, others are facultative anaerobes or microaerophilic, but all are catalase positive (see Chapter 2). *Bacillus* species also produce oval or cylindrical endospores that are resistant to adverse environmental conditions and provide a selective advantage for survival and dissemination (Fig. 1.3). Several members of the genus have important industrial roles, particularly as sources of enzymes, antibiotics (bacitracin, gramicidin and polymyxin) and insecticides.

B. subtilis is a common soil microorganism that is often recovered from water, air and decomposing plant residues. This bacterium is considered to be benign as it does not possess any disease traits, unlike some of its relatives, notably *B. anthracis*, the cause of anthrax. The range of extracellular enzymes produced by this microorganism enables it to degrade a variety of natural substrates and contributes to nutrient cycling. Many of these enzymes are exploited commercially and *B. subtilis* has become widely used for the production of industrial enzymes, particularly amylases and proteases. Some of these amylases are used extensively in starch modification processes and the proteases are employed as cleaning aids in biological detergents. *B. subtilis* has also proved very useful for the manufacture of fine chemicals, especially nucleosides, vitamins and amino acids, and some strains are used in crop protection against fungal pathogens. This bacterium is also a valuable cloning host for the production of heterologous proteins.

The main features of *B. subtilis* that distinguish it from *E. coli* are the cell wall structure and the ability to produce spores. *B. subtilis* cell walls are typical of Gram-positive bacteria, being much less complex than those of Gram-negative bacteria such as *E. coli*. They are 20–50 nm thick and simply composed of 20–25 layers of peptidoglycan, associated with some lipid, protein and teichoic acid (Fig. 1.2b). Teichoic acid is a distinctive anionic polymer of glycerol phosphate, ribitol phosphate and other sugar phosphates. It is covalently linked to the *N*-acetylmuramic acid units of the peptidoglycan or to lipids of the underlying cell membrane. This component is not found in Gram-negative bacteria. Outside the cell wall, *B. subtilis* produces a polypeptide capsule that contains both D- and L-glutamic acid units. Flagellae may be present and flagellated forms are capable of chemotaxis.

The *B. subtilis* chromosome is a little smaller than that of *E. coli* at 4188 kbp, but other intracellular features are similar. However, as mentioned above, a major difference is the ability of members of this genus to form spores (Fig. 1.3). Vegetative cells continue to grow and divide by binary fission until nutrients become limiting. Deprivation of nutrients initiates sporulation through an as yet unknown chemical signal. Unequal cell division produces a smaller forespore cell and a larger mother cell by the formation of an asymmetric septum near one pole of the cell. The mother cell goes on to engulf the forespore. A primordial wall of peptidoglycan is then formed around the forespore, which later becomes the cell wall of the vegetative cell that emerges when the spore germinates. This primordial wall is then overlain by a much thicker layer of complex peptidoglycan, known as the spore cortex. This cortex has a unique spore-specific composition, where over 50% of the muramic acid residues are present as muramic acid δ-lactam. Following cortex deposition, the structure is finally enclosed within a proteinaceous coat, and the surrounding mother cell dies and lyses to release the spore. These spores are extremely dormant, exhibiting a lack of metabolism, and are highly resistant to desiccation, heat, radiation and harsh chemical treatments. They can remain viable for long periods. Under favourable growth conditions the spore germinates to form a vegetative cell.

Fungi

Fungi are a diverse group of eukaryotic microorganisms that occupy a variety of habitats. The majority of fungal species are composed of filamentous hyphae and are often referred to as moulds, whereas the yeasts, which will be described later, are unicellular fungi. Of the thousands of species known, relatively few filamentous fungi are used for industrial purposes (Table 1.1). Filamentous fungi are non-photosynthetic and have strict chemoheterotrophic absorptive nutrition (see Chapter 2). Many secrete a range of hydrolytic enzymes to degrade the complex polymeric molecules encountered into smaller units that can be absorbed. Most are saprophytic, utilizing dead animal or plant remains. Some are facultative parasites of plants or animals, and several form symbiotic and mutualistic relationships with other organisms, e.g. with an alga to form a lichen (a fungus, the mycobiont, in symbiotic association with an alga, the phycobiont).

Filamentous fungi originate from either fragments of hyphae or dispersed spores that germinate under suitable environmental conditions. Hyphae can grow rapidly in length, at rates of up to several micrometres per minute. However, there is generally little increase in girth, which maximizes the surface area for absorption of nutrients. They branch to form intertwining mats that are usually established around and within their food source, and are structurally organized into macroscopic vegetative mycelium. In higher fungi, hyphae can be aggregated together to produce large complex solid structures, such as the fruiting bodies of mushrooms, rhizomorphs, sclerotia and stromata.

Individual hyphae are 1–15 μm in diameter depending upon the species. Their cell walls are composed of 80–90% polysaccharide, along with some lipid and protein constituents. Except in certain lower fungi, the polysaccharide is primarily microfibrils of chitin, a linear polymer of β-1,4-linked *N*-acetylglucosamine units, which is a strong and flexible material, similar to that found in arthropod exoskeletons. Unlike yeast cell walls (see p. 16), glucan and mannan are relatively minor components. Glucans are mostly laid down at the growing hyphal tips. Here the cell walls are only 50 nm thick, but behind the tip wall thickness increases up to 125 nm through the further deposition of chitin.

Fungal hyphae may be aseptate or septate. Hyphae of aseptate fungi lack cross-walls and such cells are referred to as **coenocytic**. This is a multinucleate state resulting from repeated nuclear division without cytokinesis (cell division). Hyphae of septate fungi are divided into cells by cross-walls called septa, containing pores that allow organelles and other cellular materials to move from cell to cell (Fig. 1.4). Cellular organelles of fungi are typically eukaryotic and are particularly associated with the growing hyphal tips, which are often packed with mitochondria, ribosomes and small vesicles. In addition, the hyphae tend to have numerous vacuoles containing storage products, normally glycogen, lipid and volutin (a polymer of metaphosphate).

Fungal chromosomes and nuclei are relatively small, and nuclear division is somewhat different from that in most other eukaryotes. During mitosis, the nuclear envelope remains intact with the spindle located within. After separation of replicated chromosomes, the nuclear envelope constricts to form two new nuclei, during which the spindle disappears.

Although fungal hyphae and spores are normally haploid, except for transient diploid stages in the sexual life-cycles, some mycelia may be genetically heterogeneous, resulting from the fusion of hyphae of separate origins. In such cases, each hypha contributes nuclei that may remain segregated into different parts of the same mycelium, or may mingle and even exchange genes in a process similar to crossing-over events during meiosis.

Most fungi reproduce by releasing vast quantities of

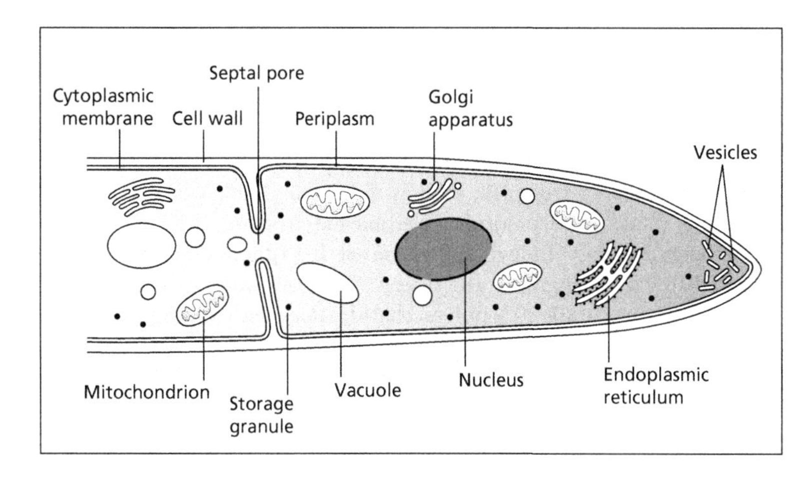

Fig. 1.4 Diagrammatic representation of a septate fungal hypha.

unicellular spores. These propagules are dispersed over a wide area by wind or water. They are stable for long periods, but will germinate if the environmental conditions and substratum are suitable. There are various ways by which spores may be formed asexually, depending upon the fungus (Fig. 1.5). Their mode of formation determines whether they are blastospores, sporangiospores, conidiospores, and so on. Variations in shape and arrangement of spore-bearing structures are often used as a basis for fungal identification.

The sexual cycle in fungi differs from that in other eukaryotic organisms in that syngamy, sexual union of haploid cells of opposite mating strains, + and –, occurs in two stages separated in time: first, plasmogamy, the fusion of cytoplasm, followed some time later by karyogamy, the fusion of nuclei. Following plasmogamy, a dikaryon is formed through the pairing of haploid nuclei from each parental strain, but the pairs of nuclei do not fuse immediately. These nuclear pairs within a dikaryon may remain separate and go on dividing synchronously for long periods. Eventually they fuse to form a diploid cell that directly undergoes meiosis, ultimately producing genetically diverse haploid spores.

Traditionally, the 100 000 known species of fungi have been divided into four main classes, primarily based upon how and where they specifically generate their sexual spores.

Phycomycetes

Phycomycetes are the lower fungi, which are subdivided into *Mastigomycotina* (zoosporic, motile spores) and *Zygomycotina* (zygosporic). They are the simplest fungi and their vegetative hyphae are aseptate. The *Mastigomycotina* contains the water moulds, *Saprolegnia*, and the important plant pathogens *Pythium* and *Phytophthora*. Unusually, their cell walls are composed of cellulose, along with other glucans or chitin. Industrially important members of the *Zygomycotina* include *Mucor*, *Rhizomucor* and *Rhizopus* species, which are used in some traditional food fermentations, and whole cell and enzyme bioconversions.

Asexual reproduction results in the formation of sac-like sporangia at the tips of upright hyphae or sporangiophores (Fig. 1.5a). Hundreds of haploid spores are produced by mitosis within the sporangium, which are then wind dispersed. In sexual reproduction mycelia of opposite mating types each form gametangia containing several haploid nuclei. Plasmogamy of + and – gametangia, and pairing of haploid nuclei, results in the formation of a dikaryotic zygosporangium. This structure is metabolically inactive and resistant to desiccation. When conditions become favourable, karyogamy occurs between paired nuclei and the resulting diploid nuclei undergo meiosis to produce haploid spores.

Ascomycetes or *Ascomycotina* (sac fungi)

This is the largest class of fungi and includes the yeasts that are utilized in many industrial fermentation processes. Other filamentous members that have industrial and commercial roles are *Neurospora* species, *Claviceps* species, and important edible fungi, including *Morchella* species (morels) and *Tuber* species

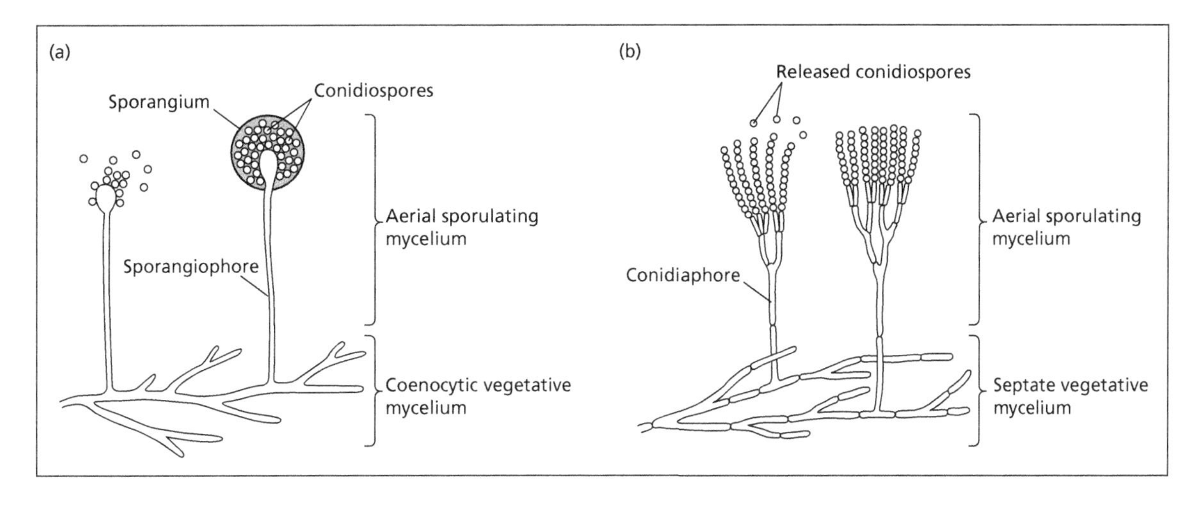

Fig. 1.5 Asexual spore formation (a) in the sporangia of *Mucor*, a coenocytic filamentous fungus and (b) conidia of *Penicillium*, a septate filamentous fungus.

(truffles). Their hyphae are septate, and in asexual reproduction, the tips of specialized hyphae (conidiophores) form chains of haploid conidiospores that are usually wind dispersed. Sexual spores or ascospores are contained within a sac-like ascus, which may be enclosed within a fruiting body or ascocarp. The yeasts, such as *Saccharomyces cerevisiae*, produce the equivalent of an ascus during sexual reproduction and most bud during asexual reproduction in a manner similar to the formation of conidiospores (see below). However, certain yeasts, notably *Schizosaccharomyces* species, do not bud, but undergo binary fission in a similar manner to bacteria.

Basidiomycetes or *Basidiomycotina* (club fungi)

Members of this division produce septate hyphae and their sexual haploid basidiospores are borne on club-shaped structures called basidia. In some cases, these are contained within large sexual fruiting bodies, the basidiocarps, as in the mushrooms, e.g. *Agaricus bisporus* (button mushroom) and *Lycoperdon* species (puff balls). Other industrially important basidiomycetes are certain wood-rotting fungi involved in biodeterioration and biodegradation processes, e.g. *Phanerochaete chrysosporium* (white rot). Also included in this division are the rusts and smuts, which are important plant pathogens.

Deuteromycetes or *Deuteromycotina* ('imperfect' fungi)

This division contains a diverse group of around 25 000 species, which have been grouped together simply because they lack a defined sexual (perfect) stage, or one has not been observed. It is assumed that this group may represent the conidial stages of ascomycetes whose ascus stage has not yet been discovered or has been lost in the course of evolution. Parasexuality has been demonstrated in some species, which has proved important for genetic study and strain development. Many industrially important fungi are classified as deuteromycetes, including species of *Aspergillus, Cephalosporium, Fusarium, Penicillium* (Fig. 1.5b) and *Trichoderma.*

Yeasts

Yeasts, notably *S. cerevisiae*, are worthy of special attention, due to their major contribution as industrial microorganisms. The term 'yeast' refers to a unicellular form rather than to a phylogenetic classification. It can be used to describe a unicellular phase of the life-cycles of fungi that may be predominantly filamentous, but exhibit yeast–mould dimorphism. Changes from one form to the other are often influenced by the prevailing nutritional conditions. However, this term is used more commonly as a generic name for fungi that have only a unicellular phase, such as baking or brewing yeasts. These true yeasts and many others of industrial importance are strains of *S. cerevisiae*, a member of the ascomycetes.

Yeasts are heterotrophic and are found in a wide range of natural habitats, being particularly common on aerial plant organs, especially flowers, and plant debris in the surface layers of soil. Unlike most fungi, which are obligate aerobes, many yeasts are facultative anaerobes, able to grow in the absence of oxygen. They generally have relatively simple nutritional needs that include a reduced carbon source, which can be compounds as simple as acetate, and a variety of minerals. Inorganic nitrogen sources, such as ammonium salts, are readily assimilated, although a variety of organic nitrogen compounds can also be used, including urea and various amino acids. Often the only other complex compounds or growth factors required are vitamins, e.g. biotin, pantothenic acid and thiamine.

Several yeasts have industrial and food uses, and some, notably *S. cerevisiae*, have GRAS (generally regarded as safe) status. Strains of *S. cerevisiae* are the most well-known and commercially important yeasts, and have long been employed to produce alcoholic beverages and leaven bread. They are now also used in the production of several other fermentation products, particularly fuel ethanol, single cell protein, enzymes and heterologous proteins, e.g. human insulin. In addition, *S. cerevisiae* has been the model system in molecular genetics research because its basic mechanisms of replication, recombination, cell division and metabolism are very similar to those of higher eukaryotes, including mammals.

Very few yeasts are animal pathogens. *Cryptococcus neoformans* causes a relatively rare form of meningitis, but the best known is *Candida albicans*. This organism is carried by most people in a benign form, but it may commonly infect the skin, and mucosal membranes of the mouth, vagina and alimentary tract. *C. albicans* can become a serious opportunistic pathogen, particularly in individuals whose immune response is weakened. Such infections are difficult to treat, due to the similarity between the host's and the pathogen's metabolism. This yeast exhibits dimorphism, growing predominantly as a yeast form, but develops into branching hyphae, with accompanying changes in cell wall composition, under certain cultural conditions.

CELL GROWTH AND THE YEAST LIFE-CYCLE

Members of the genus *Saccharomyces* produce cells that are spherical to ellipsoidal and vary in size from approximately 1–7 μm wide and 5–10 μm long. Polyploid industrial strains, containing multiple sets of chromosomes, tend to be at the larger end of this range. The yeast cell is surrounded by a thick wall, within which is the cell membrane, enclosing the cytoplasmic matrix that contains enzymes, storage granules, several different types of organelle and membrane systems.

Their asexual cell division involves budding of a daughter cell from a mother cell and cell growth is largely associated with bud development (Fig. 1.6).

A mature cell initiates a bud at a site where the cell wall has been weakened by lytic enzymes and the bud grows to approximately the same size as the mother cell. Wall growth is polarized, mainly by deposition of new cell wall material at its tip, which is associated with microvesicles in the underlying cytoplasmic matrix. During this period mitosis occurs. The chromosomes replicate, and the mitotic spindle forms and aligns across the junction between the two cells. It then elongates and facilitates the formation of two separated sets of chromosomes that become enclosed within separate nuclei, one in each cell. The cell wall between the mother cell and bud becomes completed, and the two cells may then separate. This cycle can then begin again for both cells, each may bud 10–20 times.

Sexual reproduction involves haploid cells of two mating types (**a** and **α**) and mating is mediated by the secretion of small peptide pheromones. Each mating type responds by halting bud formation, then each cell elongates and differentiates to become a specialized pear-shaped gamete. Cells of opposite mating types that are in contact or close proximity undergo plasmogamy and karyogamy to form a diploid cell. These diploid forms are stable and can perform repeated cell division. However, if the physiological conditions are suitable meiosis occurs. This results in sporulation, normally producing four haploid nuclei, which are incorporated into four stress-resistant ascospores, encapsulated within an ascus.

CELL WALL STRUCTURE

Some yeasts develop slimy polysaccharide capsules outside the cell wall that may have a protective function or aid in attachment to surfaces. The cell wall itself is approximately 100–200 nm thick and comprises 15–25% of the dry weight of a cell. Its major components, some 80–90% of wall material, are the polysaccharides glucan, phosphomannan and chitin, along with some proteins (Fig. 1.7). Glucan is a highly branched polymer of β-linked glucose units, predominantly β-1,3 linkages, in the form of microfibrils that provide strength in the inner wall adjacent to the cell membrane. Phosphomannan, a branched polymer of the hexose sugar mannose,

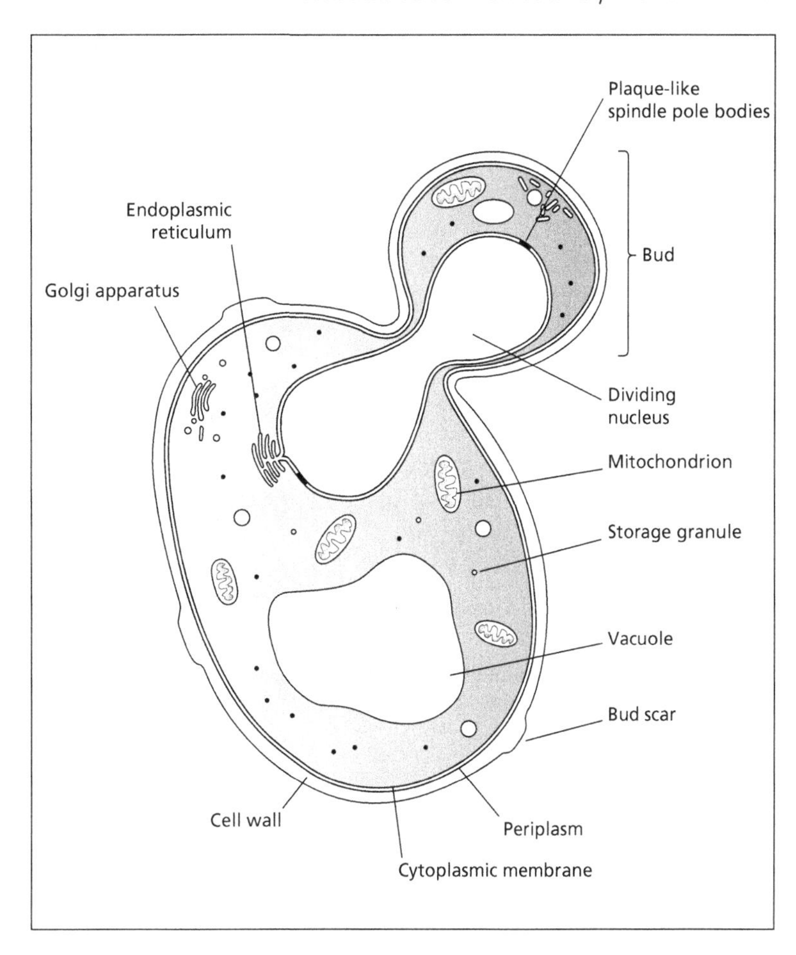

Fig. 1.6 Diagrammatic representation of a budding yeast cell.

is located towards the outside of the cell wall. Chitin may form less than 1% of the yeast wall and is primarily found around bud scars. The number of bud scars on the wall indicates the number of times that a cell has budded and as they never occur twice in the same place, they can be used to estimate the 'age' of a cell. Protein wall components are both structural and enzymic; the former are largely associated with mannan. Enzymes that have been found here and in the periplasm include those associated with wall biosynthesis and the hydrolysis of substrates unable to cross the cell membrane. They include glucanase, mannanase, lipase, alkaline phosphatase and invertase, a mannoprotein involved in sucrose utilization.

Flocculation of cells is a common feature in some yeasts and is particularly important for certain industrial strains. Flocs are aggregates of cells, not chains of unseparated buds. There are two main theories of floc formation. The first concerns calcium bridging, where a calcium ion links two cells via negatively charged cell wall components. Support for this theory comes from the observation that flocs are dispersed by the chelating agent ethylenediamine tetraacetic acid (EDTA). Secondly, the lectin hypothesis, involving protein–carbohydrate binding, proposes that a surface protein of one cell links to a mannose residue on an adjacent cell. This hypothesis is supported by the fact that sugars, particularly mannose, inhibit floc formation. In addition, proteinaceous cell wall protrusions of 0.1–10 μm, termed fimbriae, are present in many strains and may be associated with cell–cell interactions, including flocculation.

CELL MEMBRANE STRUCTURE

Between the cell wall and underlying membrane, there is

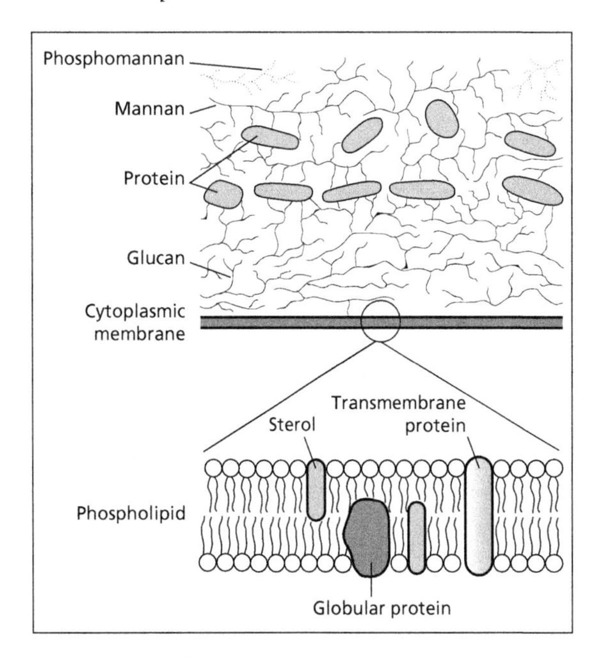

Fig. 1.7 Diagrammatic representation of the structure of a yeast cell wall and cell membrane.

a periplasmic space of 3.5–4.5 nm containing secreted proteins that cannot escape through the wall, including the enzymes mentioned above. The cell membrane controls all materials entering and exiting the cell, and is thought to manage cell wall biosynthesis. Yeast cell membranes, like those of other cells, are primarily composed of a lipid bilayer some 7.5 nm thick that arises from the tail-to-tail alignment of lipid molecules (Fig. 1.7). This produces inner and outer hydrophilic regions that sandwich the hydrophobic lipid 'tails'. The lipid constituents are composed of mono-, di- and triglycerides, glycerophosphatides and sterols. Unsaturated sterol components include ergosterol and zymosterol, which are not found in prokaryotic cell membranes. They are vital for membrane stabilization and membrane rigidity, and molecular oxygen is required for their biosynthesis. Membrane lipid composition varies depending upon growth conditions and also affects the cell's tolerance to ethanol. For example, brewing strains have been found to contain higher levels of phosphatidyl choline than baking strains.

Located on, within and across the lipid bilayer are both structural and functional protein molecules, together with a small portion of carbohydrate. Protein components may be involved with solute transport, or are signal transduction components that respond to external stimuli and ultimately initiate an internal response. Membrane-associated enzymes include a range of permeases for the transportation of compounds, such as sugars and amino acids; along with adenosine triphosphatases (ATPases), in association with systems responsible for the generation of proton-motive force across the membrane, which provide energy for solute transport. There are also wall synthesizing enzymes. For example, chitin synthetase is present in an inactive form, which is activated by proteolytic cleavage when required.

THE NUCLEUS

The nucleus is surrounded by a double membrane that has regularly spaced pores. This organelle contains the majority (80–85%) of the cellular DNA; the remainder is present as circular molecules in the mitochondria or cytoplasm (see below). Haploid cells of *S. cerevisiae* contain approximately 12 000 kbp of DNA, which is 3–10 times more than a typical bacterium and about 100–150 times less than mammalian cells. Nuclear DNA is associated with basic histone proteins and is in the form of 16 linear chromosomes made up of 150–2500 kbp. Over 6000 genes have been identified, but around 50% have unknown function. Associated with the nucleus is a structure referred to as a plaque, which has microtubules that pass into both the nucleus and cytoplasmic matrix. It functions in a similar way to the centrioles of animal cells, forming the spindle upon which the replicated chromosomes separate during mitosis.

THE MITOCHONDRIA

Fully developed mitochondria are present only in yeasts that are growing aerobically. Under anaerobic conditions they are simple structures, referred to as promitochondria, as anaerobic growth induces their dedifferentiation. Fully developed organelles are spherical to rod-shaped, enclosed by a double membrane. The inner membrane possesses proteins involved in electron transport and oxidative phosphorylation, and is folded into the lumen of the organelle to form finger-like structures called cristae. Located within the lumen is a fluid matrix containing most of the enzymes of the tricarboxylic acid (TCA) cycle, ribosomes and 75 kbp circular DNA molecules that code for some 10% of mitochondrial proteins.

'Petite' yeast mutants lack mitochondria and are therefore incapable of respiration. They can grow only fermentatively and on solid media their colonies are very small, compared with wild-type colonies, hence their name.

OTHER ORGANELLES AND CYTOPLASMIC STRUCTURES

The cytoplasmic matrix also contains ribosomes, various single membrane-bound organelles and vacuoles, a cytoskeleton composed of microtubules and microfilaments, and an elaborate membrane system, the endoplasmic reticulum. Endoplasmic reticulum is continuous with the cell membrane and outer membrane of mitochondria. It is associated with vesicle production in a similar manner to the Golgi bodies of other eukaryotes, but the yeast apparatus is less well defined. At maturity the cell has a large vacuole that arises from the fusion of smaller vacuoles. This may contain hydrolytic enzymes and also acts as a store for certain nutrients.

Microbodies, including peroxisomes and glyoxysomes, are membrane-bound organelles containing various specific enzymes. Peroxisomes contain catalase and various oxidases, and glyoxysomes possess catalase and the enzymes of the glyoxylate cycle. The number and size of peroxisomes varies depending upon the environment and available nutrients. For example, when methylotrophic yeasts (e.g. *Pichia pastoris*) are grown on C_1 substrates such as methanol, they develop several large peroxisomes, containing methanol oxidase, catalase and other enzymes necessary for the metabolism of this substrate.

Cytoplasmic storage granules of yeasts mostly contain lipids or carbohydrates. The latter include the polysaccharide glycogen and the disaccharide trehalose, which comprise two glucose residues linked by their reducing end (1–1-α-glucopyranosido-α-glucopyranoside). These carbohydrates may constitute up to 40% of the dry weight of a yeast cell. Glycogen is the predominant storage material, but the proportion of trehalose increases under aerobic conditions and under certain stresses. It is considered that glycogen has a role during periods of starvation and respiratory adaptation, whereas trehalose may be involved in only the former.

The cytoplasmic matrix of *S. cerevisiae* may also contain 50–60 copies of the extrachromosomal circular 2 μm plasmid. However, the only information that these plasmids appear to contain is for their own maintenance. They seem to provide no obvious advantage to the cells that possess them. Nevertheless, they have proved to be very useful vectors in the genetic modification of yeasts via transformation processes (see Chapter 4). Other extrachromasomal elements have also been found in some strains and include virus-like double-stranded linear RNA molecules.

Further reading

Papers and reviews

Gooday, G. W. (1993) Cell envelope diversity and dynamics in yeasts and filamentous fungi. *Journal of Applied Bacteriology* Supplement **74**, 12S–20S.

Johnstone, K. (1994) The trigger mechanisms for spore germination: current concepts. *Journal of Applied Bacteriology* Supplement **76**, 17S–24S.

Madden, K. & Snyder, M. (1998) Cell polarity and morphogenesis in budding yeast. *Annual Review of Microbiology* **52**, 687–744.

Poxton, I. R. (1993) Prokaryote envelope diversity. *Journal of Applied Bacteriology* Supplement **74**, 1S–11S.

Rose, A. H. (1993) Composition of the envelope layers of *Saccharomyces cerevisiae* in relation to flocculation. *Journal of Applied Bacteriology* Supplement **74**, 110S–118S.

Wipat, A. & Harwood, C. R. (1999) The *Bacillus subtilis* genome sequence: the molecular blueprint of a soil bacterium. *FEMS Microbiology Ecology* **28**, 1–9.

Books

Bennett, J. W. & Klich, M. A. (eds) (1992) Aspergillus—*Biology and Industrial Applications.* Butterworth-Heinemann, Boston.

Danson, M. J., Hough, D. W. & Lunt, G. G. (1992) *The Archaebacteria: Biochemistry and Biotechnology.* Portland Press, London.

Dawes, I. W. & Sutherland, I. W. (1992) *Microbial Physiology*, 2nd edition. Blackwell Scientific Publications, Oxford.

Dickinson, J. R. & Schweizer, M. (eds) (1998) *The Metabolism and Molecular Physiology of* Saccharomyces cerevisiae. Taylor & Francis, London.

Doi, R. H. & McGloughlin, M. (eds) (1992) *Biology of Bacilli.* Butterworth-Heinemann, Boston.

Ghuysen, J. M. & Hakenbeck, R. (1994) *Bacterial Cell Wall.* Elsevier, New York.

Griffin, D. H. (1994) *Fungal Physiology*, 2nd edition. Wiley, New York.

Holt, J. G. (ed.) (1989) *Bergey's Manual of Systematic Bacteriology*, Vols 1–4. Lippincott, Williams & Wilkins, Baltimore.

Holt, J. G. (ed.) (1993) *Bergey's Manual of Determinative Bacteriology*, 9th edition. Lippincott, Williams & Wilkins, Baltimore.

Lengeler, J. W., Drews, G. & Schlegel, H. G. (1998) *Biology of Prokaryotes*. Blackwell Science, Oxford.

Logan, N. A. (1994) *Bacterial Systematics*. Blackwell Science, Oxford.

Sonenshein, A. L., Hoch, J. & Losick, R. (1992) Bacillus subtilis *and other Gram-positive Bacteria: Biochemistry, Physiology and Molecular Genetics*. American Society for Microbiology, New York.

Walker, G. M. (1998) *Yeast Physiology and Biotechnology*. Wiley, Chichester.

2 Microbial growth and nutrition

Microbial nutrition

The biosynthesis of cellular components necessary for growth, reproduction and maintenance requires a supply of basic nutrients and an energy source. Nutritional classification is established on the basis of specific sources of energy, electrons/hydrogen and carbon used by an organism (Table 2.1). Microorganisms have evolved a wide range of mechanisms to gain energy, but are essentially divided into two categories. **Chemotrophs** obtain energy by the oxidation of organic or inorganic compounds, whereas **phototrophs** use energy derived from light. Both have two possible sources of hydrogen atoms or electrons. **Organotrophs** oxidize preformed organic molecules, such as sugars, to obtain electrons or hydrogen, whereas **lithotrophs** acquire electrons from reduced inorganic molecules, including hydrogen sulphide and ammonia. Carbon nutrition is divided into two classes. **Autotrophs** utilize CO_2 as their sole or primary source of carbon, whereas various **heterotrophs** use a wide range of reduced organic molecules, including hydrocarbons, lipids, organic acids, simple sugars and polysaccharides.

Microbial cells must obtain a range of chemical elements to fulfil their nutritional requirements. Four of these, the macronutrients carbon, hydrogen, oxygen and nitrogen, must be available in gram quantities per litre of growth medium. These elements, along with phosphorus and sulphur, are the principal components of major cellular polymers: lipids, nucleic acids, polysaccharides and proteins (Table 2.2). Other minor elements, including calcium, iron, potassium and magnesium, are required at levels of a few milligrams per litre; the trace elements, primarily cobalt, copper, manganese, molybdenum, nickel, selenium and zinc, are needed in only microgram quantities.

Macronutrients

Autotrophic fermentations that utilize CO_2 are rarely operated on an industrial scale: almost all involve heterotrophic growth. In heterotrophic fermentations, carbon sources are required at relatively high media concentrations, often around 10–20 g/L or greater, as they provide carbon 'skeletons' for biosynthesis and many also serve as an energy source. Generally, sugars are good carbon and energy sources, particularly glucose, the preferred substrate of most microorganisms.

Hydrogen and oxygen can be obtained from water and organic compounds. However, many organisms are also dependent on molecular oxygen as the terminal acceptor in aerobic respiration and for the synthesis of specific compounds, such as unsaturated sterols (see p. 34, Effects of oxygen on growth).

Microorganisms may contain more than 15% (w/w) nitrogen, mostly within structural and functional proteins, and nucleic acids. To fulfil these requirements a nitrogen source is normally supplied in growth media at concentrations of 1–2 g/L. Ammonium salts are often the preferred nitrogen source, but nitrate, amino acids or nitrogen-rich compounds, such as urea, may sometimes be used. In addition, certain specialized nitrogen-fixing bacteria, notably *Azotobacter* and *Rhizobium* species, can utilize molecular nitrogen.

Minor elements

Phosphorus is generally provided as inorganic phosphate ions, often as a pH buffer. This element is essential for the synthesis of nucleic acids, intermediates of carbohydrate metabolism and compounds involved in energy transduction, e.g. adenosine triphosphate (ATP) and nicotinamide adenine dinucleotide phosphate (NADP). Media concentrations normally need be no greater than 100 mg/L.

Sulphur is required for the production of the sulphur-containing amino acids cystine, cysteine and methionine, and some vitamins. It is often supplied as an inorganic sulphate or sulphide salt at a concentration of 20–30 mg/L.

Table 2.1 Nutritional categories of microorganisms

Physiological type	Source of Energy	Electrons	Carbon
Chemotroph	Chemical		
Phototroph	Light		
Organotroph		Organic compound	
Lithotroph		Inorganic molecule	
Autotroph			CO_2
Heterotroph			Organic compounds
Chemoorgano (hetero) troph (animals, fungi, protozoa, many bacteria)	Organic compound	Organic compound	Organic compound
Chemolitho (auto) troph (some bacteria)	Inorganic molecule	Inorganic molecule	CO_2
Photolitho (auto) troph (plants, most algae, some bacteria)	Light	Inorganic molecule	CO_2
Photoorgano (hetero) troph (algae, some bacteria)	Light	Organic compound	Organic compound

Table 2.2 Composition of a typical microorganism

Water	70–80%
Protein	15–18%
Polysaccharide	1–3%
Lipid	1–2%
Nucleic acids	3–7%
Inorganic salts	1–2%
Main elemental cell composition = $C_4H_7O_2N$	

Other minor elements, mostly calcium, iron, potassium and magnesium, must be provided in relatively small but significant quantities, usually less than 10–20 mg/L. Several minor elements are essential for specific enzyme activities. For instance, iron is essential for several oxidation–reduction enzymes, particularly cytochromes, and potassium is required by enzymes involved in protein synthesis and as counter-ion for the DNA phosphate backbone. Several minor elements also have structural roles: magnesium ions are involved in the stabilization of ribosomes, and some are necessary for maintaining cell wall and membrane integrity. Sodium and potassium are used in chemiosmotic energy pumps, but the former does not appear to have other specific roles, except in salt 'loving' halophilic microorganisms.

Trace elements

Trace elements include cobalt, copper, manganese, molybdenum, nickel, selenium and zinc. These elements are usually required at concentrations of 0.1–1 mg/L, or less, for a number of specific enzymes. Their normal concentrations in water supplies, or as contaminants in other media ingredients, often fulfil this requirement.

Overall nutrient demands vary for different microorganisms. Many bacteria can grow on media merely containing carbon and energy sources, and basic mineral elements. Microorganisms that are able to grow on this **minimal medium** are referred to as **prototrophs.** However, other microorganisms must be given specific compounds in a part or fully constructed form. Those that are unable to grow without additional organic substances, such as amino acids or vitamins, are called **auxotrophs.** Culture media for many organotrophic microorganisms often require certain vitamin and growth factor supplements, especially B vitamins (thiamin, B_1; riboflavin, B_2; pyridoxine, B_6; cobalamin, B_{12}; biotin; nicotinic acid; and pantothenic acid). Few microorganisms require fat-soluble vitamins (A, D, E and K); and vitamin C, although often improving growth, is not a true microbial growth factor.

Nutrient uptake

Nutrients from the environment must be transported across the cell membrane into the cell. This is often the rate-limiting step in the conversion of raw materials to products and therefore is of major importance to industrial fermentation processes. Uptake of a few nutrients is by **passive diffusion**, which does not require carriers. Such nutrients are usually soluble in lipids and can enter hydrophobic membranes, e.g. glycerol and urea. However, it is an inefficient mechanism, as the rate of uptake is dependent on the magnitude of the concentration gradient across the membrane. Most solutes must be transported via specific active mechanisms, because membranes are selectively permeable. Also, microorganisms usually inhabit natural environments where nutrient concentrations are low. Consequently, it is essential that they can accumulate solutes against concentration gradients, as intracellular concentrations of compounds are often higher than environmental levels.

Most solute uptake involves carrier proteins (permeases) that span the membrane. Their participation in **facilitated diffusion** requires no direct input of energy. It is driven solely by the concentration gradient across the membrane and is reversible. However, nutrient uptake into the cell continues, because their intracellular concentration does not increase, as they are immediately metabolized on entry. This mechanism, rarely seen in prokaryotes, is used to transport sugars and amino acids, and their selective carriers usually function for a group of related solutes. They greatly increase diffusion rates, at least up to concentrations at which the carrier protein becomes saturated with its nutrient.

Active transport mechanisms allow the accumulation of materials against a concentration gradient, which is important in environments where nutrient levels are low. Some systems allow accumulation to 100–1000 times greater than the external concentration. However, these mechanisms require the direct input of substantial amounts of metabolic energy, ATP or proton gradients, to drive transportation. As in facilitated diffusion, protein carriers are involved; many are highly specific, whereas others function with groups of related compounds. Where proton gradients are involved (see Chapter 3, Electron transport), the nutrient entering the cell, such as a sugar, amino acid or organic acid molecule, is simultaneously transported with a proton, and is referred to as **symport**. Proton gradients may also be used to establish a sodium ion gradient across the membrane. Here sodium ions leave the cell in exchange for the entry of protons, which is termed **antiport**. The sodium gradient established drives uptake of nutrients, e.g. sugars and amino acids.

Some compounds may be modified during uptake. Sugars, for example, can be phosphorylated using phosphoenolpyruvate (PEP) as the phosphate donor. This is referred to as **group translocation**, which is performed by many prokaryotic cells.

During periods when nutrient concentrations are very low, as during the stationary phase of a batch culture (see below, Microbial growth kinetics), some organisms produce metabolites capable of scavaging for remaining metal ions. Sideramines, which include the hydroxamate ferrichrome, are well-known examples. Alternatively, some microorganisms may produce compounds whose role *in vivo* is to render their membranes more permeable to certain metal ions, e.g. macrotetralides (ionophore antibiotics; see Chapter 3, Secondary metabolism).

Utilization of high molecular weight materials

Utilization of polymeric substrates (polysaccharides, proteins and lipids) requires additional activities. Protozoa and other eukaryotic organisms without cell walls can ingest relatively large pieces of food materials from their environment by phagocytosis (engulfment) into what becomes a membrane-bound food vacuole. Hydrolytic enzymes are then secreted into the vacuole to break down the polymers to their constituent monomers. For organisms with a rigid cell wall this is not usually possible. They must secrete extracellular hydrolytic enzymes—proteases, amylases, cellulases, etc.—directly into the environment and then take up the resultant hydrolysis products. Many of these extracellular enzymes have major industrial applications (see Chapter 9).

Microbial growth kinetics

Microbial growth can be defined as an orderly increase in cellular components, resulting in cell enlargement and eventually leading to cell division. This definition is not strictly accurate as it implies that a consequence of growth is always an increase in cell numbers. However, under certain conditions growth can occur without cell division, for example, when cells are synthesizing storage compounds, e.g. glycogen or poly β-hydroxybutyrate. In this situation the cell numbers remain constant, but the concentration of biomass continues to increase. This is also true for coenocytic organisms, such as some fungi, that are not divided into separate cells. Their growth results only in increased size.

Growth kinetics of homogeneous unicellular suspension cultures can be modelled using differential equations in a continuum model. However, filamentous growth and growth in heterogeneous cell aggregates and assemblages, particularly biofilms, colonies, flocs, mats and pellicles, is much more complex. In fact, heterogeneous systems require a very different approach using cellular automaton and Swarm system models, e.g. BacSim. The growth kinetics of filamentous organisms and heterogeneous systems will not be discussed here.

The growth model that will be examined is bacterial binary fission in homogeneous suspension cultures, where cell division produces identical daughter cells. Each time a cell divides is called a **generation** and the time taken for the cell to divide is referred to as the **generation time**. Therefore, the generation time or doubling time (t_d) is the time required for a microbial population to double. Theoretically, after one generation, both the microbial cell population and biomass concentration have doubled. However, as previously stated, under certain conditions growth can be associated with an increase in biomass and not cell numbers. Also, the generation time recorded during microbial growth is in reality an average value, as the cells will not be dividing at exactly the same rate. At any one time there are cells at different stages of their cell cycle. This is termed asynchronous growth. However, under certain conditions synchronous growth can be induced so that all cells divide simultaneously, which is a useful research tool in the study of microbial physiology.

Microbial fermentations in liquid media can be carried out under different operating conditions, i.e. batch growth, fed-batch growth or continuous growth. Batch growth involves a closed system where all nutrients are present at the start of the fermentation within a fixed volume. The only further additions may be acids or bases for pH control, or gases (e.g. aeration, if required). In fed-batch systems fresh medium or medium components are fed continuously, intermittently or are added as a single supplement and the volume of the batch increases with time. Continuous fermentations are open systems where fresh medium is continuously fed into the fermentation vessel, but the volume remains constant as spent medium and cells are removed at the same rate.

Batch growth

During batch fermentations the population of microorganisms goes through several distinct growth phases: lag, acceleration, exponential growth, deceleration, stationary and death (Fig. 2.1). In the **lag phase** virtually no growth occurs and the microbial population remains relatively constant. Nevertheless, it is a period of intense metabolic activity as the microbial inoculum adapts to the new environment. When cells are inoculated into fresh medium they may be deficient in essential enzymes, vitamins or cofactors, etc., that must be synthesized in order to utilize available nutrients, prior to cell division taking place. The chemical composition of the fermen-

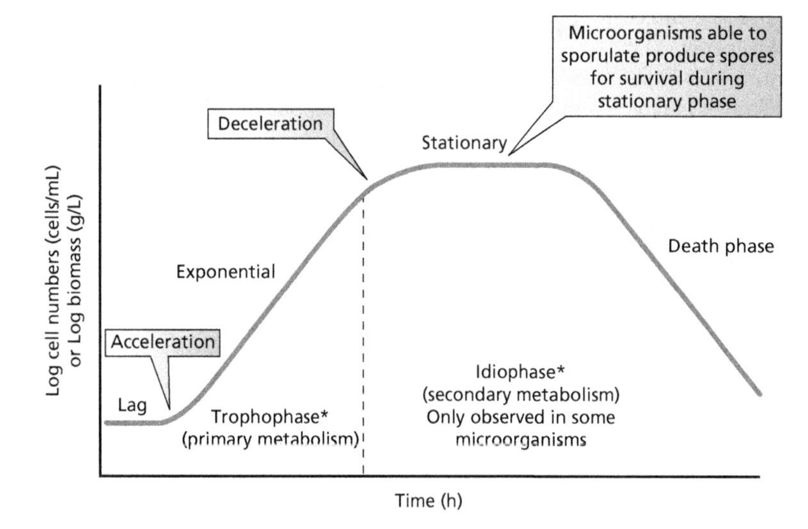

Fig. 2.1 Growth of a microorganism in a batch culture. *Trophophase and idiophase: see Chapter 3, Secondary metabolism.

tation media influences the length of the lag phase. It is usually longer if the inoculum was grown up using a carbon source different from that in the fresh medium, because the cells must synthesize enzymes required to catabolize the new substrate. Physiological stress may also have an effect, especially as cells are often transferred from an inoculum medium of low osmotic pressure (low solute concentration) to fresh medium of higher osmotic pressure (high solute concentration). Other factors influencing the length of the lag phase are the age, concentration, viability and morphology of the inoculum. Generally, inocula prepared from cells harvested in the exponential growth phase (period of most rapid growth) exhibit shorter lag phases than those harvested from subsequent stages.

Once the cells have adapted to their new environment they enter the **acceleration phase**. Cell division occurs with increasing frequency until the maximum growth rate (μ_{max}) for the specific conditions of the batch fermentation is reached. At this point **exponential growth** begins and cell numbers/biomass increase at a constant rate. Mathematically, this exponential growth can be described by two methods; one is related to biomass (x) and the other to cell numbers (N). For cell biomass, growth can be considered as an autocatalytic reaction. Therefore, the rate of growth is dependent on the biomass concentration, i.e. catalyst, that is present at any given time. This can be described as follows:

rate of change of biomass is $dx/dt = \mu x$ 2.1

where x = concentration of biomass (g/L), μ = **specific growth rate** (per hour) and t = time (h).

When a graph is plotted of cell biomass against time, the product is a curve with a constantly increasing slope (Fig. 2.2a). Equation 2.1 can also be rearranged to estimate the specific growth rate (μ):

$\mu = 1/x * dx/dt$ 2.2

During any period of true exponential growth, equation 2.1 can be integrated to provide the following equation:

$x_t = x_0 e^{\mu t}$ 2.3

where x_t = biomass concentration after time t, x_0 = biomass concentration at the start exponential growth, and e = base of the natural logarithm.

Taking natural logarithms, $\log_e$ (ln), gives

$\ln x_t = \ln x_0 + \mu t$ 2.4

This equation is of the form $y = c$ (intercept on y axis) $+ mx$ where m = gradient (μ in equation 2.4), which is the general equation for a straight-line graph. For cells in exponential phase, a plot of natural log of biomass concentration against time, a semilog plot, should yield a straight line with the slope (gradient) equal to μ (Fig. 2.2b), or

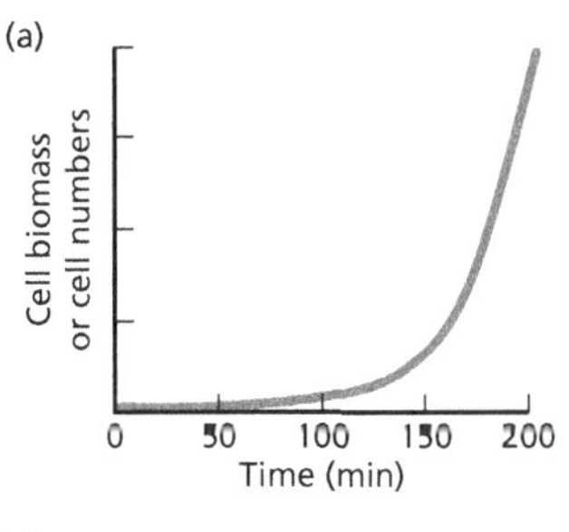

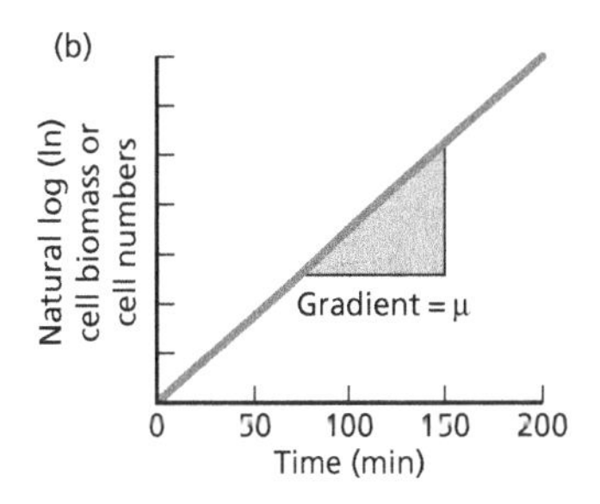

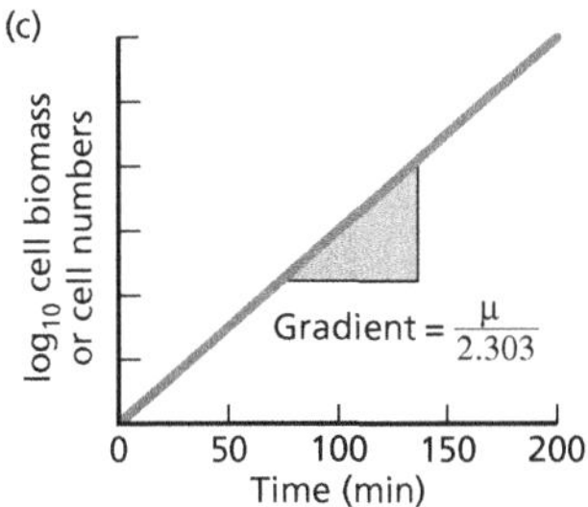

Fig. 2.2 Exponential growth of a unicellular microorganism with a doubling time of 20 min (a) arithmetic plot, (b) semilog plot using natural logarithms and (c) semilog plot using logarithms to the base 10 (μ = specific growth rate).

$$\mu = (\ln x_t - \ln x_o)/t \qquad 2.5$$

(note: when plotting $\log_{10}$ values instead of the natural log, the gradient of the semilog plot is equal to $\mu/2.303$; see Fig. 2.2c).

A second approach is to examine growth in relation to cell number, where the number of cells at the start of **exponential growth** is N_o. If we consider a hypothetical case where the microbial cell population at this point of exponential growth is one ($N_o = 1$), we can describe binary fission as follows:

No. of divisions	0	1	2	3	n
No. of cells	1	2	4	8	2^n
Mathematically	N_o	N_o2	N_o2*2	N_o2*2*2	–
	–	N_o2^1	N_o2^2	N_o2^3	N_o2^n

Consequently, after a period of exponential growth, time (t), the number of cells (N_t) is given by

$$N_t = N_o 2^n \qquad 2.6$$

where n = the number of divisions, N_o = initial cell number. Taking natural logarithms gives

$$\ln N_t = \ln N_o + n \ln 2 \qquad 2.7$$

Therefore, the number of divisions (n) that have taken place is given by

$$n = \frac{\ln N_t - \ln N_o}{\ln 2} \qquad 2.8$$

The number of divisions per unit time during this period of exponential growth is determined by dividing by the time period (t):

$$\frac{n}{t} = \frac{\ln N_t - \ln N_o}{t \ln 2} \qquad 2.9$$

where n/t = **division rate constant** (average number of generations per hour).

Often, we are not really interested in the number of divisions per hour, unit time, but rather in the **mean generation time** or **doubling time** (t_d), that is, the time required to undergo a single generation that doubles the population. Thus,

$$t_d = \frac{t}{n} = \frac{t \ln 2}{\ln N_t - \ln N_o} \qquad 2.10$$

During exponential growth, when all nutrients are supplied in excess and are therefore non-limiting, there is a direct relationship between cell numbers and biomass concentration, assuming that mean cell size is constant. This is balanced growth, and a direct relationship between specific growth rate and doubling time can also be established. However, under conditions where an essential nutrient becomes limiting, unbalanced growth arises, and variations in cell numbers (N) and biomass (x) concentration occur, as during the synthesis of cell storage compounds.

If we consider a situation where at time zero, the cell biomass is x_o, then after a fixed period of time (t) of exponential growth, equivalent to one doubling time (t_d), the microbial biomass will double to $2x_o$, i.e. $x_t = 2x_o$, when $t = t_d$. Substituting these parameters into equation 2.3 gives

$$2x_o = x_o e^{\mu t_d} \qquad 2.11$$

Taking natural logarithms produces

$$\ln 2x_o = \ln x_o + \mu t_d \qquad 2.12$$

or

$$\mu t_d = \ln 2 \qquad 2.13$$

Therefore, in this case

$$t_d = \frac{0.693}{\mu} \qquad 2.14$$

Equation 2.1, rate of change of biomass ($dx/dt = \mu x$), predicts that growth will occur indefinitely. However, during batch growth the microorganisms are continuously metabolizing the finite supply of nutrients available in the fermentation broth. After a certain time the growth rate decreases and eventually stops. This cessation of growth can be due to depletion of essential nutrients (carbon source, essential amino acids, etc.) or the build-up of toxic metabolites, such as ethanol and lactic acid, or a combination of nutrient depletion and toxin accumulation. Monod showed that growth rate is an approximate hyperbolic function of the concentration of the growth-limiting nutrient(s) (Fig. 2.3). This impact of essential nutrient depletion on growth can be described mathematically by the Monod equation, in a form similar to that used in biochemistry, where Michaelis–Menten kinetics define the rate of an enzyme-catalysed reaction in relation to its substrate concentration:

$$\mu = \frac{\mu_{max} S}{K_s + S} \qquad 2.15$$

where μ_{max} = maximum specific growth (per hour) of the cells, i.e. when substrate concentration is not limiting;

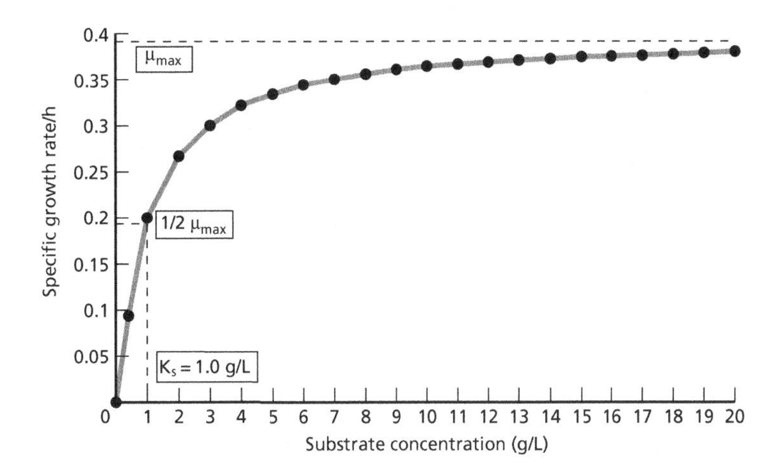

Fig. 2.3 The relationship between substrate concentration and specific growth rate.

S = concentration of limiting substrate (g/L); K_s = **saturation constant**, concentration (g/L) of limiting nutrient enabling growth at half the maximum specific growth rate, i.e. $\mu = {}^1/_2\ \mu_{max}$, and is a measure of the affinity of the cells for this nutrient.

When a microorganism is provided with the limiting substrate at a concentration much greater than the K_s, and with all other nutrients in excess, the microorganism will grow exponentially at its maximum rate, i.e. when $S \gg K_s$, then $\mu = \mu_{max}$. However, as the level of this substrate decreases, it eventually becomes limiting and can no longer sustain μ_{max}. This is the beginning of the **deceleration phase.** As the residual concentration of the limiting substrate approaches K_s and then falls below this concentration, there is an accompanying gradual decrease in growth rate (μ). The growth rate of a microorganism with a very high affinity for a rate-limiting substrate (i.e. a low K_s) will not be affected until the substrate concentration becomes very low. However, where there is a low affinity for the limiting substrate (i.e. a high K_s), the growth rate will begin to fall even at relatively high substrate concentrations and the organism exhibits a longer deceleration phase.

The specific growth rate of the microorganism continues decelerating until all of the available limiting substrate is metabolized. Growth is no longer sustainable and the cells enter the **stationary phase.** At this point, the overall growth rate has declined to zero and there is no net change in cell numbers/biomass (rate of cell division equals rate of cell death). However, the microorganisms are still metabolically active, involved in metabolizing intracellular storage compounds, utilizing nutrients released from lysed cells, and in some cases producing secondary metabolites. The duration of the stationary phase varies depending on the microorganisms involved and the environmental conditions. For cells unable to survive by forming spores, this is followed by an exponential **death phase** when the cells die at a constant rate and often undergo lysis.

APPLICATION OF BATCH FERMENTATIONS

During batch fermentations certain environmental conditions continually change, particularly nutrient and product concentrations, as does the specific growth rate, because the cells must pass through the sequence of growth phases described above. Consequently, the system never achieves steady-state conditions. A further disadvantage is that several distinct practical stages are associated with the operation of a batch fermentation:

1 charging of the fermenter with fresh medium;
2 sterilization of the fermenter and medium;
3 inoculation of the fermenter;
4 production of microbial products;
5 harvesting of biomass and spent fermentation broth; and finally
6 cleaning of the vessel.

This has major economic implications for industrial processes. For a considerable period of time, the fermentation vessel is not producing microbial products, but is being cleaned, filled, sterilized, etc. The non-productive period is referred to as the down-time of the fermenter.

OPTIMIZATION OF BATCH FERMENTATIONS

During the development of a batch process, key growth parameters can be determined that enable the production of a given microbial product to be optimized, whether it is the biomass itself or a specific metabolite. An important parameter is the **yield coefficient** (Y), which is determined on the basis of the quantity of rate-limiting nutrient, normally the carbohydrate source, converted into the microbial product. In respect of biomass, it is defined as

$$x = Y_{x/S}(S - S_r) \qquad 2.16$$

where x = biomass concentration (g/L), $Y_{x/S}$ = yield coefficient (g biomass/g substrate utilized), S = initial substrate concentration (g/L), and S_r = residual substrate concentration (g/L).

In the case of biomass production, the yield coefficient relates to the quantity of biomass produced per gram of substrate utilized. Therefore, the higher the yield coefficient, the greater the percentage of the original substrate converted into microbial biomass. For microbial metabolic products (p) the yield coefficient is related to the quantity of metabolite produced in relation to the quantity of substrate used ($Y_{p/S}$). Determination of yield coefficients is vitally important because the cost of the fermentation medium, particularly the carbon source, can be a significant proportion of the overall production cost (see Chapter 5).

By performing a range of experiments under different operating conditions, varying medium constituents and component concentrations, pH, temperature, etc., optimum growth/production conditions can be established. It is also important to determine the **maximum specific growth rate** (μ_{max}) of the production organism. This is particularly true for primary metabolites, where product formation is related to growth. To optimize the overall productivity of the system, the microorganism must usually be grown at its μ_{max}. As previously stated, the operating substrate concentration has a major effect on the growth rate of a microorganism. By performing a series of batch fermentations, each with a different initial concentration of the limiting substrate, the specific growth rate (μ) for each experiment can be determined. These data can then be used to estimate both μ_{max} and the **saturation constant** (K_s) by simply taking the reciprocal values in the Monod equation and rearranging equation 2.15 to give

$$\frac{1}{\mu} = \frac{K_s + S}{\mu_{max} S} \qquad 2.17$$

or

$$\frac{1}{\mu} = \frac{K_s}{\mu_{max} S} + \frac{S}{\mu_{max} S}$$

and

$$\frac{1}{\mu} = \frac{K_s}{\mu_{max}} * \frac{1}{S} + \frac{1}{\mu_{max}} \qquad 2.18$$

A plot of $1/\mu$ against $1/S$ should produce a straight line with the intercept on the y-axis at $1/\mu_{max}$ and a gradient equal to K_s/μ_{max}. Therefore, all the key kinetic parameters can be readily determined and when the values of K_s, μ_{max} and Y are known, a complete quantitative description can be given of the growth events occurring during a batch culture.

Continuous growth kinetics

Continuous culture fermentations have many applications for both laboratory research and industrial-scale processes. Studies can be performed on all aspects of cell growth, physiology and biochemistry. They are useful for ecological studies and as a genetics tool for the examination of mutation rates, mutagenic effects, etc. Application in industrial fermentations overcomes many limitations of batch processes.

Initially, continuous fermentations start as batch cultures, but exponential growth can then be extended indefinitely, in theory, through the continuous addition of fresh fermentation medium. The reactor is continuously stirred and a constant volume is maintained by incorporating an overflow weir or other levelling device (Fig. 2.4). Fresh medium is continuously added and displaces an equal volume of spent fermentation broth and cells at the same rate as fresh medium is introduced. Steady-state conditions prevail, where the rate of microbial cell growth equals the rate at which the cells are displaced from the vessel.

As with batch fermentations, the specific rate at which the microorganism grows in continuous culture is controlled by the availability of the rate-limiting nutrient. Therefore, the rate of addition of fresh medium controls the rate at which the microorganisms grow. However, the actual rate of growth depends not only on the volumetric flow rate of the medium into the reactor, but also on the **dilution rate** (D). This equals the number of reactor volumes passing through the reactor per unit time and is expressed in units of reciprocal time, per hour.

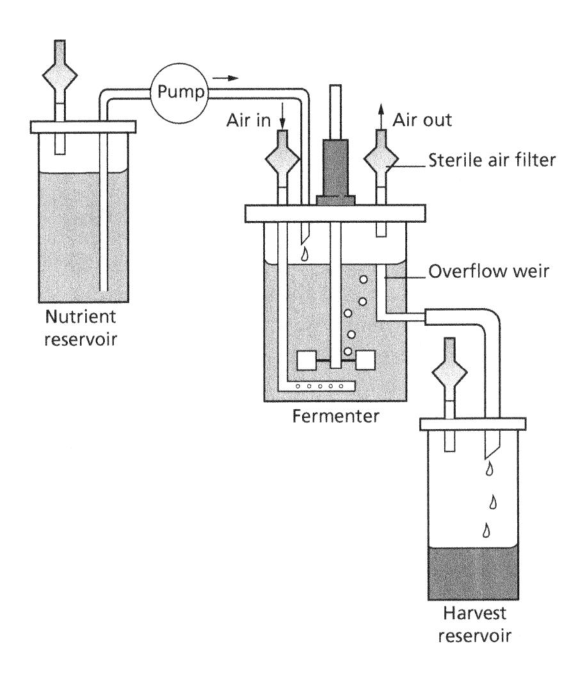

Fig. 2.4 Continuous culture apparatus.

$$D = \frac{F}{V} \qquad 2.19$$

where D = dilution rate (per hour), F = flow (L/h) and V = reactor volume (L).

The term D is the reciprocal of the mean residence time or hydraulic retention time, as used in waste-water treatment (see Chapter 15). Addition of fresh medium into the reactor can be controlled at a fixed value, therefore the rate of addition of the rate-limiting nutrient is constant. Within certain limits, the growth rate and the rate of loss of cells from the fermenter will be determined by the rate of medium input. Therefore, under steady-state conditions the net biomass balance can be described as

$$\frac{dx}{dt} = \text{rate of growth in reactor} - \text{rate of loss from reactor (wash-out)} \qquad 2.20$$

or

$$\frac{dx}{dt} = \mu x - Dx \qquad 2.21$$

Under steady-state conditions the rate of growth = rate of loss, hence $dx/dt = 0$ and therefore

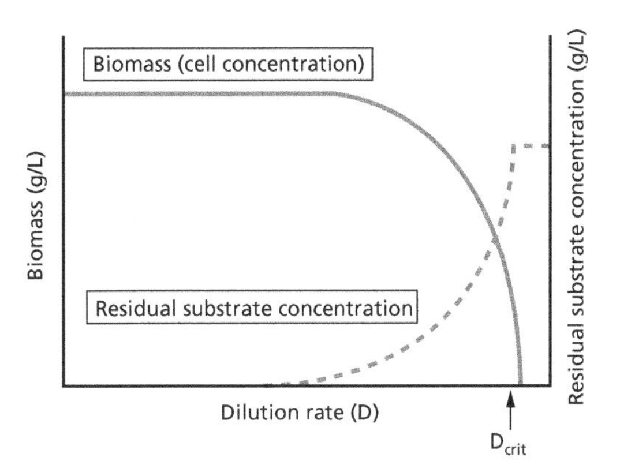

Fig. 2.5 Growth of a microorganism in continuous chemostat culture.

$$\mu x = Dx \qquad 2.22$$

and

$$\mu = D \qquad 2.23$$

Consequently, at fixed flow rates and dilution rates under constant physical and chemical operating conditions, i.e. under steady-state conditions, the specific growth rate of the microorganism is dependent on the operating dilution rate, up to a maximum value equal to μ_{max} (Fig. 2.5). If the dilution rate is increased above μ_{max}, complete wash-out of the cells occurs, as the cells have insufficient time to 'double' before being washed out of the reactor via the overflow. The point at which this is just avoided is referred to as the **critical dilution rate** (D_{crit}).

For any given dilution rate, under steady-state conditions, the residual substrate concentration in the reactor can be predicted by substituting D for μ in the Monod equation (equation 2.15):

$$D = \frac{\mu_{max} S_r}{K_s + S_r} \qquad 2.24$$

where S_r is the steady-state residual substrate concentration in the reactor at a fixed dilution rate. Rearrangement gives

$$D(K_s + S_r) = \mu_{max} S_r \quad \text{or} \quad DK_s + DS_r = \mu_{max} S_r$$

dividing by S_r then gives

$$\frac{DK_s}{S_r} + D = \mu_{max}$$

Hence,

$$S_r = \frac{DK_s}{\mu_{max} - D} \qquad 2.25$$

Consequently, the residual substrate concentration in the reactor is controlled by the dilution rate. Any alteration to this dilution rate results in a change in the growth rate of the cells that will be dependent on substrate availability at the new dilution rate. Thus, growth is controlled by the availability of a rate-limiting nutrient. This system, where the concentration of the rate-limiting nutrient entering the system is fixed, is often described as a **chemostat**, as opposed to operation as a **turbidostat**, where nutrients in the medium are not limiting. In this case turbidity or absorbance of the culture is monitored and maintained at a constant value by regulating the dilution rate, i.e. cell concentration is held constant.

At low dilution rates with fixed substrate concentrations, the residual substrate concentration will be low (Fig. 2.5). However, as D approaches μ_{max} the residual substrate concentration increases along with the growth rate of the microorganism. Beyond D_{crit}, input substrate concentration will equal output concentration, as all the cells have been lost from the system. Thus, this continuous reactor can be described as a self-regulating nutrient-limited chemostat.

The concentration of biomass or microbial metabolite in a continuous fermenter under steady-state conditions can be related to the yield coefficient, as described in the batch fermentation section. Inserting the equation for residual substrate (equation 2.25) into the biomass or metabolic product yield coefficient equation (equation 2.16) gives, in this case for steady-state biomass ($\bar{x}$),

$$\bar{x} = Y_{x/S}\left(S_R - \frac{DK_s}{\mu_{max} - D}\right) \qquad 2.26$$

where S_R is the substrate concentration of inflowing medium or

$$\bar{x} = Y_{x/S}(S_R - S_r) \qquad 2.27$$

Therefore, the biomass concentration under steady-state conditions is controlled by the substrate feed concentration and the operating dilution rate. Under non-inhibitory conditions, where there is no substrate or product inhibition, the higher the feed concentration, the greater the biomass concentration and residual substrate concentration remains constant. However, the higher the dilution rate, the faster the cells grow, which results in a simultaneous increase in the residual substrate concentration and a consequent reduction in the steady-state biomass concentration. As D approaches μ_{max}, the biomass concentration becomes even lower, yet the cells grow faster and there is a concurrent increase in the residual substrate concentration (Fig. 2.5).

Monitoring microbial growth in culture

During a fermentation, methods are required for the routine determination of the microbial population, cell number and/or biomass, in order to monitor its progress. Numerous direct and indirect methods are available for this purpose. Direct procedures involve dry weight determination, cell counting by microscopy and plate counting methods. Indirect methods include turbidimetry, spectrophotometry, estimation of cell components (protein, DNA, RNA or ATP), and on-line monitoring of carbon dioxide production or oxygen utilization. The method adopted in any given situation depends upon the fermentation and any specific requirements. Several factors must be considered, such as the degree of accuracy and sensitivity needed, and the duration of the analysis. Estimation of unicellular organisms, provided that they are not prone to flocculation, is relatively straightforward, but filamentous organisms, fungi and actinomycetes present additional problems. Also, culture media vary in viscosity, colour and the quantity of particulate solids, all of which may influence the choice of monitoring method. The speed of analysis may be critical, as an instant indication of biomass or cell concentration is often required. However, faster methods are frequently less reliable, and longer procedures are generally more accurate and reproducible.

Direct microscopic counting methods

Cell numbers in a suspension can be measured, except for filamentous organisms, by direct microscopic counts, using **Petroff–Hauser** or **Neubauer**-type counting chambers. The former is more suitable for counting bacteria. The counting grids, when the chamber is covered with a glass coverslip, hold a known volume of culture. By counting the number of cells within a proportion of the grid, the number of cells per millilitre can be determined. Direct microscopic counts are rapid, but limited by their inability to distinguish living from dead cells, unless differentiated by use of a vital staining technique. Also, samples must contain relatively high cell concentrations, normally a minimum of 5×10^6 cells/mL.

Electronic cell counters

Electronic cell counting equipment is also rapid, but most methods are more suitable for larger unicellular microbes, such as yeasts, protozoa and some algae, and less useful for estimating bacteria. The **Coulter Counter**, for example, is used to count and size particles, and is based on the measurement of changes in electrical resistance produced by non-conductive particles suspended in an electrolyte. This method involves drawing a suspension of cells through a small aperture across which an electrical current is maintained. As a cell passes through the aperture it displaces its own volume of electrolyte and changes the electrical resistance. These changes are detected and converted to a countable pulse. However, it essentially counts particles, and is consequently prone to errors due to cells clumping and the presence of particulate debris.

Plate counting techniques

Plate counting methods detect viable cells, i.e. those able to form colonies on an appropriate solid nutrient medium. The two methods routinely used are spread plating and pour plating. Prior to plating, it is usually necessary to prepare a serial dilution series in sterile diluent for concentrated cell suspensions. Alternatively, for samples with low cell numbers, as in water analysis, a concentration step is required.

In spread plating, samples, usually 0.1 mL, are spread on the surface of a suitable agar-based nutrient medium using a sterile spreading device, e.g. a bent glass rod. The plates are then inverted and incubated at the optimum growth temperature. All resultant colonies should be well separated and easily counted. This also enables the isolation of pure cultures where required. However, microorganisms with low tolerance to oxygen do not grow (unless incubated under appropriate conditions; see p. 35) and pour plating may be preferred. Here a suspension of microorganisms, normally a 1 mL sample, is placed in a Petri dish and thoroughly mixed with molten agar media at 48–50°C, and allowed to solidify prior to incubation. Pour plates result in the development of microbial colonies throughout the agar. Those organisms with lower oxygen tolerance grow within the agar. Colonies of aerobic organisms often have variable sizes, as those nearer the surface have a better oxygen supply. Consequently, it is often harder to see similarity in colonial morphology between colonies on the surface and those within the agar. Also, counting may be more difficult than for spread plates. Care has to be taken to ensure that the molten agar is not too hot, otherwise some microbial cells may be injured and slow to form visible colonies or even killed.

For statistical reliability, results are recorded only for plates containing 30–300 colonies. Calculation of the cell concentration in the original sample is then carried out, taking into account the dilution and volume plated. Both methods measure **colony-forming units** (CFU). This may or may not be the same as number of cells depending upon the extent of clumping and the cellular morphology of the microorganism. These techniques are accurate, but a minimum of 1–2 days incubation is usually necessary before the colonies are countable. However, more rapid microcolony counting techniques are also available.

Turbidimetric and spectrophotometric techniques

These methods provide a simple, rapid and convenient means of total biomass estimation. They are usually performed at a specific optimum wavelength for each microorganism. Turbidimetric methods measure the light scattered by a suspension of cells, which is proportional to the cell concentration. Alternatively, spectroscopy may be employed, using absorbance or transmittance of a cell suspension. Some modern fermentation monitoring systems now employ methods based on near-infrared spectroscopy.

Turbidimetric and spectrophotometric methods require the construction of appropriate calibration curves, prepared using standard cell suspensions containing known concentrations of cells. Also, care must be taken when interpreting the results if the fermentation broth contains particulate matter or is highly coloured.

Dry weight estimation

This method determines the weight of total cells, both living and dead, in liquid culture samples. It involves separating the biomass from a known volume of a homogeneous cell suspension. This is usually achieved by filtration under vacuum, through a preweighed membrane filter with a pore size of 0.2 μm or 0.45 μm. The filter with collected cells is 'washed' with water to remove any residual growth medium and dried to a constant weight in an oven at 105°C. Results are normally expressed as milligrams of cells per millilitre of culture. Obviously, any other suspended non-cellular materials above the size of the filter pores is also collected and can lead to errors. Further limitations are the time needed to obtain the results and the relatively large volume of sample required to obtain sufficient biomass for accurate weighing, as an individual bacterium weighs only about 10^{-12} g.

ATP bioluminometry

ATP is rapidly lost from dead cells; consequently, it is very useful for determining the concentration of viable microorganisms. The amount of ATP in a sample can be quantified using ATP bioluminometry. This technique utilizes an enzyme–substrate complex, luciferase–luciferin, obtained from the firefly, *Photinus pyralis*, which generates a photon of light for each molecule of ATP. When an aliquot of luciferase–luciferin is added to ATP extracted from a sample of cell suspension, the light generated can be detected in a bioluminometer. The resulting signal is amplified and then expressed as a digital or analogue data output.

$$\text{ATP} + \text{O}_2 + \text{luciferin} \xrightarrow{\textit{luciferase}} \text{oxyluciferin} + \text{AMP} + \text{CO}_2 + \text{PP}_i + \text{photon of light} \quad (562\,\text{nm})$$

This procedure is very rapid and sensitive. Under optimum conditions as little as 10 femtomoles (1 femtomole $= 10^{-15}$ mol) of ATP can be detected, which is approximately equivalent to 1 yeast cell or about 10 bacteria. ATP bioluminometry is most suitable for direct measurement of samples that are not coloured, as quenching of light may be a problem. However, as the method is very sensitive, samples may require considerable dilution, which often overcomes any colour quenching problem.

On-line estimation

On-line monitoring of fermenters can provide real-time estimation of biomass, and minimizes the requirement for repeated sampling and off-line analysis. Monitoring systems may involve optical density or capacitance-based probes. Also, the microorganisms involved in most fermentation processes will either have a requirement for oxygen and/or will produce carbon dioxide. In such cases it is theoretically possible to establish a mathematical relationship between factors, such as carbon dioxide evolution or oxygen utilization, and the biomass concentration within the bioreactor. Therefore, estimation of biomass concentrations and even product formation can be made by measuring these parameters on-line using carbon dioxide or oxygen detectors and biosensors attached to a computer. This can give an accurate estimate of the biomass concentration, provided that the mathematical algorithms developed are reliable.

Effects of environmental conditions on microbial growth

Growth and development of microorganisms are greatly affected by the chemical and physical conditions of their environment. Nevertheless, microorganisms have evolved to occupy niches throughout the range of environments on Earth, some of which are very hostile. As shown in later sections, microorganisms from some of these extreme environments have useful properties that can be exploited.

Effects of temperature on growth

As the temperature rises, the rate of chemical reactions increases. This rate doubles for every 10°C rise in temperature. Thus, cells should grow faster as the temperature is raised. However, there are maximum limits beyond which some temperature-sensitive macromolecules (proteins, nucleic acids and lipids) will become denatured, and hence, non-functional. There is also a minimum temperature for growth, below which the lipid membrane is not fluid enough to function properly.

All organisms have an optimum temperature for growth and different groups of microorganisms have evolved to grow over different temperature ranges. A typical microorganism, referred to as **stenothermal**, can grow over a temperature range of approximately 30°C, and **eurythermal** organisms grow over even wider ranges.

As for temperature optima, **mesophiles** have optimum growth temperatures in the range 20–45°C and a minimum around 15–20°C (Fig. 2.6). **Psychrophiles** usually have their optimum temperatures below 15°C and these organisms are often killed by exposure to room temperature. They are able to function at low temperature because their membranes contain a high proportion of unsaturated fatty acids. These molecules remain fluid at the lower temperatures, whereas membranes containing saturated fatty acids become non-functional. **Psychrotrophs** are sometimes referred to as facultative psychrophiles. They grow most

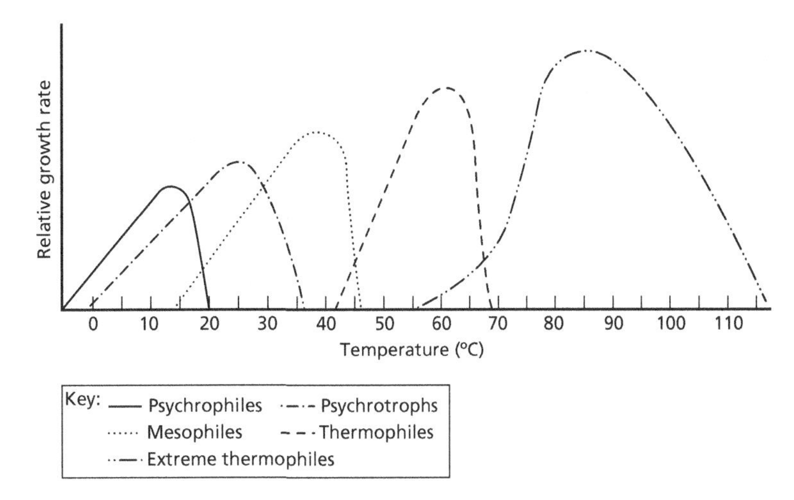

Fig. 2.6 Temperature ranges for microbial growth.

rapidly at temperatures above 20°C, but are capable of slow growth at temperatures down to around 0°C. Several are responsible for the spoilage of refrigerated products.

Organisms with optima above 50°C are termed **thermophiles.** Several algal, protozoal and fungal species have temperature maxima up to 55°C. However, it is only certain prokaryotes that are truly thermophilic, as no eukaryotic microbes grow at such high temperatures. Some extreme (hyper-) thermophiles can grow at 100°C or higher. For example, *Pyrolobus fumarii* has an optimum of 106°C and continues growing up to 113°C. All of these are archaeans (Table 2.3), which are mainly anaerobic sulphate reducers or have other metabolism with lower requirements for thermolabile cofactors such as NADH and NADPH. They have potential industrial uses in coal desulphurization, metal leaching, methane generation and the production of commercial enzymes, especially DNA polymerases and restriction endonucleases.

Exposure of non-thermophiles to high temperatures normally causes damage to cytoplasmic membranes, breakdown of ribosomes, irreversible enzyme denaturation and DNA strand breakage. It is difficult to determine which of these heat 'injuries' is the most deleterious to microorganisms, as all will reduce growth rate and contribute to inhibition and ultimately death. The actual mechanisms of thermotolerance are not fully elucidated. However, ribosomes of thermophiles work effectively at higher temperatures and their membranes also remain intact. Enzymes from thermophiles, such as

Table 2.3 Examples of thermophilic prokaryotes and their temperature optima

	Optimum temperature (°C)
Archaeans	
Methanobacterium thermoautotrophicum (methanogen)	65
Methanococcus jannaschii (methanogen)	85
Pyrococcus furiosis (accumulates H_2)	100
Pyrolobus fumarii (obligate H_2 chemolithotroph)	106
Sulfolobus acidocaldarius (acidophilic sulphur oxidizer)	75
Eubacteria	
Bacillus stearothermophilus	60–65
Clostridium thermocellum (cellulolytic anaerobe)	60
Synechococcus lividans (cyanobacterium)	67
Thermoanaerobacter ethanolicus (ethanol producer)	70

Bacillus stearothermophilus, are generally more heat stable, which may be the result of a few amino acid changes within a protein that can markedly increase thermostability. Maintenance of DNA integrity under these conditions is obviously crucial and some extreme thermophiles have DNA with a high guanine–cytosine ratio, giving increased interchain hydrogen bonds that provide greater thermostability. They also maintain higher intracellular potassium ion levels, often with the unusual counter-ion, 2,3-diphosphoglycerate, which helps stabilize proteins.

Microbial spores are particularly heat resistant. These dormant structures have thick spore coats and low water activity (see below). Bacterial endospores (Fig. 1.3) contain high levels of calcium dipicolinate, which may constitute up to 15% of their dry weight. Its role is thought to be in stabilizing nucleic acids and proteins. However, some mutants lacking this compound appear to retain heat resistance.

Effects of pH on growth

As with temperature, every microorganism has a pH optimum and a pH range over which it grows. Generally, fungi tend to grow at lower pHs (pH 4–6) than most bacteria. True acidophiles have pH optima between 1 and 5.5, and often have mechanisms for the exclusion of protons in order to maintain their internal pH at a higher level. However, the majority of microorganisms are neutrophiles, growing between pH 5 and 9, as most natural environments fall within this range. Alkalophiles, such as species of *Bacillus* and *Micrococcus,* have pH optima between 8.5 and 11.5, but take up protons to maintain their internal pH at a lower value.

Effects of solutes and water activity on growth

Most microorganisms contain approximately 70–80% water and require a certain amount of free water for the performance of specific metabolic activities. When microbial cells with rigid cell walls are placed in hypotonic solutions (lower osmotic pressure than cellular contents), they take up water to become turgid, whereas those without walls swell and ultimately burst. In dry and hypertonic environments water uptake and retention is problematical. Hypertonic environments may contain considerable amounts of water, but it is not necessarily available. In fact, the high external levels of solutes, such as salt or sugars, causes water to pass out of most cells and growth is halted.

Actual water availability to a microorganism is indicated by the term **water activity** (A_w). This is the ratio of the vapour pressure of water in the solution surrounding the microorganism, P_{soln}, to the vapour pressure of pure water:

$$A_w = \frac{P_{soln}}{P_{water}}$$

i.e. pure water has an A_w of 1.

Microbial species vary in their tolerance to dry conditions and environments of high osmotic strength. Most bacteria, with a few exceptions, cannot grow below an A_w of 0.9, which can provide a valuable means of food preservation. Many fungi that cause biodeterioration of stored grain (A_w = 0.7), including *Aspergillus restrictus*, are **xerotolerant** and can grow at low moisture levels. However, truly **xerophilic** filamentous fungi and **osmophilic** yeasts, e.g. *Zygosaccharomyces rouxii*, have evolved to inhabit environments where the A_w can be as low as 0.6. Some of these species may not grow under conditions of high water availability.

Marine bacteria are optimally adapted to the salt concentration in sea water (A_w = 0.98) and possess enzymes that have a specific requirement for sodium ions. True **halophiles**, e.g. *Halobacterium halobium*, found in areas such as the Dead Sea and Salt Lakes, will not grow in salt concentrations less than 3 mol/L. These are archaeans that have highly modified cell walls and unusual membrane lipids.

Osmophilic, xerophilic and halophilic organisms accumulate compensating solutes, which balance the osmotic strength of the external solute. Examples include polyols; arabitol is accumulated by several yeasts and filamentous fungi, and high levels of glycerol are found in the halophilic alga *Dunaliella salina*. Many halophiles also store large amounts of potassium ions and some bacteria amass amino acids, notably proline and glutamic acid. Although not specialists, *Escherichia coli* synthesizes osmoprotective betaine (*N,N,N*-trimethyl glycine) and *Penicillium chrysogenum* can accumulate glucose, fructose and potassium chloride.

Effects of oxygen on growth

Microorganisms are classified into five main groups on the basis of their requirements for oxygen.

1 **Obligate aerobes** grow only in the presence of oxygen, as it is required as the terminal electron acceptor for electron transport in aerobic respiration (see Chapter 3).

2 **Facultative anaerobes** function with or without oxygen, but grow more efficiently when oxygen is available.

3 Microaerophiles require some oxygen for the biosynthesis of certain compounds, but cannot grow at normal atmospheric oxygen concentrations of 21% (v/v). They must have lower oxygen levels of 2–10% (v/v).

4 Aerotolerant anaerobes essentially ignore oxygen and grow equally well in its presence or absence.

5 Obligate anaerobes cannot tolerate oxygen; exposure to it results in their death.

On exposure to oxygen, most organisms interact with it to produce highly reactive toxic products. These reduction products of atmospheric oxygen include superoxide ($O_2^{\bullet-}$), hydrogen peroxide (H_2O_2) and hydroxyl radicals ($OH^{\bullet}$). Unless detoxified, these products react destructively with any organic molecules that they encounter, including lipids, proteins and nucleic acids. Three groups of enzymes catalyse free radicals to help to render them harmless: **superoxide dismutase**, **catalase** and **peroxidases**:

$$\text{superoxide } (2O_2^{\bullet-}) + 2H^+ \xrightarrow{\text{superoxide dismutase}} O_2 + H_2O_2$$

$$\text{hydrogen peroxide } (2H_2O_2) \xrightarrow{\text{catalase}} O_2 + 2H_2O$$

$$\underset{}{H_2O_2} + \underset{\text{glutathione}}{2GSH} \xrightarrow{\text{glutathione peroxidase}} 2H_2O + \underset{\text{oxidized glutathione}}{GSSG}$$

Aerobes and all other organisms able to tolerate oxygen have superoxide dismutase, which dissipates any superoxide. Many aerobes, but not aerotolerant organisms, also possess catalase for eliminating hydrogen peroxide. The presence of ascorbic acid, vitamin E and β-carotenes may also be helpful in mopping up free radicals non-enzymatically. Strict obligate anaerobes lack superoxide dismutase and catalase, or produce inadequate quantities. Consequently, they have very limited ability to detoxify oxygen radicals and are unable to survive in the presence of oxygen.

Anaerobic habitats are rather more common than expected. They may arise from the utilization of oxygen by facultative anaerobes to create an anaerobic microenvironment where obligate anaerobes can grow. In general, wherever organic matter accumulates, microorganisms will use oxygen faster than it can be replaced, creating an anaerobic environment.

In order to grow obligate anaerobes in the laboratory, procedures to exclude oxygen from the culture must be used and all manipulations must be performed in an anaerobic atmosphere. Suitable procedures include:

1 the use of media containing reducing agents, such as thioglycollate, which chemically combine with oxygen;

2 physical removal of oxygen from a growth chamber, by pumping out the air with a vacuum pump, and then flushing the vessel with nitrogen gas ± carbon dioxide;

3 the use of GasPak jars, which involves sealing the microbial cultures inside the jar, along with a palladium catalyst and a hydrogen gas generating system; the oxygen is removed through its reaction with the hydrogen to form water; and

4 anaerobic cabinets for the growth and manipulation of the organisms, which are flushed with a gas mixture of nitrogen : carbon dioxide : hydrogen (80 : 10 : 10) and have a 'scrubber system' to remove residual oxygen.

Effects of radiation on growth

Visible and ultraviolet (UV) light are parts of the electromagnetic spectrum, which extends from strong radiation (γ-rays) to very weak radiation (heat and radio waves). Even visible light, particularly the more energetic violet and blue regions, can be damaging. It may generate singlet oxygen, a very powerful oxidizing agent, which can cause damage to cellular components and may kill some microorganisms. UV light, which is most harmful at 260 nm, causes specific damage to DNA, notably the formation of thymine dimers. Ionizing radiation, such as γ-rays emitted from the excited nucleus of ^{60}Co, generates free radicals, $OH^{\bullet}$ and $H^{\bullet}$, and hydrated electrons. Resulting damage includes breakage of hydrogen bonds in functional molecules, oxidation of many chemical groups and breakage of DNA strands.

Certain microorganisms, notably the bacterium *Deinococcus radiodurans*, exhibit considerable natural resistance to radiation, as do bacterial endospores. Many air-borne bacteria are afforded some protection by pigments that absorb radiation, and in other microorganisms resistance involves effective repair mechanisms, especially for damaged DNA. In both *Saccharomyces cerevisiae* and *E. coli*, for example, there are at least three systems for repairing DNA; mutants lacking these repair enzymes exhibit greater sensitivity to radiation. Some microorganisms, including *E. coli*, are also less susceptible to radiation damage in anaerobic environments than when respiring aerobically, and may be partially protected by reducing agents and sulphydryl compounds.

Effects of hydrostatic pressure on growth

In nature, many microorganisms are never subjected to pressures in excess of 1 atm (101.3 kPa or 1.013 bar) and higher pressures generally inhibit microbial growth. For example, in cider production, yeast performance is impaired where fermenter depths are greater than 14.5 m, which produce hydrostatic pressures above 1.5 atm. However, it is difficult to determine which process or function is most damaged and the order of inhibition. Interference with both protein synthesis and energy transfer from catabolic to biosynthetic systems is often observed, as is the inability of microorganisms to adapt to new substrates. However, many marine organisms must be **barotolerant**, as they live at depths in oceans where they may be subjected to pressures of up to several hundred atmospheres, and only fail to grow at above 200–600 atm. Some may even grow better under these conditions and are truly **barophilic**, capable of living under 700–1000 atm.

The control (inhibition) of microbial growth

The control or prevention of microbial growth is necessary in many practical situations, particularly in health care, food processing and preparation, and in the preservation of materials. Control may be achieved using physical or chemical agents that either kill microorganisms or inhibit their further growth. Agents which kill cells are called '**-cidal**' agents, whereas '**-static**' agents inhibit the growth of cells without killing them. Some agents affect groups of organisms; the term bactericidal, for example, refers to killing bacteria and bacteriostatic refers to inhibiting the growth of bacterial cells. **Sterilization** procedures completely destroy or eliminate all viable organisms, including spores, and may be performed using heat, radiation and chemicals, or by the physical removal of cells.

Microorganisms are not killed instantly on exposure to a lethal agent. The population decreases exponentially, by a constant fraction at constant intervals, and several factors influence the effectiveness of any antimicrobial treatment. These include:

1 population size: the larger the microbial population, the longer the time required to kill all the microorganisms present;
2 population composition: different microorganisms vary in their sensitivity to a specific lethal agent, and its effectiveness may be influenced by their age, morphology and physiological condition;
3 concentration of the antimicrobial agent or intensity of the treatment: higher concentrations or greater intensities are generally more efficient, but the relationship is rarely linear;
4 period of exposure to the lethal agent: the longer the exposure, the greater the number of organisms killed;
5 temperature: normally a higher temperature increases the effectiveness of the agent; and
6 environmental conditions, such as pH, medium viscosity and the concentration of organic matter, can have major influences on the effectiveness of most antimicrobial agents.

Control of microbial growth by physical agents

HEAT

Whenever heat is used to control microbial growth both temperature and exposure time must be considered. The temperature required to kill a specific microorganism (the lethal temperature) varies quite widely. Also, the time needed to kill depends on several factors (see above). Heat is the most important and widely used means of sterilization (see terms relating to heat sterilization in Table 2.4), and may be achieved through:
1 **incineration**, where burning at 500°C physically destroys the organisms and is particularly useful for some solid wastes;
2 **moist heat**, which is suitable for sterilizing most items, except heat-labile substances that would be denatured or destroyed. It is carried out using steam under pressure to achieve 121°C for 15 min, and is extensively used in fermentation processes for the sterilization of vessels, connecting pipe work and culture media (see Chapter 6); or
3 **dry heat** is less efficient than moist heat, requiring higher temperatures and much longer exposure times. It is performed in hot air ovens at 160°C for 2 h, and can be used for glassware, metal objects and moisture-sensitive materials such as powders and oils.

Boiling of aqueous solutions or immersion of

Table 2.4 Terms relating to heat sterilization used in the food and fermentation industries

Thermal death time (TDT) is the shortest time required to kill all microorganisms in a sample at a specific temperature and under defined conditions

Decimal reduction time (*D*-value) is the time required to kill 90% of the microorganisms in a sample at a specific temperature. This term is used extensively in the food industry

Z-value is the rise in temperature required to reduce *D* to $^1/_{10}$ of its previous value

***F*-value** is the time in minutes at a specific temperature (usually 250°F or 121.1°C) necessary to kill a population of cells or spores

Del factor (∇) = ln (N_0/N_t), where N_0 is the number of organisms at the start of sterilization and N_t is the number remaining after time *t*. Therefore, the Del factor is a measure of the fractional reduction in viable organism count produced by a certain heat and time regime. The larger the Del factor, the greater the sterilization regime required. This term is mostly used in the fermentation industry

solid objects in boiling water at 100°C for 30 min does not guarantee sterility. It kills most vegetative cells, but not all endospores. To kill endospores, intermittent boiling or tyndallization is required. This involves exposure of the material to elevated temperatures to kill the vegetative cells, followed by incubation at 37°C to allow any spores to germinate and form new vegetative cells. A second exposure to elevated temperatures kills the newly germinated vegetative cells.

Heat sterilization is commonly employed in canning and bottling, and ultra-high temperature treatments (UHT) are used in some sterile packaging procedures. Pasteurization is a milder heat treatment, used to reduce the number of microorganisms in products or foods that are heat-sensitive and unable to withstand prolonged exposure to high temperatures. Batch pasteurization at 63°C for 30 min can be used for filled bottles; however, flash pasteurization at 72°C for 15 s may be preferred for certain foods and beverages. For example, it is performed on milk and beer as it has fewer undesirable effects on quality or flavour. In the case of pasteurization of milk, the time and temperature used are targeted at killing milk-transmitted pathogens, notably staphylococci, streptococci, *Brucella abortus*, *Mycobacterium bovis* and *Mycobacterium tuberculosis*.

LOW TEMPERATURE

These treatments involve refrigeration or freezing. Organisms are not usually killed, but the majority do not grow or grow very slowly at temperatures below 5°C. Perishable foods are stored at low temperatures to slow the rate of microbial growth and consequent spoilage. Often, it is psychrotrophs, rather than true psychrophiles, that are the cause of food spoilage in refrigerated foods.

LOW WATER ACTIVITY

This is used extensively to preserve foods, especially fruits, grains and some meat products. Methods involve removal of water from the product by heating or freeze-drying; alternatively, water activity may be reduced by the addition of solutes, usually salt or sugar.

IRRADIATION

Microwave, UV, X and γ radiation can be used to destroy microorganisms. UV radiation is effective, but its use is limited to surface sterilization because it does not penetrate glass, dirt films, water and other substances. Ionizing radiation treatment mostly involves X-rays or γ-rays. This is particularly effective due to its ability to penetrate materials.

Irradiation is used commercially to sterilize items such as Petri dishes, and in some countries, spices, fruits and vegetables are irradiated to increase their storage life up to 500% by destroying spoilage microorganisms. The process is relatively expensive, and due to the nature of food materials and doses given, not all organisms are killed. Irradiation may also destroy some vitamins. In some foods the levels of vitamin C and E may be reduced by 5–10% and up to 25%, respectively. Such treatments are not suitable for all products; for example, when beer and some shampoos are irradiated, hydrogen sulphide may be generated.

Irradiation of food has not been accepted worldwide. Currently, about 40 countries have passed legislation allowing its use for specified materials. However, it is now gaining greater acceptance, particularly with the increased incidence of food-borne diseases such as *E. coli* 0157:H7. In the USA, use of these techniques has been extended to the treatment of beef, lamb and pork products following approval by the US Food and Drug Administration (FDA).

HIGH-INTENSITY PULSED ELECTRIC FIELD TREATMENT

This is another non-thermal treatment and involves exposing the material to an electric field of 15–20 kV/cm for just a few milliseconds or less. The underlying mechanism of cell inactivation has not been fully elucidated, but is thought to involve the breakdown of membranes through electroporation (i.e. the formation of pores within the membranes). It is considered that high-intensity pulsed electric field treatments could, in the future, have several applications, including food preservation.

Sterile filtration

Sterile filtration involves the physical removal of all cells from a liquid or gas. It is especially useful for sterilizing heat labile solutions of antibiotics and other drugs, amino acids, sugars, vitamins, animal cell culture media and some beverages. However, the resultant products may not be virus-free and ultrafiltration is necessary to remove viruses. There are three main types of sterile filter:

1 **depth filters** are thick fibrous or granular filters that remove microorganisms by physical screening, entrapment and adsorption;
2 **membrane filters** are thin filters with defined pore sizes, usually of 0.2 or 0.45 μm, through which the microorganisms cannot pass; and
3 **high-efficiency particulate air (HEPA) filters** are used in laminar flow biological safety cabinets and containment rooms to sterilize the air circulating in the enclosure.

Control of microbial growth by chemical agents

Chemical antimicrobial agents can be used to kill microorganisms or inhibit their growth and may be of natural or synthetic origin. They include chemical preservatives, disinfectants, antiseptics, and drugs used in antimicrobial chemotherapy. Some are also used as liquid and gaseous sterilants. These are highly toxic and kill all forms of life, e.g. formaldehyde, glutaraldehyde, ethylene oxide and diethyl pyrocarbonate.

ANTISEPTICS AND DISINFECTANTS

Antisepsis is the prevention of infection of living tissue. It is achieved using antiseptic microbiocidal agents that do not harm skin and mucous membranes, but should not be taken internally. Examples include iodine solution, alcohols and quaternary ammonium compounds (Table 2.5). Disinfectants are agents that kill microorganisms, but not necessarily their spores. They are not safe enough for direct application to living tissues and are used on inanimate objects such as tables, floors, utensils, etc. Chlorine, hypochlorites, chlorine compounds and phenols are extensively used as disinfectants. Disinfectants and antiseptics are distinguished on the basis of whether they are safe for application to mucous membranes. Often, safety depends on the concentration of the compound. Some phenolic compounds, for example, are used as disinfectants at high concentration, but at low concentrations they are suitable for use as antiseptics (Fig. 2.7).

ANTIMICROBIAL PRESERVATIVES FOR FOOD AND RELATED PRODUCTS

These compounds are mostly static agents used to inhibit the growth of microorganisms in food and those pharmaceutical or cosmetic products that may be ingested (Table 2.6 and Fig. 2.7). There are relatively few compounds approved for these purposes. Most countries use three categories of classification for 'added food preservatives', based on that issued by the FDA. Group 1 are natural organic acids (lactic, citric, etc.); whereas group 2 has substances classified as generally regarded as safe (GRAS), which includes other organic acids (benzoic, propionic, sorbic, etc.), the parabens (esters of para-hydroxybenzoic acid) and SO_2. Other compounds proved safe for man or animals that cannot be placed within group 1 or 2 are assigned to group 3. Common salt, sugars, vinegars, spices and oils have some preservative action, but are not classified as 'added preservatives'. Obviously, any preservative used in food, apart from being safe, should ideally be free from flavour and aroma. The minimum concentration should be added and preservatives must not be used to disguise poor manufacturing practice.

In the case of the organic acids and the parabens, it is their undissociated form that exhibits the antimicrobial properties. Their pK_a values (the pH at which there is 50% dissociation of the molecules) lie in the range pH 3–5. As the pH falls the concentration of the undissociated form increases, thus elevating the antimicrobial activity. However, above pH 5.5–5.8 most are ineffective, apart from sorbic acid and the parabens. They retain antimicrobial activity to pH 6.5 and pH 7.0–8.0,

Table 2.5 Common chemical antimicrobial agents used as antiseptics, disinfectants, non-food preservative and sterilants

Chemical	Action	Uses
Alcohols Ethanol (50–70%, v/v) Isopropanol (50–70%, v/v)	Denature proteins and solubilize lipids	Widely used disinfectants and antiseptics; will not kill endospores Antiseptic used on skin
Formaldehyde (8%) and **Glutaraldehyde** (5%) (cold sterilization)	React with NH_2, SH and COOH groups. Highly reactive	Disinfectants, kill endospores at high conc., may irritate the skin; possible carcinogens
Formaldehyde donors (e.g. Germall 115-imidazolidinyl urea effective at 0.3%)	See formaldehyde	Advantages over free formaldehyde, used in some cosmetics
Tincture of iodine (2% in 70%, v/v alcohol)	Inactivates proteins	Antiseptic used on skin
Chlorine (Cl_2) gas	Forms hypochlorous acid (HClO), a strong oxidizing agent $Cl_2 + H_2O \rightarrow HCl + HClO$ $HClO \rightarrow HCl + O$ (nascent oxygen, powerful oxidizing agent)	Disinfection of drinking water; general disinfectant
Silver nitrate ($AgNO_3$)	Precipitates proteins	General antiseptic and used in pharmaceutical products for the eyes
Mercuric chloride	Inactivates proteins by reacting with sulphydryl groups	Disinfectant, although occasionally used as an antiseptic on skin
Quaternary ammonium compounds e.g. cetrimide and benzalkonium chloride	Disrupts cell membranes Cationic surfactants Activity reduced by organic matter, more effective against Gram (+) bacteria than Gram (–) bacteria, yeasts or moulds. Rapid effect, and activity enhanced by EDTA, benzyl alcohol and certain other compounds	Skin antiseptics and disinfectants for food utensils, low toxicity Used in ophthalmic preparations at 0.001% (w/v)
Phenolic compounds	Denature proteins and disrupt cell membranes.	Antiseptics at low concentrations; disinfectant at higher concentrations
Phenol	More effective against Gram (+) bacteria	0.5% (w/v) in cosmetics and toiletries; note: 0.5% (w/v) solution used as a standard against which other antimicrobial agents are measured
Ortho-cresol	Similar to phenol	0.3% (w/v) used in injectables, preservative of creams and ointments
Chlorocresol	More effective than phenol and ortho-cresol	0.1% (w/v) used in eye drops and injectables, skin creams and ointments
Chloroxylenol	Activity against Gram (–) bacteria enhanced by adding EDTA	Used in many cosmetic preparations
Sterilizing gases e.g. ethylene oxide, diethyl pyrocarbonate	Alkylating agent notably of sulphydryl groups	Disinfectant; used to sterilize heat-sensitive objects such as rubber and plastics. Kills spores and viruses
Bronopol (2-bromo-2-nitropropan-1,3-diol)	Broad spectrum against Gram (+) and (–) bacteria, fungi, wide pH range of 3–8, affects membranes and dehydrogenase enzymes	Preservative for fuels, paints, textiles and and particularly cosmetics and pharmaceuticals, used at 0.01–0.02% (w/v)

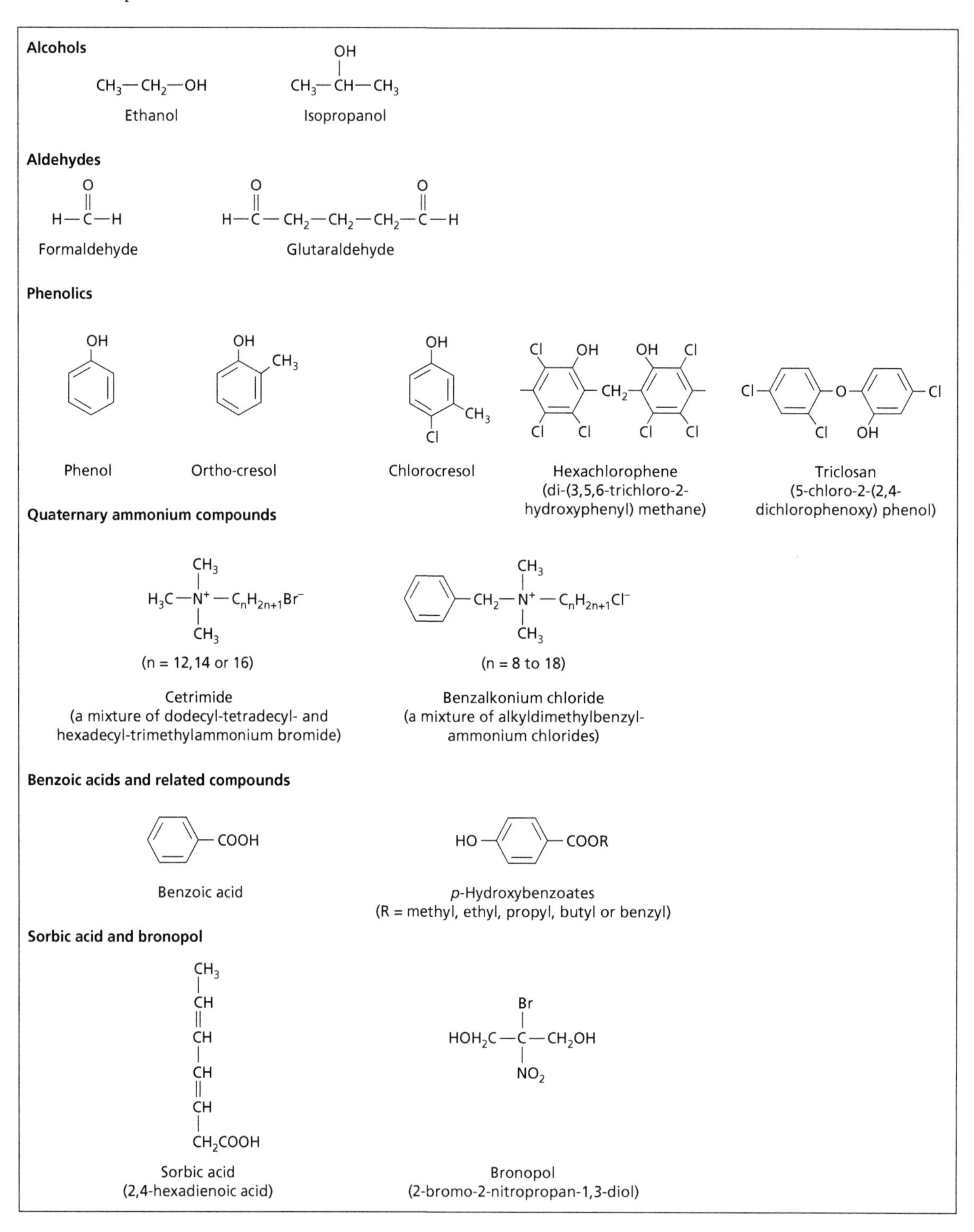

Fig. 2.7 Examples of chemical antiseptics, disinfectants and food preservatives.

Table 2.6 Common food preservatives and their uses

Preservative	Effective concentration	Uses
Acetic acid	0.4% (w/v)	Used in baked goods
Benzoic acid (E210), benzoates and parabens*	0.1% (w/v) 0.01–0.05% (w/v)*	Antimicrobial agents in margarine, cider, soft drinks, cosmetics, etc.
Lactic acid	0.3% (w/v)	Antimicrobial agent in cheeses, buttermilk, yoghurt and pickled foods
Nisin (E234)	50–500 IU/g effective against Gram (+) bacteria; particularly useful against spore formers	Used in processed cheese, other dairy produce and canned products at acid pH
Propionic acid (E280) and propionates	0.3% (w/v)	Antifungal agent in breads, cake, cheeses, dried fruits
Sodium nitrite (E250)	0.02% (w/v) more effective at pH 5.0–5.5	Antibacterial agent in cured meats and fish. Retards growth of *Clostridium botulinum*. Problems due to formation of nitrosamines on cooking
Sorbic acid (E200) and sorbates	0.2% (w/v)	Antifungal agent in cheeses, syrups, cakes, fruit juices, etc.
Sulphur dioxide (E220), sulphites	0.02–0.03% (w/v) active form is undissociated sulphurous acid	Antimicrobial agent in dried fruits, grapes, molasses and beverages. Inhibits fungi and bacteria. Used extensively in the past, now reductions in use due to induction of asthmatic attacks in sensitive individuals

respectively. Consequently, the parabens have widespread use in food, pharmaceuticals and cosmetics, particularly as they are tasteless, colourless and stable. Parabens have a high LD_{50} of 8 g/kg and have European Community and FDA approval (note: LD_{50} is the lethal dose of a substance, in grams per kilogram of body weight of test animals, such as mice, at which 50% of a population are killed). The only apparent problems with the parabens are low-grade sensitization exhibited by some individuals, their tendency to undergo partition into oil, particularly vegetable oils, and their reduced efficacy in the presence of glycerol and ethylene glycol. Parabens inhibit many bacteria, yeasts and fungi, but are much less effective against Gram-negative bacteria. Their action is thought to be on microbial membranes, but they may also affect nucleic acid metabolism. Effective in-use concentrations of parabens are normally 0.01–0.05% (w/v). Mixtures are often used because they appear to have synergistic action, e.g. two parts methyl paraben plus one part propyl paraben.

Very few microbial products, apart from organic acids, are used as food preservatives. Normally, antibiotic agents are not acceptable in human food due to the potential development of resistant strains, and possible disturbance of the microbial ecology of the gut or allergic reactions. Only nisin, and to a lesser extent natamycin, currently have food applications (see Chapter 13).

Other possible biological preservatives include the iron-binding lactoferrin; avidin, a chelator of biotin; ovoinhibitor, a protease inhibitor; and lactoperoxidase, an oxidizer of SH groups. Lysozyme, which causes bacterial cell lysis, can be used as a preservative, but has a limited bacteriolytic spectrum. Similar bacteriolytic enzymes, such as those from the cellular slime mould, *Polysphondylium* species, may be ultimately more useful.

Antimicrobial chemotherapy

Antimicrobial chemotherapy involves the control or the destruction of disease-causing microorganisms once within the tissues of humans and other animals. Probably the most important property of a clinically useful antimicrobial agent is its **selective toxicity**, i.e. the agent acts to inhibit or kill the pathogenic microorganism, but has little or no toxic effect on the host. Selective toxicity is more readily achieved in treating bacterial disease, because of the structural and functional differences between prokaryotic bacterial cells and the host, eukaryotic animal cells. Major differences are the absence of cell walls in animal cells; the size of their ribosomes, except mitochondrial ribosomes; and specific details of metabolism.

Chemical antimicrobial agents are of synthetic origin, whereas **antibiotics** are often defined as 'low molecular weight organic compounds, produced by microorganisms, that kill or inhibit other microorganisms'. However, a broader definition of an antibiotic includes natural chemical products from any type of cell that kill or inhibit the growth of other cells. Some antibiotics can now be completely chemically synthesized or natural antibiotics may be chemically modified to improve their properties (see Chapter 11).

SYNTHETIC CHEMICALS USED IN ANTIMICROBIAL CHEMOTHERAPY

Many synthetic chemotherapeutic agents are competitive inhibitors, often referred to as **antimetabolites**, and may be bacteriostatic or bactericidal. Most are 'growth factor analogues', which are structurally similar to a microbial growth factor but cannot fulfil its metabolic function. Some are analogues of purine and pyrimidine bases that become incorporated into nucleic acids, but are unable to form functional base pairs. Consequently, replication and transcription processes are disrupted.

The **sulphonamides** were the first compounds found to suppress bacterial infections selectively. They inhibit the synthesis of tetrahydrofolic acid (THF), which is essential for one-carbon transfer reactions. Folic acid also functions as a coenzyme for the synthesis of purine and pyrimidine bases of nucleic acids. Sulphonamides are structurally similar to para-aminobenzoic acid (PABA) or its derivatives, which serve as substrates for enzymes in the THF pathway. They competitively inhibit folic acid synthesis, which halts microbial growth. Selectivity is provided as animal cells do not synthesize their own folic acid: they must obtain the preformed vitamin from their diet.

Quinolones, such as nalidixic acid, are synthetic chemotherapeutic agents that are bacteriocidal, killing mainly Gram-negative bacteria. They bind to, and inhibit, the DNA gyrase (a topoisomerase). This enzyme is essential during DNA replication, as it enables DNA supercoils to be relaxed and re-formed.

ANTIBIOTICS

Currently, most clinically useful antibiotics are produced by microorganisms and are used to treat bacterial infections. Antibiotics may have cidal or static effects on a range of microorganisms. They are classified according to their range of effectiveness and the general microbial group they act against, i.e. antibacterial, antifungal or antiprotozoal. The range of bacteria or other microorganisms affected by a specific antibiotic is expressed as its **spectrum of action**. Antibiotics effective against a wide range of Gram-positive and Gram-negative bacteria are said to be **broad spectrum**. If they are predominantly effective against Gram-positive or Gram-negative bacteria, they are **narrow spectrum**; and **limited spectrum** antibiotics are effective against only a single organism or disease.

Antibiotics are low molecular weight compounds that are produced as secondary metabolites by mainly soil microorganisms. They are non-protein molecules, although there are some peptide antibiotics. Most producer microorganisms form a spore or other dormant cell. Also, there is thought to be a relationship, besides temporal, between antibiotic production and the processes of sporulation (see Chapter 3, Secondary metabolism). Among the filamentous fungi, the notable antibiotic producers are *Penicillium* and *Cephalosporium* (*Acremonium*), which are the main source of β-lactam antibiotics, e.g. penicillin and related compounds. Within the bacteria, the filamentous actinomycetes, especially many *Streptomyces* species, produce a variety of antibiotics, including aminoglycosides, macrolides and tetracyclines. Endospore-forming *Bacillus* species and relatives produce polypeptide antibiotics, such as polymyxin and bacitracin.

Antibiotics exhibit four primary modes of action.

Inhibition of cell wall synthesis

Inhibitors of bacterial cell wall/envelope synthesis generally target some step in the formation of peptidoglycan. The β-lactam antibiotics, penicillins and cephalosporins, are the best known examples. Penicillin was the first antibiotic to be discovered and used therapeutically. Natural penicillins, such as penicillin G (Fig. 2.8) or penicillin V, are commercially produced by fermentation of *Penicillium chrysogenum* (see Chapter 11). Most penicillins are derivatives of 6-aminopenicillanic acid and like all β-lactam antibiotics contain a β-lactam ring, which is essential for activity. The side chains attached may provide other characteristics, including resistance to enzymic degradation and the mechanism of cell wall penetration. Penicillin G and V

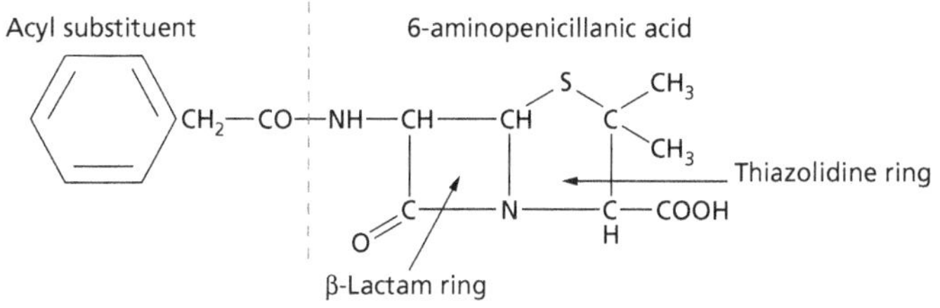

Fig. 2.8 The structure of Penicillin G.

are considered to be narrow spectrum, as they are not effective against Gram-negative rods. Also, penicillin G is unstable in acid conditions, therefore oral administration is ineffective due to the low pH within the stomach. In an effort to overcome these shortcomings, semisynthetic forms were developed with improved properties over natural penicillins. The main portion of the molecule, 6-aminopenicillanic acid, is produced by fermentation and is then chemically modified by the addition of various side chains (see Chapter 11). Improved properties exhibited by these semisynthetic penicillins include resistance to penicillinase, suitability for oral administration and increased spectrum of activity, e.g. effectiveness against Gram-negative rods.

Penicillins inhibit peptidoglycan synthesis and have no effect on established cell walls. Consequently, they are bactericidal to actively growing cells, which through the action of penicillin become sensitive to osmotic stress. Cell wall peptidoglycan is made in three stages (see Chapter 3). The third step involves carboxypeptidases and transpeptidases in the final cross-linking between peptide side chains, to form a rigid matrix. Penicillins have a structural resemblance to the end D-alanyl-D-alanine residues of the small peptides involved in the peptidoglycan cross-links. Thus, penicillins appear to act as substrate analogues. They covalently bind to transpeptidase via the β-lactam ring and prevent further functioning of the enzyme. As a result, a cell wall containing this loosely formed peptidoglycan is much weaker and the cell undergoes lysis. Also, penicillins may induce certain autolytic events.

Cephalosporins are produced by species of *Cephalosporium* (*Acremonium*) and are β-lactam antibiotics with essentially the same mode of action as penicillins. They have a low toxicity and a rather broader spectrum than natural penicillins, and are particularly useful for patients allergic to penicillins.

Vancomycin, a product of *Streptomyces orientalis*, has become a clinically important antibiotic since the emergence of methicillin-resistant strains of *Staphylococcus aureus* (MRSAs). It too inhibits peptidoglycan synthesis, but does so by binding to the D-alanyl-D-alanine group on the side chain of the *N*-acetyl glucosamine-*N*-acetyl muramic acid-pentapeptide subunits, inhibiting their incorporation into new peptidoglycan (see Chapter 3). Bacitracin, a peptide antibiotic produced by *Bacillus* species, also prevents peptidoglycan synthesis. It acts by inhibiting the release of peptidoglycan subunits from the lipid-carrier molecule that transports them across the cytoplasmic membrane. Teichoic acid synthesis, which requires the same carrier, is also inhibited. Bacitracin has a high toxicity and is primarily used for topical (external) treatments, rather than for systemic applications.

Cell membrane inhibitors

In bacteria, the integrity of the cytoplasmic membrane is vital for both nutrient transport and energy generation. Outer membranes of Gram-negative bacteria also have a protective role. Any agents that disrupt these membranes kill the microbial cells. However, eubacterial and eukaryotic membranes have similar structures and constituents. Therefore, antimicrobial compounds with this mode of action are rarely specific enough to permit their systemic use. **Polymyxin** is the only membrane-targeting antibacterial antibiotic of clinical importance. It acts by binding to membrane phospholipids to interfere with membrane function. This antibiotic is mainly effective against Gram-negative bacteria, but is normally restricted to topical use due to damaging side-effects.

Fungal infections are generally much more difficult to treat than bacterial diseases. The similarity between fungi and animal host (both are eukaryotic) limits the ability of a drug to have a selective point of attack. However, one potential target is ergosterol, a component of fungal cell membranes, which is replaced by cholesterol in higher eukaryotes. Polyene macrolide antibiotics, such as nystatin, and the synthetic azoles, utilize this difference. Azoles inhibit ergosterol biosynthesis, whereas nystatin binds to ergosterols to destabilize the fungal membrane.

Protein synthesis inhibitors

Protein synthesis is a complex process involving DNA transcription to mRNA, which is then translated to form polypeptides (linear assembly of amino acids) through interaction with the ribosomes and transfer (t) RNAs (see Chapter 3). This provides a number of specific targets for different groups of antibiotics. Several antibiotics target the translation step where their attack is invariably at the ribosome, rather than the stage of amino acid activation or their attachment to tRNAs. As for selective toxicity, it is fortunate that there is a major difference between prokaryotic and eukaryotic ribosomal structure. Apart from some organellar ribosomes, eukaryotic ribosomes are 80 S, consisting of a 60 S and a 40 S subunit (note: Svedberg units are not additive); whereas prokaryotic 70 S ribosomes are composed of a 50 S and a 30 S subunit. This difference accounts for the selective toxicity of several antibiotics that target this point in protein synthesis. The majority have an affinity or specificity for 70 S, as opposed to 80 S, ribosomes. However, as mitochondria of eukaryotic cells contain 70 S ribosomes, very similar to those of bacteria, antibiotics targeting 70 S ribosomes may have adverse effects on host cells.

Important antibiotics with this mode of action include aminoglycosides, macrolides and tetracyclines. The aminoglycosides, notably streptomycin, kanamycin, tobramycin and gentamicin, are products of *Streptomyces* species. They bind to bacterial ribosomes and prevent the initiation of protein synthesis. Macrolides, such as erythromycin and oleandomycin, contain large lactone rings linked with amino acids through glycoside bonds. They bind to the bacterial 50 S ribosomal subunit, which interferes with 'elongation' of the protein by peptidyl transferase or prevents translocation of the ribosome.

The tetracyclines are also products of *Streptomyces*, although some semisynthetic derivatives are now produced. They include tetracycline, chlortetracycline, oxytetracycline and doxycycline, which are broad-spectrum antibiotics, active against both Gram-positive and Gram-negative bacteria. These compounds block the binding of aminoacyl-tRNA to the A site on the ribosome. Tetracyclines inhibit protein synthesis by both isolated 70 S and 80 S ribosomes. However, despite this apparent general action, they exhibit very low toxicity in animals. This is because most bacteria possess an active transport system for tetracycline that results in intracellular accumulation of the antibiotic to concentrations 50-fold higher than in the external medium. As a consequence, a blood level of tetracycline that is harmless to animal tissues, as they do not accumulate inhibitory levels, can prevent protein synthesis in invading bacteria.

Effects on nucleic acids

Several antibiotics can block the growth of cells by interfering with the synthesis of DNA or RNA. However, the majority have no therapeutic use because they are unselective, affecting both microbial cells and animal cells. One group that does exhibit selectivity are the rifamycins of *Streptomyces* species. The most widely known is rifampicin, a semisynthetic derivative of rifamycin. It is relatively specific towards bacterial DNA-dependent RNA polymerase, exhibiting no effect on RNA polymerase from animal cells or DNA polymerase. Rifampicin blocks mRNA synthesis by binding to the β-subunit of the polymerase. This appears to block the entry of the first nucleotide, which is necessary to activate the enzyme.

Novobiocin, like quinolones (see p. 42), is an inhibitor of DNA gyrase. This antibiotic is produced by *Streptomyces niveus* and has some clinical applications. However, it is usually employed as a reserve drug, when other antibiotics have failed to cure an infection.

Further reading

Papers and reviews

Adams, M. W. W. (1993) Enzymes and proteins from organisms that grow near and above 100°C. *Annual Review of Microbiology* 47, 627–658.

Battista, J. R. (1997) Against all odds: the survival strategies of *Deinococcus radiodurans*. *Annual Review of Microbiology* 51, 203–224.

Debono, M. & Gordee, R. S. (1994) Antibiotics that inhibit fungal cell wall development. *Annual Review of Microbiology* 48, 471–497.

Kreft, J.-U., Booth, G. & Wimpenny, J. W. T. (1998) BacSim, a simulator for individual-based modelling of bacterial colony growth. *Microbiology* 144, 3275–3287.

Madigan, M. T. & Marrs, B. L. (1997) Extremophiles. *Scientific American* 276, 82–87.

Morris, J. G. (1994) Obligately anaerobic bacteria in biotechnology. *Applied Biochemistry and Biotechnology* **48**, 75–106.

Prosser, J. L. & Tough, A. J. (1991) Growth mechanisms and growth kinetics of filamentous microorganisms. *Critical Reviews in Biotechnology* **10**, 253–274.

Scheper, T.-H. & Lammers, F. (1994) Fermentation monitoring and process control. *Current Opinion in Biotechnology* **5**, 187–191.

Setlow, P. (1994) Mechanisms which contribute to the long-term survival of spores. *Journal of Applied Bacteriology* Supplement **76**, 49S–61S.

Welch, W. J. (1993) How cells respond to stress. *Scientific American* **268**, 56–64.

Wimpenny, J. W. T. & Colasanti, R. (1997) A unifying hypothesis for the structure of microbial biofilms based on cellular automaton models. *FEMS Microbiology Ecology* **22**, 1–16.

Yayanos, A. A. (1995) Microbiology to 10 500 meters in the deep sea. *Annual Review of Microbiology* **49**, 777–805.

Books

Berry, D. R. (ed.) (1988) *Physiology of Industrial Fungi*. Blackwell Scientific Publications, Oxford.

Dawes, I. W. & Sutherland, I. W. (1992) *Microbial Physiology*, 2nd edition. Blackwell Scientific Publications, Oxford.

Dickinson, J. R. & Schweizer, M. (eds) (1998) *The Metabolism and Molecular Physiology of* Saccharomyces cerevisiae. Taylor & Francis, London.

Griffin, D. H. (1994) *Fungal Physiology*, 2nd edition. Wiley, New York.

Isaac, S. & Jennings, D. (1995) *Microbial Culture*. BIOS Scientific Publishers, Oxford.

Jennings, D. M. (1995) *The Physiology of Fungal Nutrition*. Cambridge University Press, Cambridge.

Moat, A. G. & Foster, J. W. (1995) *Microbial Physiology*, 3rd edition. John Wiley, New York.

Pirt, S. J. (1975) *The Principles of Microbe and Cell Cultivation*. Blackwell Scientific Publications, Oxford.

Russell, A. D., Hugo, W. B. & Ayliffe, G. A. J. (1992) *Principles and Practice of Disinfection, Preservation and Sterilization*, 2nd edition. Blackwell Scientific Publications, Oxford.

Wainwright, M. (1990) *Miracle Cure—the Story of Antibiotics*. Blackwell Scientific Publications, Oxford.

Walker, G. M. (1998) *Yeast Physiology and Biotechnology*. Wiley, Chichester.

3 Microbial metabolism

Metabolism encompasses all enzyme-catalysed reactions of a cell, and can be divided into **primary** and **secondary** processes. Primary metabolic pathways are largely common to most organisms. They involve both energy-generating metabolism, referred to as **catabolism**, and **anabolism**, which utilizes this energy in the biosynthesis of cellular components for growth and maintenance. Catabolism and anabolism are considered separately for convenience, but are highly integrated processes. Products of primary metabolism that are of particular industrial importance include alcohols, amino acids, organic acids, nucleotides, enzymes and microbial cells (biomass).

Some microorganisms also perform secondary metabolism, which involves pathways that are not used during rapid growth. Secondary metabolism produces diverse, often species-specific end-products; the many industrially important secondary metabolites include alkaloids, antibiotics, toxins and some pigments. Certain secondary metabolites may confer an ecological advantage, whereas others have no apparent value to the producer organism.

Catabolism

All vegetative microbial cells require a continuous supply of energy for processes associated with growth, transport, movement and maintenance. In chemoheterotrophic microorganisms, the organic energy sources are acquired from the environment and then transformed by an ordered series of enzyme-controlled reactions within specific metabolic pathways. This breakdown metabolism (catabolism) leads to the generation of potential energy in the form of adenosine 5′-triphosphate (ATP) and reduced coenzymes, such as nicotinamide adenine dinucleotide (NADH), nicotinamide adenine dinucleotide phosphate (NADPH) and flavin adenine dinucleotide ($FADH_2$), and heat. Microorganisms have a very wide diversity of metabolism for the generation of ATP and reduced coenzymes, that in many cases also provides precursors for the biosynthesis of cellular components (see p. 54, Anabolism).

Most microorganisms can utilize carbohydrates, which may act as sources of both carbon and energy. The hexose, six carbon (C_6) sugar, glucose is the preferred substrate for a great many microorganisms and there are very few organisms that cannot use it. In nature, free glucose is not normally available, but can be obtained via various routes. It may be derived from the interconversion of other hexoses, the hydrolysis of disaccharides, oligosaccharides and polysaccharides from the environment, or from cell storage materials, such as starch, glycogen and trehalose.

Generation of energy from glucose is preceded by its phosphorylation, followed initially by oxidation through to mostly pyruvate (C_3). Metabolism of glucose to pyruvate, broadly termed glycolysis, provides ATP, precursors and reducing power (primarily NADH). However, only a limited amount of ATP is produced, which is formed via **substrate-level phosphorylation**. A maximum of two ATP molecules are generated for each glucose unit oxidized to this point. Resulting pyruvate occupies a pivotal position in intermediary metabolism and is the starting point for further catabolism. The various routes to this key intermediate operated by different organisms include the following.

1 The **Embden–Meyerhof–Parnas (EMP) pathway**, which is the most common route and is found in all major groups of organisms, including filamentous fungi, yeasts and many bacteria (Fig. 3.1). This pathway can operate under anaerobic or aerobic conditions and consists of a sequence of 10 enzyme-catalysed reactions located within the cytoplasmic matrix. The three key regulatory enzymes of the pathway (hexokinase, phosphofructokinase and pyruvate kinase) act irreversibly (Fig. 3.1). All other steps are freely reversible, which is important for the biosynthetic role of the pathway during glucose synthesis (gluconeogenesis, see p. 55). The irreversible steps have bypasses to enable the pathway to operate in this anabolic mode.

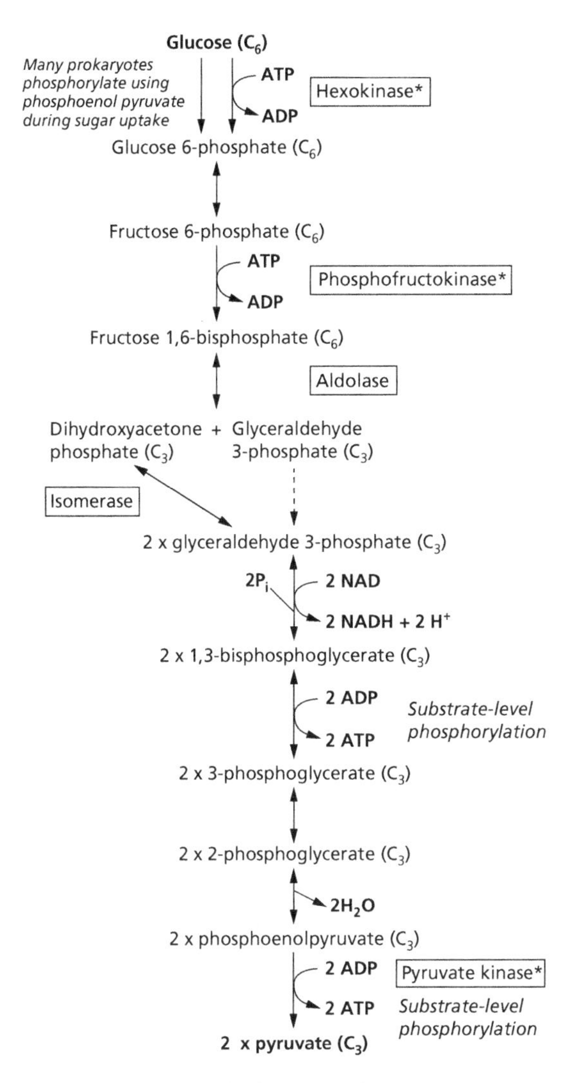

Fig. 3.1 Embden–Meyerhof–Parnas pathway (*irreversible steps).

Early stages of glucose breakdown actually consume two ATP molecules in the three-stage formation of fructose 1,6-bisphosphate. This molecule is then cleaved through the action of aldolase to form two different triose (C_3) phosphates, glyceraldehyde 3-phosphate (GAP) and dihydroxyacetone phosphate (DAP). Only GAP is directly processed in this pathway and DAP must be isomerized to GAP before it can be used. Oxidation of the resultant two GAP molecules to pyruvate generates energy in the form of four ATP molecules via two substrate-level phosphorylation reactions. However, for each glucose molecule oxidized to two pyruvate molecules, the net gain is only two ATP, due to its consumption in the earlier reactions.

$$\text{Glucose}\ (C_6) + 2\text{ADP} + 2\,P_i + 2\text{NAD}^+ \rightarrow 2\ \text{pyruvate}\ (C_3) + 2\text{ATP} + 2\text{NADH} + 2\text{H}^+$$

2 The **pentose phosphate (PP) pathway** or **hexose monophosphate pathway** is found in many bacteria and the majority of eukaryotic organisms (Fig. 3.2). This pathway often operates at the same time as the EMP pathway. In yeasts, for example, 10–20% of glucose (more during rapid growth) is degraded via the PP pathway, and the remainder is catabolized via the EMP pathway. The PP pathway functions under aerobic or anaerobic conditions, and has both catabolic and anabolic roles. This pathway is important in the provision of NADPH, mainly for use in reductive steps in anabolic processes; intermediates for aromatic amino acid synthesis, particularly erythrose-4-phosphate; pentoses, especially ribose for nucleic acid biosynthesis; and other biosynthetic intermediates. Pentose sugars such as xylose can also be catabolized via this route.

The PP pathway is cyclic and like all these glycolytic pathways, its enzymes are located in the cytoplasmic matrix. It begins with the two-step oxidation of glucose 6-phosphate (G6P) to the pentose (C_5) phosphate, ribulose 5-phosphate (RuMP), via 6-phosphogluconate. This involves the loss of one carbon as CO_2 and the formation of two NADPH. Following this oxidative phase, RuMP undergoes rearrangement in a series of two-carbon and three-carbon fragment exchanges, catalysed by the key enzymes transketolase and transaldolase. For every three glucose units processed, one GAP, six NADPH and two fructose 6-phosphate (F6P) molecules are generated. F6P molecules are converted back to G6P to maintain the operation of the cycle. The GAP may be oxidized to pyruvate by EMP pathway enzymes or it too may be returned to the start of the pathway via conversion of two GAP to one G6P.

$$3\ \text{glucose 6-phosphate}\ (C_6) + 6\text{NADP}^+ + 3\text{H}_2\text{O} \rightarrow 2\ \text{fructose 6-phosphate}\ (C_6) + \text{glyceraldehyde 3-phosphate}\ (C_3) + 3\text{CO}_2 + 6\text{NADPH} + 6\text{H}^+$$

3 The **Entner–Doudoroff (ED) pathway** (Fig. 3.3) is used by a relatively few microorganisms that lack the EMP pathway. Most are Gram-negative bacteria, including species of *Azotobacter*, *Pseudomonas*, *Rhizobium*, *Xanthomonas* and *Zymomonas*, but it is rare in fungi. The pathway begins with the formation of 6-phosphogluconate, as in the PP pathway. However, it

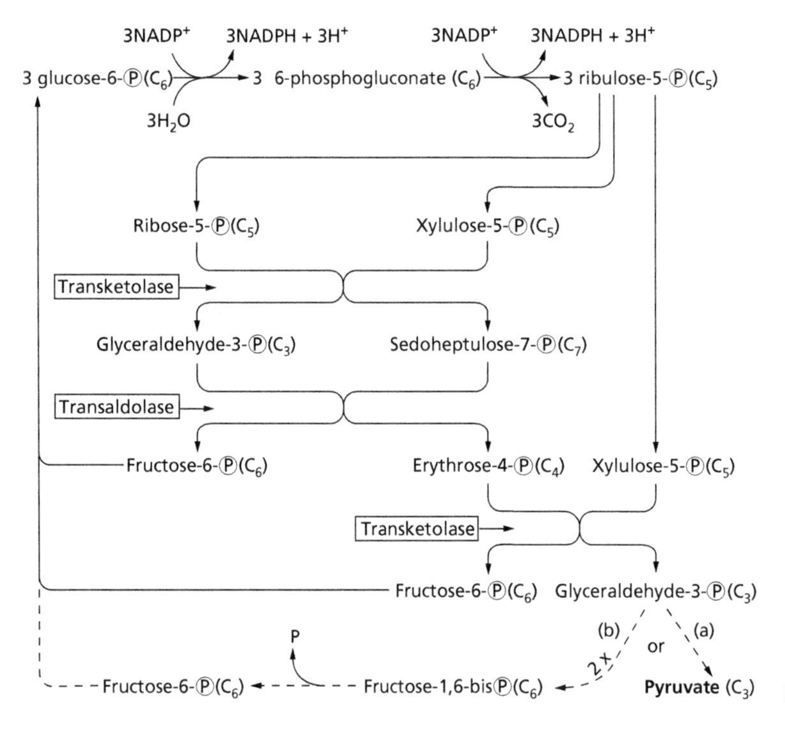

Fig. 3.2 The pentose phosphate pathway.

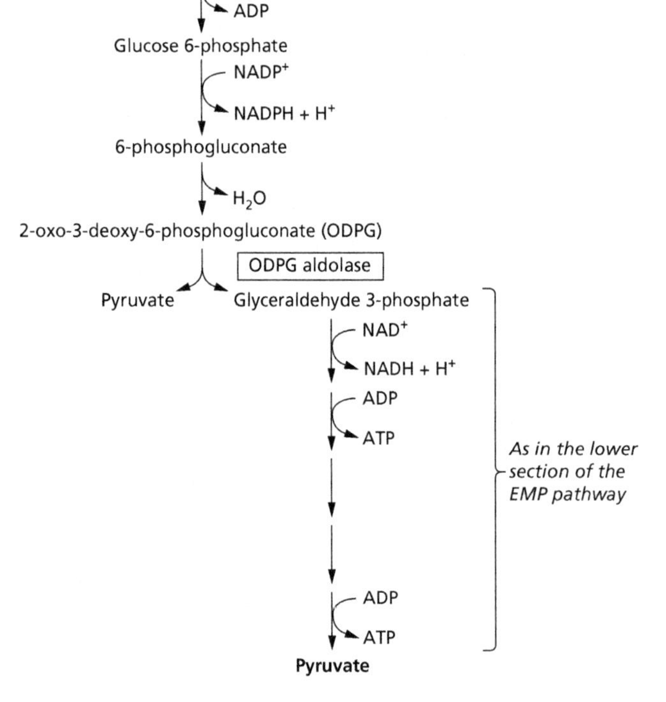

Fig. 3.3 The Entner–Doudoroff pathway.

is then dehydrated, rather than oxidized, to form 2-oxo-3-deoxy-6-phosphogluconate. This six-carbon molecule is cleaved by an aldolase to form two C_3 compounds, pyruvate and GAP, and the latter may also be converted to pyruvate. Overall, from each glucose molecule metabolized, the pathway can generate two pyruvate molecules, one ATP, one NADH and one NADPH, which is a lower energy yield than for the EMP pathway.

4 The **phosphoketolase (PK) pathway** or **Warburg–Dickens pathway** (Fig. 3.4) is operated by some lactic acid bacteria, notably species of *Lactobacillus* and *Leuconostoc*. It involves the oxidation and decarboxylation of glucose 6-phosphate to RuMP, as in the PP pathway. RuMP is isomerized to xylulose 5-phosphate (C_5) and then cleaved by a phosphoketolase to GAP (C_3) and acetyl phosphate (C_2). The former is ultimately converted to lactate and the latter to ethanol (see Heterolactic fermentation p. 52). This pathway produces only half the yield of ATP compared with the EMP pathway.

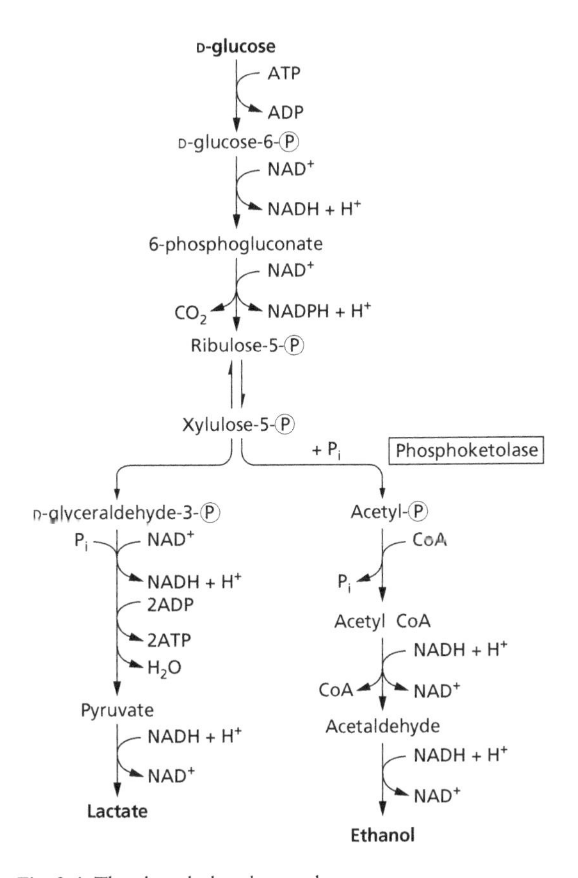

Fig. 3.4 The phosphoketolase pathway.

However, it does allow pentose formation from hexose sugars for nucleic acid synthesis and the catabolism of pentoses.

The tricarboxylic acid cycle

In many bacteria, yeasts, filamentous fungi, algae and protozoa, further catabolism of pyruvate under aerobic conditions involves its direction into the tricarboxylic acid (TCA) cycle (Fig. 3.5). TCA cycle enzymes are located within the mitochondrial matrix in eukaryotes, whereas in prokaryotes they are cytoplasmic. The step immediately before the cycle involves the oxidative decarboxylation of pyruvate to form the C_2 unit acetyl coenzyme A (acetyl CoA), catalysed by the multienzyme complex, pyruvate dehydrogenase. Acetyl CoA can also be formed via breakdown of lipids and some amino acids, and from the oxidation of alkanes, which certain microorganisms can utilize.

$$\text{Pyruvate } (C_3) + NAD^+ + CoA \rightarrow \text{acetyl CoA } (C_2) + CO_2 + NADH + H^+$$

The cycle proper commences with the condensation of this two-carbon compound with oxaloacetate (C_4) to form citrate (C_6). During the following eight steps of the complete TCA cycle, the two-carbon fragment is oxidized to two CO_2 molecules and oxaloacetate is regenerated to accept a further two-carbon unit. Three reactions within the cycle result in NADH formation and one generates $FADH_2$, and a single ATP molecule is formed indirectly by a substrate-level phosphorylation. Net yield is as follows:

$$\text{acetyl CoA } (C_2) + 3NAD^+ + FAD + ADP \rightarrow 2CO_2 + 3NADH + 3H^+ + FADH_2 + ATP$$

In terms of catabolism, the TCA cycle completes the oxidation of pyruvate to CO_2, and reduces electron carriers to produce NADH and $FADH_2$. These reduced coenzymes may then be used for further ATP synthesis in respiration (see p. 50), and importantly, are thereby reoxidized to participate in further catabolism.

The TCA cycle is not only a catabolic pathway, but also provides intermediates, C_4 and C_5 compounds, for the biosynthesis of amino acids, purines and pyrimidines. Under anaerobic conditions it does not function as a cycle. However, as several intermediates are still required for biosynthesis, it operates as a branched biosynthetic pathway. This facility to produce biosynthetic intermediates is also present in other microorganisms which lack a complete TCA cycle.

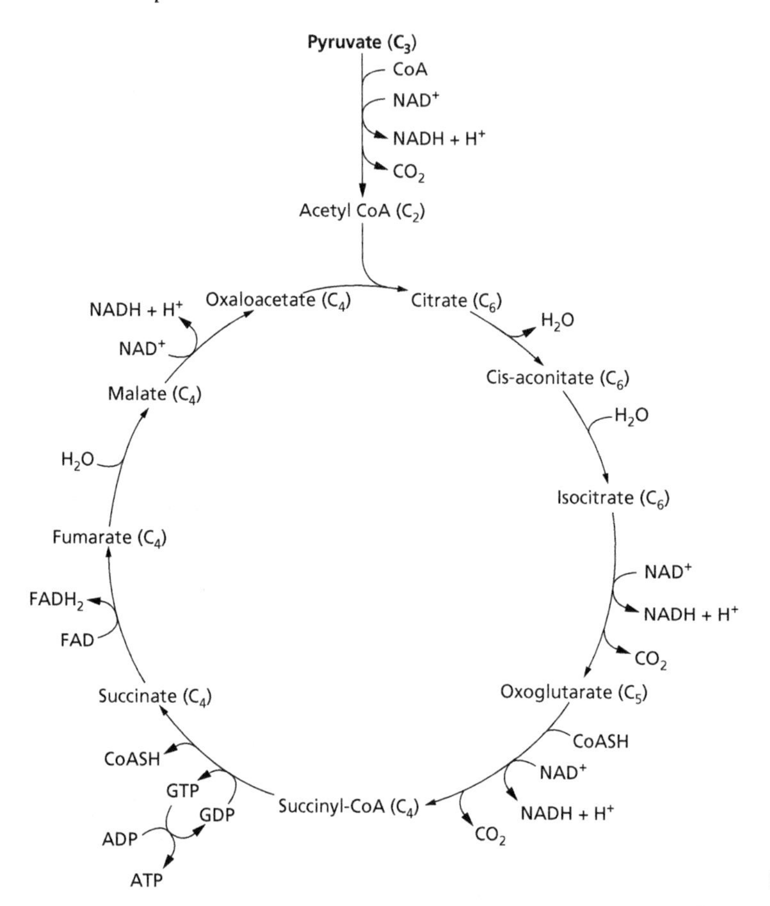

Fig. 3.5 The tricarboxylic acid cycle.

Respiration

All ATP formation from glucose oxidation has, to this point, been formed via substrate-level phosphorylation. Respiration leads to the generation of much more ATP from the oxidation of NADH and $FADH_2$. It requires oxygen or, as in some facultative and obligate anaerobic prokaryotes, another compound or ion that has a low redox potential, to act as the terminal electron acceptor. Oxygen is ideal for this, as it has a low redox potential and becomes reduced to water, whereas some of the final products of anaerobic respiration are toxic, e.g. nitrate reduced to nitrite.

In order to transfer electrons and protons (H^+) to oxygen or an alternate final electron acceptor, a special series of intermediate electron carriers, constituting the electron transport system (ETS), is required. Eukaryotes have their ETS located in the inner mitochondrial membrane, whereas in prokaryotes it is bound within the cytoplasmic membrane. The series of reactions involved can be coupled to energy-requiring processes, such as solute transport, or to the production of ATP via **oxidative phosphorylation.**

Electrons derived from the oxidation of substrates are passed from NADH or $FADH_2$, through a series of redox or reduction–oxidation reactions, to the terminal acceptor (Fig. 3.6). In this process, the energy released during electron flow is used to create a gradient of protons across a membrane. This proton gradient is used to drive ATP generation by a chemiosmotic oxidative phosphorylation mechanism. The **chemiosmotic hypothesis** proposed by Mitchell (1961) explains the process as follows: as electrons flow through the transport system, protons (H^+) are moved from inside to outside the membrane (excretion). Because the membrane is essentially impermeable to protons they cannot pass back and a proton gradient is established. As positive charges are removed from inside the membrane,

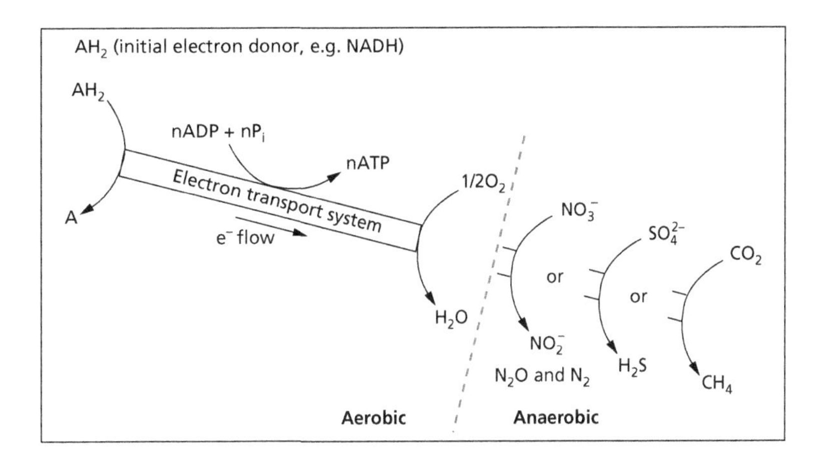

Fig. 3.6 Aerobic and anaerobic respiration.

negative charge remains within, mostly in the form of hydroxyl (OH^-) ions. Consequently, the pH immediately outside the membrane can reach 5.5, whereas within it approaches 8.5. This represents potential energy stored as a proton gradient or proton-motive force, which drives the synthesis of ATP via chemiosmotic phosphorylation, mediated by ATP synthase (ADP + P_i → ATP).

A molecule of NADH contains approximately 218 kJ of potential energy and it takes approximately 30 kJ to generate one ATP molecule. Hence, a maximum of seven ATP molecules could, in theory, be generated per NADH if energy conversion were 100% efficient. However, these systems are only up to 40% efficient and consequently produce a maximum of only three ATP molecules per NADH.

Aerobic respiration is performed by strict aerobes and facultative anaerobes in the presence of oxygen. Electrons enter the ETS from carriers, notably NADH or $FADH_2$. When these carriers donate their electrons they are reoxidized to allow their further participation in metabolism. The electrons are then passed through a series of carriers within the membrane, whereas protons are moved from inside to outside of the membrane, as described above. The protons ultimately become attached to oxygen, forming water. In eukaryotes, the ETS within the inner mitochondrial membrane generates three ATP molecules per NADH, and is composed of the following chain of electron carriers:

NADH → flavoprotein → iron sulphur proteins → quinone → cytochrome b → cytochrome c → cytochrome a → cytochrome a_3 → oxygen

In prokaryotes the ETS within the cytoplasmic membrane has somewhat different carriers, with often shorter incomplete chains that generate only one or two ATP molecules per NADH. Cytochromes involved include: b_{558}, b_{595}, b_{562}, d, and o. Certain facultative anaerobes, such as *Escherichia coli*, have at least two terminal respiratory systems, each with a different affinity for oxygen. Consequently, they can adapt to varying oxygen concentrations, continuing to use oxygen as the terminal electron acceptor even at low oxygen levels.

Some bacteria that are facultative anaerobes or obligate anaerobes perform **anaerobic respiration**. This involves a similar ETS, but has terminal electron acceptors other than oxygen. Examples of anaerobic respiration include the following.

1 Nitrate respiration, which is conducted by facultative anaerobic bacteria. The redox potential of nitrate is +0.42 volts, compared with +0.82 volts for oxygen. Consequently, less energy is gained than with oxygen as terminal electron acceptor and fewer molecules of ATP can be formed. The process may have several steps, where nitrate is reduced via nitrite and nitrous oxide ultimately to dinitrogen, which is referred to as **dissimilatory nitrate reduction** or **denitrification**. Denitrifiers include species of *Pseudomonas*, *Paracoccus denitrificans* and *Thiobacillus denitrificans*. Other facultative anaerobes, including *E. coli* and relatives, merely reduce nitrate to nitrite, and the enzyme responsible, nitrate reductase, is repressed in the presence of oxygen.

2 Sulphate respiration is practised by a small group of heterotrophic bacteria, which are all obligate anaerobes, e.g. *Desulfovibrio* species. Their main product is hydrogen sulphide, formed via several intermediates.

3 **Carbonate respiration** is performed by archaeans such as *Methanococcus* and *Methanobacterium*. They are obligate anaerobes that reduce CO_2, and sometimes carbon monoxide, to methane. These methanogenic bacteria commonly use hydrogen as their energy source and are found in anoxic environments low in nitrate and sulphate, e.g. the gut of some animals, marshes, rice fields and sewage sludge digesters.

$$CO_2 + 4H_2 \rightarrow CH_4 + 2H_2O$$

In addition to nitrate, sulphate and carbon dioxide, ferric iron (Fe^{3+}), manganese (Mn^{4+}) and several organic compounds (dimethyl sulphoxide, fumarate, glycine and trimethylamine oxide) can serve as terminal electron acceptors for anaerobic respiration in certain bacteria.

Fermentations

When respiration is not an option, organisms must use an alternative mechanism for the regeneration of the limited supply of coenzymes, reduced during the oxidation of glucose to pyruvate. If NAD(P)H is not oxidized back to $NAD(P)^+$, further catabolism will cease. Consequently, a suitable terminal electron acceptor must be found to take the electrons. Fermentations use an organic molecule, pyruvate or a derivative, as this final electron acceptor, thereby regenerating $NAD(P)^+$ and allowing catabolism to continue. This results in the formation of reduced 'waste' products, such as alcohols and acids, that are then excreted from the cell (Fig. 3.7). However, this is very wasteful in terms of recovery of potential energy from glucose, as little or no further ATP is formed. Conversely, complete oxidation to CO_2, in conjunction with oxidative phosphorylation during respiration, can generate a possible further 36 molecules of ATP per glucose molecule.

Alcoholic fermentation is performed by yeasts and certain filamentous fungi and bacteria. It is a two-step process, where pyruvate from the EMP pathway, or via the ED pathway in the case of *Zymomonas*, is first decarboxylated to acetaldehyde; NAD^+ is then regenerated during reduction of acetaldehyde to ethanol (Fig. 3.7a).

Lactic acid fermentations are carried out by a number of bacteria, including *Streptococcus*, *Lactobacillus*, *Lactococcus* and *Leuconostoc*, along with some fungi, algae and protozoa. Here pyruvate, rather than a derivative of pyruvate, is the electron acceptor and forms lactate. There are two forms of this fermentation:

1 **homolactic** fermentation is operated by bacteria such as *Lactobacillus acidophilus* and *Lactobacillus casei*, which reduce virtually all pyruvate generated by glycolysis to lactic acid (Fig. 3.7b); this also occurs in animal muscle deprived of oxygen; and

2 **heterolactic** fermentation generates other products along with the lactic acid (Fig. 3.7c). The organisms that perform it include *Leuconostoc mesenteroides* and *Lactobacillus brevis*, which operate the PK pathway. They form lactate from pyruvate and ethanol from acetyl phosphate, and in certain cases some acetate may be produced.

Mixed acid fermentation is carried out by *E. coli* and related facultative anaerobes. The products include lactate, acetate, small quantities of ethanol and formate. Some organisms have the ability to further reduce formate to hydrogen and CO_2 (Fig. 3.7d).

2,3-Butanediol fermentation is performed by *Enterobacter*, *Erwinia*, *Klebsiella* and *Serratia*. It is similar to the mixed acid fermentation, but generates butanediol, along with ethanol and acids (Fig. 3.7e).

Propionic acid fermentation is conducted by several gut bacteria, such as species *Propionibacterium*, some of which are involved in the commercial production of certain Swiss-type cheeses and vitamin B_{12} (cobalamin). The propionate is formed from pyruvate via the methylmalonyl CoA pathway, where pyruvate is carboxylated to oxaloacetate, and then reduced to propionate via malate, fumarate and succinate (Fig. 3.7f). This requires the important cofactors, biotin, CoA and CoB_{12}.

Butyric acid fermentation is carried out by species of *Clostridium*. These anaerobic spore formers also produce acetone, butanol, propanol, other alcohols and acids (Fig. 3.7g). They also ferment amino acids and other nitrogenous compounds, as well as carbohydrates.

Catabolism of lipids and proteins

Carbon sources other than carbohydrates must be processed before they can enter catabolism. Lipids such as di- and triglycerides are hydrolysed by lipases to produce free fatty acids and glycerol. Fatty acids may then be broken down by the β-oxidation pathway. Here FAD and NAD^+ are used to accept electrons, and two-carbon units are successfully removed, in the form of acetyl CoA, which can then feed directly into the TCA cycle. The glycerol may be phosphorylated to glycerol phosphate, followed by its oxidization to DAP and isomer-

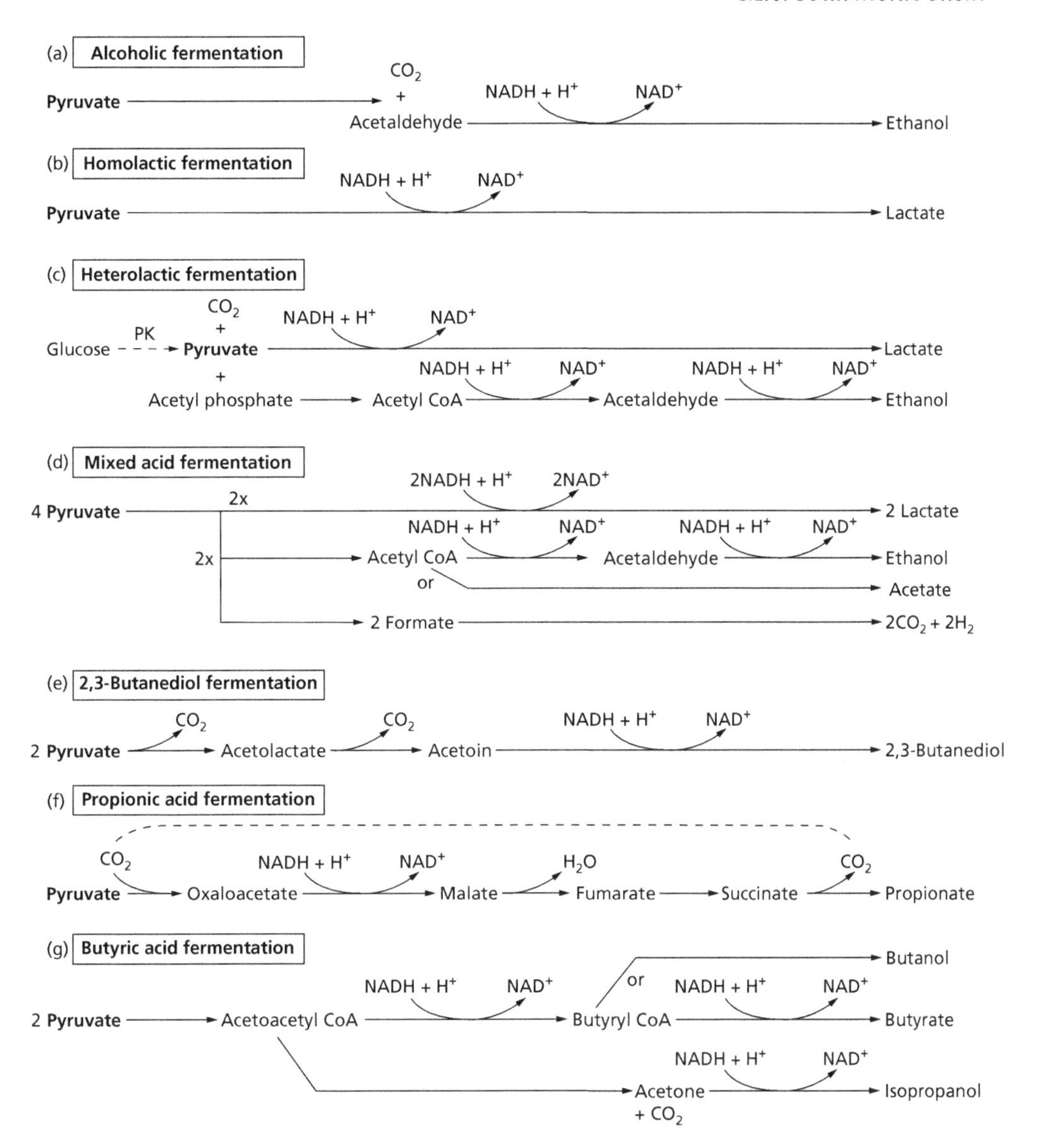

Fig. 3.7 Common fermentation products derived from pyruvate.

ization to GAP. This C_3 unit can then enter the EMP pathway.

Extracellular proteins are hydrolysed by proteases to produce free amino acids, which can be transported into the cell. Amino acid catabolism initially involves removal of their amino group(s). This is usually achieved via transamination, where the amino group is donated to a keto host, e.g. amination of pyruvate to form alanine. The keto acid resulting from transamination can then be oxidized within the TCA cycle. Excess amino groups may accumulate and are often excreted as ammonium ions, which accounts for the increasing pH of media during the growth of some bacteria.

Energy storage

When excess nutrients are available, microorganisms normally synthesize compounds that can be stored for later use during periods of nutrient shortages. **Storage carbohydrates** include the polysaccharides, glycogen and starch, and the disaccharide trehalose, all of which act as both carbon and energy sources. Trehalose is produced by filamentous fungi, yeasts and some bacteria, and glycogen is a major storage carbohydrate in many groups of microorganisms. When required, these polymeric reserves are hydrolysed by phosphorylases, forming glucose 1-phosphate, which can directly enter catabolism.

Many microorganisms store energy-rich lipids. Poly β-hydroxybutyrate is commonly found in bacteria, but is not produced by eukaryotes. Filamentous fungi and yeasts often store neutral lipids (triglycerides) within vacuoles. Volutin granules, composed of **polymetaphosphates**, are also stored by some prokaryotes and eukaryotes, which act as both phosphate and energy reserves. Sulphur bacteria may accumulate sulphur granules by the oxidation of H_2S. When no further media supplies of sulphide are available, this sulphur store is oxidized to sulphate, providing reducing equivalents for CO_2 fixation or oxidative phosphorylation (see p. 61, Chemolithotrophic autotrophy).

Anabolism: the synthesis of biomolecules

Anabolism is the *de novo* biosynthesis of complex organic molecules from simpler ones. These processes are endergonic, requiring an input of energy to drive them, which mostly comes from the ATP provided by catabolism. The anabolic processes often comprise two stages, involving the synthesis of small metabolic intermediates that are then assembled to form polymers.

Many catabolic processes not only generate energy, in the form of ATP and reduced coenzymes, for use in biosynthesis, but also provide carbon skeletons that can feed into anabolism. Pathways with dual catabolic and anabolic roles are referred to as **amphibolic.** Amphibolic pathways include the EMP pathway and the TCA cycle. The former provides pyruvate, hexose phosphates and triose phosphates, whereas the latter supplies oxaloacetate and oxoglutarate. These products can be utilized in the synthesis of cellular components and storage materials. Many reactions of amphibolic pathways are freely reversible or have 'bypasses' of irreversible steps to facilitate their dual function. The irreversible steps have separate enzymes for the two directions that can provide suitable control points for the independent regulation of catabolism and anabolism.

Anaplerotic (replenishment) reactions

As a consequence of removing intermediate compounds from these amphibolic pathways for biosynthesis, their levels may become depleted. For example, oxaloacetate is taken from the TCA cycle to furnish the demand for carbon skeletons in amino acid biosynthesis. Hence, these intermediates have to be replenished via an alternate route, referred to as an **anaplerotic pathway**, in order to maintain operation of this cycle. One such anaplerotic route is the **glyoxalate cycle** (Fig. 3.8). This cycle operates in many organisms for the replenishment of oxaloacetate, particularly for gluconeogenesis (see below) from lipids and during growth on C_2 compounds. It involves isocitrate lyase, which cleaves isocitrate to succinate and glyoxalate. Acetyl CoA is then condensed with glyoxylate to generate malate and coenzyme A (CoASH) by the action of malate synthase. The malate can then be transformed to oxaloacetate to allow the continued operation of the TCA cycle.

Gram-negative bacteria such as *E. coli* and other enteric bacteria can generate oxaloacetate, a four-carbon compound, by fixing CO_2 to the three-carbon compound, phosphoenolpyruvate (PEP), using PEP carboxylase. Many Gram-positive bacteria and yeasts have a similar carboxylation system, but one which utilizes the biotin-requiring enzyme, pyruvate carboxylase, to add CO_2 to pyruvate. However, this route consumes energy as it has a requirement for ATP. Both of these 'fixation' reactions are almost irreversible:

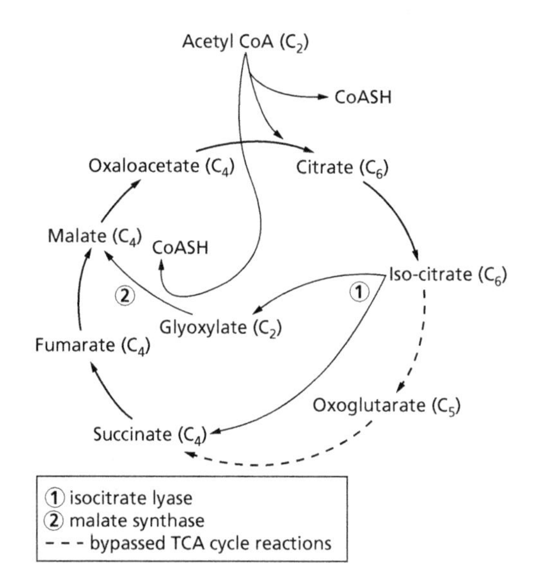

Fig. 3.8 The glyoxylate cycle.

$$\text{phosphoenolpyruvate} + CO_2 \xrightarrow{\textit{phosphoenol pyruvate carboxylase}} \text{oxaloacetate} + P_i$$

$$\text{pyruvate} + CO_2 + ATP + H_2O \xrightarrow{\textit{pyruvate carboxylase}} \text{oxaloacetate} + ADP + P_i$$

Gluconeogenesis, essentially the reversal of glycolysis, also fulfils a similar anaplerotic role. It is particularly important during growth on pyruvate, related C_3 compounds and C_2 units. However, as mentioned earlier, not all steps are reversible and appropriate bypasses are necessary. The reversal of the flow of carbon from pyruvate maintains a supply of hexoses, which would otherwise become depleted. These intermediates are mostly required for the synthesis of cell wall components and storage carbohydrates.

Biosynthesis of nucleotides and nucleic acids

Nucleotides are composed of a cyclic nitrogenous base, a pentose sugar (ribose or deoxyribose) and phosphate. The base may be a purine, either adenine or guanine, or one of three pyrimidines, cytosine, thymine or uracil. These nucleotides are the building blocks of deoxyribonucleic acid (DNA) and ribonucleic acid (RNA). They are also important constituents of several coenzymes, and have key roles in the activation and transfer of sugars, amino acids and cell wall components.

Most microorganisms can synthesize their own purines and pyrimidines. Purine synthesis is complex, the purine skeleton being derived from several components, whereas pyrimidines originate from orotic acid, a product of the condensation of carbamoyl phosphate with aspartic acid.

Ribose 5-phosphate is generated in the PP pathway (see p. 48) and is used here in the activated form of phosphoribosyl pyrophosphate (PRPP). It condenses with the base precursor orotic acid to form pyrimidines. In purine biosynthesis, the compounds are built up by several additions of fragments to the PRPP molecule. Deoxyribose forms of the nucleosides are then generated by reduction of the sugar moiety of ribonucleosides. The nucleosides are finally phosphorylated to form nucleotide triphosphates by successive phosphorylation from ATP, in order that they can participate in further metabolic processes.

NUCLEIC ACIDS

As mentioned above, there are two types of nucleic acid, DNA and RNA, and both are composed of nucleotide building blocks. The two nucleic acids differ in the sugar found in their nucleotides, either deoxyribose or ribose. Both polymers are made by the formation of covalent bonds between the sugar and phosphate groups of adjacent nucleotides. This sugar–phosphate backbone is common to all nucleic acid molecules. The unique nature of nucleic acid molecules resides in the sequence of bases. Four bases can occur in each nucleic acid. Three bases, adenine, guanine and cytosine, are common to both RNA and DNA, but thymine of DNA is replaced by uracil in RNA. The various forms of RNA, which play key roles in protein synthesis (see below), include messenger (mRNA), ribosomal (rRNA) and transfer (tRNA).

DNA is double stranded, composed of two helical chains each coiled around the same axis. Both chains follow right-handed helices, with the two chains running in opposite directions. Bases of each strand are on the inside of the helix with the phosphates on the outside. A key feature of the structure is the manner in which the two chains are held together by the purine and pyrimidine bases (Fig. 3.9). A single base from one chain is hydrogen bonded to a single base from the other chain, so that the two lie side by side. One of the pair must be a purine and the other a pyrimidine, and only specific pairs of bases can bond together. These base pairs are: adenine (purine) with thymine (pyrimidine) and guanine (purine) with cytosine (pyrimidine). Hence, the two strands of DNA have a complementary sequence of bases.

Biosynthesis of amino acids and protein

Microorganisms are normally capable of synthesizing the 20 amino acids required for protein production. The carbon skeletons for the amino acids are derived from intermediary metabolism (Table 3.1), and the amino groups are introduced by direct amination, or more usually for most amino acids, via transamination. Microorganisms readily assimilate ammonia, but if nitrate, nitrite or more rarely molecular nitrogen are the nitrogen source, they must first be reduced to ammonia. Primary assimilation of ammonia in many bacteria involves L-glutamate dehydrogenase and L-alanine dehydrogenase, which catalyse the reductive amination without the requirement for ATP. However, when ammonium ion concentrations are low, assimilation of nitrogen may be via glutamine synthetase, which has a higher affinity for ammonium ions but demands expenditure of ATP (Fig. 3.10). Glutamate is the predominant component of the free cytoplasmic amino acid pool. Consequently, it is readily available for amino group donation in transamination reactions that form other amino acids.

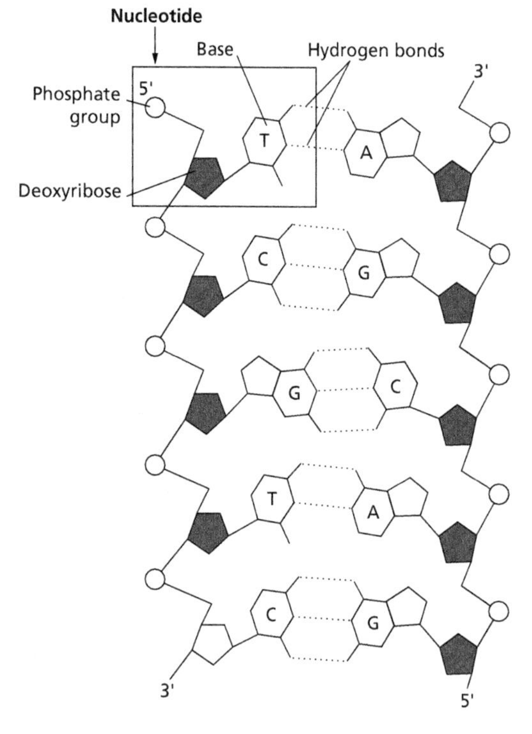

Fig. 3.9 Two-dimensional representation of a section of deoxyribonucleic acid. Bases: adenine (A), thymine (T), cytosine (C) and guanine (G).

Polypeptides consist of a chain of amino acids joined by very stable covalent peptide bonds. These bonds form between the amino group of one amino acid and the carboxylic acid group of another. The term 'protein' normally refers to the final entire functional assembly, which may be composed of one or several polypeptides. The order of amino acids within a polypeptide is determined by the base sequence of the DNA that constitutes the structural gene coding for the specific polypeptide. DNA codes are not read directly, but transcribed into mobile copies of the code, in the form of mRNA. These act as templates for polypeptide synthesis and are translated into a specific sequence of amino acids. Initiation and termination of both transcription and translation are strictly controlled. The mRNAs are synthesized by RNA polymerases that bind to the promoter site located upstream from the structural gene coding for the polypeptide. Base sequences of the template or sense strand of double-stranded DNA, composed of adenine, guanine, cytosine and thymine, are transcribed into single-stranded mRNA copies. They contain the complementary sequence of bases, except that uracil replaces thymine.

There are several differences between the modes of transcription in prokaryotes and eukaryotes, but the outcome is essentially the same. Several mRNA copies are made, with the number of mRNA copies produced and their rate of production and degradation playing a role in metabolic control, i.e. regulating specific enzyme concentrations. Prokaryotic transcription and translation processes are coupled, enabling protein synthesis to start before transcription ends. In eukaryotic cells, the mRNA must first exit the nucleus and move into the

Table 3.1 Amino acid groups with common biosynthetic pathways (the compound/skeleton from which they are derived is in parenthesis)

Glutamate family (2-oxoglutarate)	Glutamate, glutamine, proline, arginine
Aspartate family (oxaloacetate)	Aspartate, asparagine, lysine, methionine, threonine, isoleucine
Pyruvate family (pyruvate)	Alanine, leucine, valine
Serine family (3-phosphoglycerate)	Serine, glycine, cysteine
Aromatic amino acids (erythrose 4-phosphate + phosphoenolpyruvate)	Phenylalanine, tryptophan, tyrosine
Histidine (phosphoribosyl pyrophosphate (PRPP) + ATP)	Histidine is the only amino acid member of its biosynthetic route

(a) **Glutamate dehydrogenase (GDH)**

$$\text{Oxoglutarate}^* + NAD(P)H + H^+ + NH_4^+ \rightleftharpoons \text{Glutamate} + NAD(P)^+ + H_2O$$

*5-carbon compound produced in the TCA cycle
(replace with pyruvate for alanine dehydrogenase,
and the product is alanine)
Note: no ATP required

(b) **Glutamine synthetase (GS)**—direct amination

$$\text{Glutamate} + NH_4 + ATP \rightleftharpoons \text{Glutamine} + ADP + P_i$$

Glutamate synthetase (GOGAT)—a transamination

$$\text{Glutamine} + \text{oxoglutarate} + NAD(P)H + H^+ \rightleftharpoons 2\ \text{glutamate} + NAD(P)^+ + H_2O$$

Fig. 3.10 Amination mechanisms (a) direct amination (b) the route of assimilation at low ammonium ion concentrations.

cytoplasm or to rough endoplasmic reticulum, where the ribosomes are located.

The genetic code is composed of a linear series of triplets of bases (codons) that represent the amino acids of proteins, along with start and stop signals for protein synthesis. This code is virtually the same for all organisms and is degenerate for most amino acids, i.e. there are 20 amino acids and 64 possible triplet combinations from the four bases. Consequently, each amino acid is represented by more than one triplet code.

Polypeptide synthesis takes place when a mRNA molecule forms a complex with a ribosome. It involves the interaction between the small ribosomal unit and the ribosomal binding site of the message. Bacterial ribosomes have a sedimentation value of 70 S and eukaryotic ribosomes, except those contained within certain organelles, are larger 80 S structures. However, both types are composed of two subunits containing proteins and rRNA. Several ribosomes may 'read' a single mRNA molecule simultaneously, forming a polyribosome. Each ribosome holds the mRNA, moves along its length in the 5′ to 3′ direction, and constructs the polypeptide from the amino terminal end to the carboxyl terminal end. Different tRNA molecules are responsible for delivering specific amino acids to the ribosomes in an activated form as aminoacyl-tRNAs. The order of assembly is determined by the matching of codons on the mRNA to complementary triplet anticodons on the tRNAs. Amino acids sequentially bond together to form the polypeptide, which on completion spontaneously, or with the aid of chaperone proteins, folds to form secondary and tertiary structures. Constituent amino acids exhibit various chemical properties determined by their side chains. For example, some are hydrophobic and others have hydrophilic characteristics (Fig. 3.11). Thus, the properties of different regions of protein molecules are determined by the types of amino acids present. Some polypeptides may undergo post-translational modification before becoming functional. This can involve glycosylation, phosphorylation and cleavage of preproteins or signal sequences.

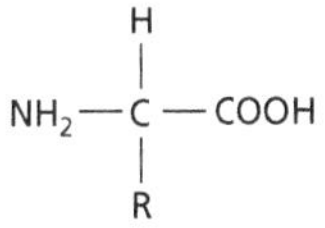

$$NH_2-\underset{\substack{|\\R}}{\overset{\substack{H\\|}}{C}}-COOH$$

Fig. 3.11 The structure of an amino acid. The side group R can be any of 20 different chemicals. Some side chains are very nonpolar, others very polar, or electrically charged.

Biosynthesis of monosaccharides and polysaccharides

Carbohydrates can exist either as single units (monosaccharides) or joined together in molecules ranging from two units (disaccharides) to thousands of units (polysaccharides). Monosaccharides containing an aldehyde group are called aldoses, whereas those containing a ketone group are ketoses. Monosaccharides are also classified by the number of carbon atoms that they contain, i.e. trioses, tetraoses, pentoses, hexoses, etc. Free sugars are almost never found in cells, rather they are present in small quantities as sugar phosphates or sugar nucleotides, e.g. uridine diphosphate (UDP)-glucose, UDP-N-acetylglucosamine, guanosine diphosphate (GDP)-mannose. The monosaccharides are derived from common sugar substrates such as glucose or hydrolysed polysaccharide substrates. As previously mentioned, during growth on non-carbohydrate carbon sources, monosaccharides must be synthesized via gluconeogenesis, requiring reversal of sections of the EMP pathway and the bypassing of irreversible steps (Fig. 3.1).

Within each cell the bulk of the monosaccharides are present in the form of polysaccharides. The component sugar units of these polymers are joined by covalent **glycosidic bonds**. They are not assembled from free sugars, but from sugar nucleotides. Different polysaccharides can be formed by varying the orientation of the glycosidic bond to form α- or β-linkages, or by changing the monomer. Homopolymers are composed of units of just one monomer, whereas heteropolymers contain more than one type of monomer. Examples of important microbial polysaccharides include the following.

1 Glycogen and starch, branched-chain homopolymers of α-linked glucose units that function as carbon and energy reserves. Glycogen, for example, is synthesized through the activities of ADP-glucose pyrophosphorylase and the branching enzyme, glycogen synthase.
2 Cellulose and other cell wall β-glucans, composed of β-linked glucose units.
3 Chitin, a polymer of *N*-acetylglucosamine.
4 Peptidoglycan, a polymer of *N*-acetylglucosamine and *N*-acetylmuramic acid, cross-linked with peptides.
5 Various gums such as dextran (mostly α-1,6 linked glucose units) and xanthan (essentially a cellulose backbone with a trisaccharide of mannose–glucuronic acid–mannose, linked to alternate glucose units).

PEPTIDOGLYCAN BIOSYNTHESIS

Peptidoglycan (PG) consists of a rigid linear polysaccharide backbone of alternating units of *N*-acetyl glucosamine (NAG) and *N*-acetyl muramic acid (NAM), with tetrapeptide side chains whose component amino acid may vary depending upon the bacterium. Each tetrapeptide is attached to a NAM residue through a lactate unit (Fig. 3.12). The structure is made rigid by cross-linking a proportion of adjacent tetrapeptide side chains either directly or through short peptide bridges to form an overall net-like structure. In *E. coli,* this is achieved by direct transpeptidation between diaminopimelic acid (DAP) on one side chain with D-alanine on an adjacent peptide chain. Generally, there is less cross-bridging in Gram-negative peptidoglycan than in Gram-positive cell walls.

Biosynthesis of peptidoglycan requires two types of nucleotide-amino sugars, uridine diphosphate-*N*-acetyl glucosamine (UDP-NAG) and UDP-*N*-acetyl muramic acid (UDP-NAM). Five amino acids are added to UDP-NAM molecules in sequence, each directed by separate ligases, forming UDP-NAM-pentapeptides. Some of these amino acids are D-forms (isomers), whereas in proteins they are always L-forms. The NAM-pentapeptide is then transferred from UDP to a bactoprenol carrier (an isoprenoid lipid phosphate) bound at the inner side of the cell membrane. UDP-NAG then adds its NAG moiety to the bactoprenol-NAM-pentapeptide, forming a complete subunit for PG assembly. Each bactoprenol-NAG-NAM-pentapeptide unit is then transported across to the outside of the cell membrane, where its NAG-NAM-pentapeptide is released. This subunit moves into the periplasm to attach to incomplete peptidoglycan, leaving the bactoprenol carrier to return back across membrane. New cross-links are then formed between peptidoglycan chains by transpeptidation to form a rigid network. Energy for this is provided by the cleavage of the terminal amino acid of the pentapeptide by a carboxypeptidase, as ATP is not available in this location.

Biosynthesis of fatty acids and lipids

Lipids are a very diverse family of compounds, grouped together on the basis of their relative insolubility in water and solubility in non-polar solvents. They are important components of membranes and may function as energy reserves in some organisms, e.g. poly β-hydroxybutyrate and triglycerides. Simple lipids are glycerides, esters of glycerol and fatty acids. Each fatty acid consists of a long non-polar chain of 12–24 carbon atoms with a polar carboxylic acid group at one end. Most natural fatty acids contain an even number of carbon atoms. They may be saturated, unsaturated or polyunsaturated, containing no double bonds, one double bond or two or more double bonds, respectively. Synthesis of saturated fatty acids involves derivatives of coenzyme A and acyl carrier proteins (ACP). The fatty acid is built up by successive addition of two-carbon units in the form of acetyl units, donated indirectly via acetyl CoA (i.e. following its carboxylation to malonyl CoA), which are then reduced by NADPH. Two routes are known for unsaturated fatty acids synthesis, one of which has a requirement for molecular oxygen.

Monoglycerides, possessing one fatty acid, are rare, but diglycerides are common membrane lipids and triglycerides are often used as energy storage materials by some yeasts. Bacteria contain few glycerides, they

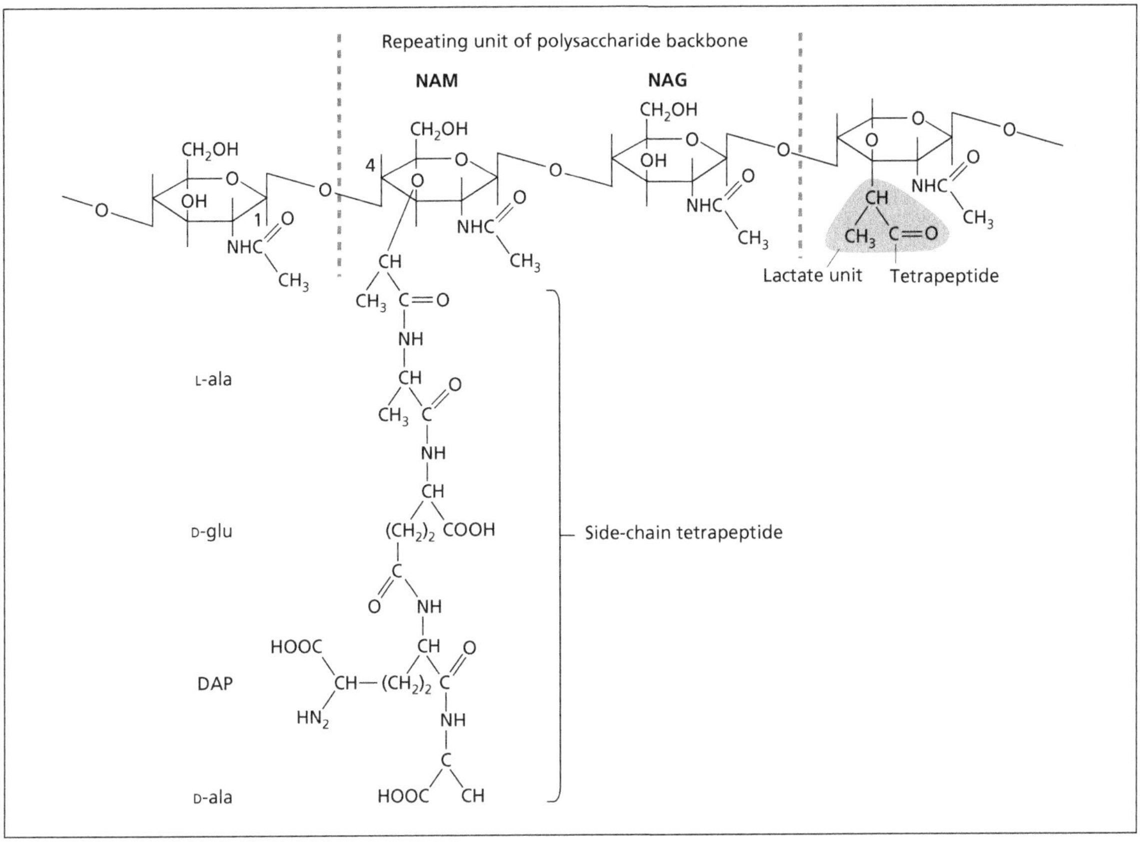

Fig. 3.12 Repeating unit of peptidoglycan (from Dawes & Sutherland (1992)).

mainly have phospholipids (phosphoglycerides), which are diglycerides containing a phosphate group. These are highly water-soluble (hydrophilic) at the phosphate end, but very water-insoluble (hydrophobic) at the fatty acid end. The phosphate may be further esterified to ethanolamine, inositol or serine.

Other important lipids include sterols, such as ergosterol, which are key cell membrane components in fungi, synthesized only under aerobic conditions. Glycolipids, including lipopolysaccharides, glycosyldiglycerides and lipoteichoic acid, are also found in many microorganisms.

Autotrophy

Autotrophs can obtain all their carbon requirements from CO_2, which is fixed (assimilated) via the Calvin cycle (see p. 62). There are two basic types of autotrophs and they are characterized by their method of obtaining energy. Phototrophs use energy derived from light and chemolithotrophs use inorganic compounds as energy sources.

Photoautotrophy

Photoautotrophs are able to absorb light energy and transform it to chemical energy (ATP and reduced coenzyme) via the 'light reactions' of photosynthesis. The chemical energy generated is used in driving the Calvin cycle to fix CO_2. Photosynthesis is initiated by the absorption of quanta of light by specific photosynthetic pigments. There are three main types employed.

1 **Bacteriochlorophyll** is found in the purple sulphur bacteria, purple non-sulphur bacteria and green bacteria, which includes all photosynthetic eubacteria except the cyanobacteria.

2 **Chlorophylls** are the photopigments of cyanobacteria, algae and higher plants.

3 **Bacteriorhodopsin**, a non-chlorophyll pigment, is used by archaeans, such as halobacteria, but will not be discussed here.

Chlorophylls (CHL) are porphyrins containing a magnesium atom at their centre. They have a hydrophobic side chain that enables the molecules to associate with membranes. The different chlorophylls may be distinguished by variations in the side chains, which also affect their absorption spectra. Most organisms have more than one pigment, allowing the absorption of a wider range of wavelengths of light. Photopigments operate in groups of a few hundred molecules situated within membranes. In prokaryotes they are associated with internal membrane systems, the chromatophores, and in eukaryotes the membranes are located within specific organelles, the chloroplasts. Pigments other than chlorophylls can serve as accessory light-capturing agents, including carotenoids and phycobiliproteins. Light energy that they absorb is transferred to chlorophylls.

ANOXYGENIC PHOTOSYNTHETIC BACTERIA

Photosynthetic units in eubacteria, other than cyanobacteria, consist of three components: the light-gathering complex, containing bacteriochlorophyll and accessory pigments; a reaction centre, composed of bacteriochlorophylls and bacteriopheophytins; and an ETS. Reduced sulphur compounds, molecular hydrogen or organic compounds are utilized as final electron donors. Consequently, oxygen is not generated during this type of photosynthesis. When the bacteriochlorophyll within a reaction centre is excited by light, an electron is donated to a bacteriopheophytin. It is passed on to a quinone and then, via a series of cytochromes, returning to the oxidized reaction centre bacteriochlorophyll. During the cyclic path that each electron travels, ATP is synthesized. This process is referred to as cyclic photophosphorylation.

In order to generate NADPH, which is needed along with ATP for CO_2 fixation, a reduced electron donor is required, e.g. H_2S or H_2. Their donation of electrons to $NADP^+$ is independent of light. However, some of the ATP synthesized via cyclic phosphorylation is required to force electrons from weak reductants to stronger ones that can then reduce $NADP^+$, i.e. reversal of the electron transport chain. Reversed electron transport is also used by lithotrophs to overcome the same problem (see p. 61).

OXYGENIC PHOTOSYNTHESIS

Cyanobacteria, algae and higher plants have two photosystems, photosystems I and II, linked through a series of electron carriers (Fig. 3.13). These systems are responsible for the generation of both ATP and NADPH. Electrons removed from the reaction centre of photosystem I, when excited by light, are passed to ferredoxin, then directly to $NADP^+$ to form NADPH. This leaves photosystem I positively charged and unable to supply any further electrons. Electrons are supplied to photosystem I from photosystem II when it too is energized by light. The electrons flow to photosystem I through an electron transport system, thereby generating ATP. This is non-cyclic photophosphorylation, as the electrons never return to photosystem II. Consequently, the

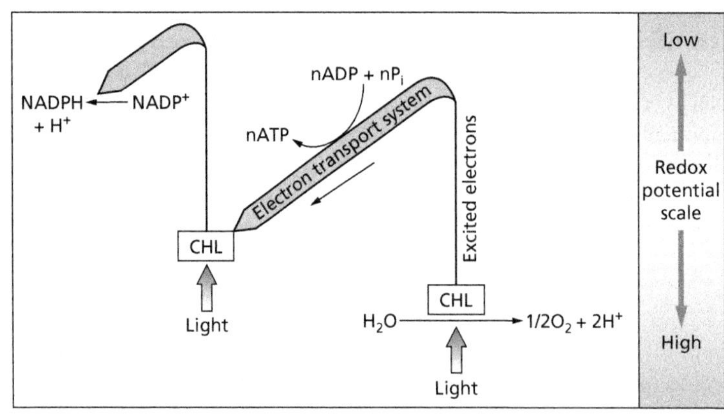

Fig. 3.13 A simplified scheme for non-cyclic photophosphorylation.

reaction centre of photosystem II requires a supply of electrons to maintain the operation of the process. They are provided by the photolysis of water and as a result oxygen is released, i.e. this process is oxygenic.

Photolysis of water:

$$H_2O\ (+\text{light energy}) \rightarrow 2e^- + 2H^+ + {}^1/_2\,O_2$$

ATP can also be produced from electrons generated by photosystem I, if they are not used in NADPH production. The primary electron acceptor of photosystem I can pass electrons through cytochromes and cycles them back to the reaction centre of photosystem I, from where they came. During this, ATP is synthesized by cyclic photophosphorylation (Fig. 3.14).

Chemolithotrophic autotrophy

Chemolithotrophs use electrons obtained from reduced inorganic compounds as an energy source and inorganic carbon, usually CO_2, as their source of carbon, which is fixed via the Calvin cycle. They possess electron transport systems, generate a proton-motive force and produce ATP by oxidative phosphorylation. Reducing power, NAD(P)H, is generated using reversed electron transport, as in bacterial photosynthesis. Reversal of electron transport is driven by ATP or the proton-motive force, and electrons are provided by available inorganic donors (Fig. 3.15). The ATP and NAD(P)H generated are then used in CO_2 fixation.

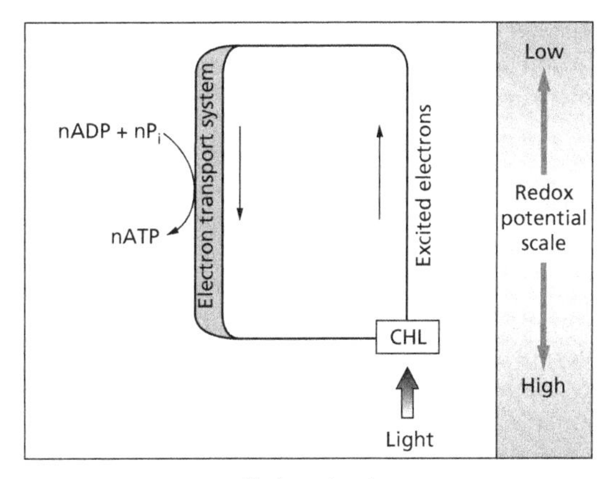

Fig. 3.14 A simplified scheme for cyclic photophosphorylation.

Inorganic energy sources utilized include reduced forms of nitrogen, sulphur, iron and hydrogen. Ammonia (NH_3) and nitrite (NO_2^-) are the most common inorganic nitrogen compounds used chemoautotrophically. They are oxidized aerobically by a limited range of bacteria, a process known as nitrification. *Nitrosomonas* oxidizes ammonia to nitrite, and *Nitrobacter* oxidizes nitrite to nitrate (NO_3^-). Aerobic nitrifying bacteria also use the Calvin cycle for CO_2 fixation. The ATP requirements for this process are considerable and place an additional burden on an inefficient energy-generating system.

Nitrosomonas:

$$2NH_4^+ + 3O_2 \rightarrow 2NO_2^-\ (\text{nitrite}) + 2H_2O + 4H^+$$

Nitrobacter: $2NO_2^- + O_2 \rightarrow 2NO_3^-$ (nitrate)

These organisms are widespread in water and particularly in soil where they have a significant impact on soil fertility. Nitrate, the end-product of nitrification, is water-soluble and rapidly leached from soils, whereas ammonia is cationic and adsorbs readily to negatively charged clay minerals. Nitrification is therefore an undesirable process in agriculture, especially in areas with high rainfall. Run-off of nitrate produced in this way and from added agricultural fertilizer, into reservoirs, rivers and groundwaters, also constitutes a major pollution problem. Nitrification occurs in well-drained soils

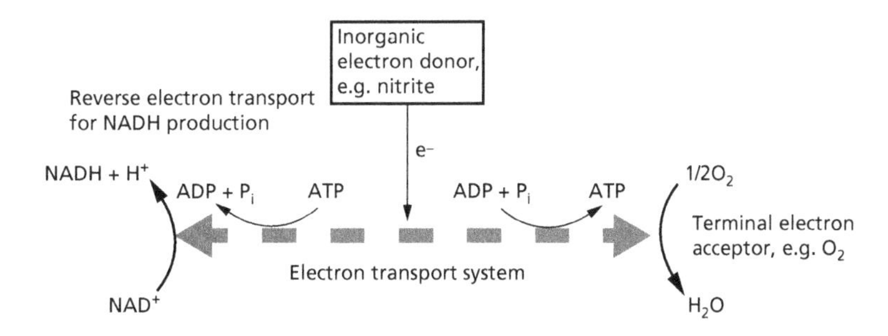

Fig. 3.15 Reverse electron transport.

at neutral pH, but is inhibited by anaerobic conditions and acidity.

The most common chemoautotrophic energy sources containing reduced forms of sulphur are hydrogen sulphide (H_2S), elemental sulphur (S^0) and thiosulphate ($S_2O_3^{2-}$). Sulphate (SO_4^{2-}) is the end-product of sulphur oxidation.

Thiosulphate: $S_2O_3^{2-} + 2O_2 + H_2O \rightarrow 2SO_4^{2-} + 2H^+$

Free sulphur: $2S^0 + 2H_2O + 3O_2 \rightarrow 2SO_4^{2-} + 2H^+$

One genus able to obtain energy in this way is *Thiobacillus*. They are Gram-negative rods approximately 2 µm in length, some of which are acidophiles that can considerably lower the pH of their environment. For example, *Thiobacillus thiooxidans* oxidizes sulphur not iron, and *T. ferrooxidans* oxidizes both. This acidophilic sulphur bacterium is one of the few microorganisms capable of the indirect generation of energy via the aerobic oxidation of ferrous (Fe^{2+}) to ferric (Fe^{3+}) ions. The energy is generated from the pH gradient across the cell membrane, created by acidic conditions. As protons (H^+) flow in, ATP is formed. Once inside, the protons need to be 'removed', to keep the internal pH near neutral. Ferrous ions are used as electron donors to oxygen as a terminal electron acceptor, combining with H^+ to form water, i.e. iron functions as electron supplier to remove protons. The cells process a large quantity of iron for very small yields of energy, which can be seen as an insoluble reddish yellow precipitate of ferric hydroxide ($Fe(OH)_3$).

T. ferrooxidans is common in acid-polluted environments such as coal mines and mining wastes. Here, one of the most common forms of iron and sulphur in nature, the mineral pyrite (FeS_2), is usually present. The problem of acid mine drainage occurs following pyrite oxidation by this organism, as it leads to formation of sulphuric acid (H_2SO_4). This process is aerobic and as long as coal is unmined no pyrite oxidation can occur. However, as coal seams are exposed by mining operations, oxygen becomes available, and they quickly become contaminated with *T. ferrooxidans*. This organism's activities generate sulphuric acid and solubilize metals, which quickly leach into surrounding waters, where both the acidity and dissolved heavy metals are toxic to aquatic life. Nevertheless, these activities can be usefully employed in commercial biomining operations, such as the leaching of copper from low-grade sulphur ores (see Chapter 15).

With regard to molecular hydrogen utilization, several chemoautotrophic bacteria, including *Ralstonia eutropha* (formerly *Alcaligenes eutrophus*) utilize it as an energy source with oxygen as a terminal electron acceptor. Such bacteria tend to be facultative chemoautotrophs, capable of chemoheterotrophic growth, although some obligatory autotrophic species have also been found. Certain bacteria incapable of autotrophy oxidize hydrogen as an energy source, but cannot utilize this reducing power to fix CO_2 and must be provided with an organic carbon source. These bacteria are sometimes referred to as **mixotrophs**. Bacteria such as sulphate reducers and methanogens use hydrogen as an electron donor and all are obligate anaerobes. Finally, some phototrophic purple sulphur bacteria can grow as hydrogen oxidizers, under aerobic conditions in darkness.

$$H_2 + NAD^+ \xrightarrow{\textit{hydrogenase}} NADH + H^+$$

(alternatively, electrons can be donated directly to the ETS).

Carbon dioxide fixation (assimilation) via the Calvin cycle

The **Calvin cycle (ribulose bisphosphate pathway)** for the assimilation of CO_2 is operated by autotrophs: algae, higher plants, most photosynthetic and chemolithotrophic bacteria, and autotrophic methylotrophs (see p. 63). In eukaryotic autotrophs this pathway is located within the stroma (matrix) of chloroplasts, and some prokaryotes have special carboxysomes that contain key Calvin cycle enzymes. This formation of multicarbon compounds from the one-carbon compound, CO_2, requires considerable quantities of ATP and NADPH.

The Calvin cycle is essentially divided into three phases: carboxylation, reduction and regeneration (Fig. 3.16). Initial carboxylation involves the addition of CO_2 to a C_5 acceptor molecule, ribulose 1,5-bisphosphate (RuBP), by RuBP carboxylase. The resulting C_6 product is immediately split into two C_3 units, forming two molecules of 3-phosphoglycerate (PGA). In the following reduction phase, a phosphate group (from ATP) and hydrogen (from NADPH) are added to each PGA molecule to produce GAP. The cycle must complete three turns, fixing a CO_2 molecule each time, in order to produce a net gain of one GAP (C_3), which can then enter anabolic metabolism.

$$3CO_2 + 9ATP + 6NADPH + 6H^+ + 6H_2O \rightarrow GAP + 9ADP + 6P_i + 6NADP^+$$

However, not all GAP is taken as product, some must be used in regenerating more RuBP acceptor to maintain the operation of the cycle. This regeneration phase involves PP pathway enzymes, including transaldolase and transketolase. Together they produce ribulose 5-phosphate, which is further phosphorylated to form the CO_2 acceptor RuBP (Fig. 3.16).

Methylotrophic metabolism

Methylotrophy is the ability of organisms to utilize one-carbon compounds, other than carbon dioxide, as their sole carbon source. This includes compounds such as methane, methanol, formate, carbon monoxide, chloromethane and cyanide. The definition is usually extended to the utilization of compounds that may contain more than one carbon atom, but no carbon–carbon bonds, such as dimethylamine, trimethylamine and dimethylsulphide. Methylotrophs are found within bacteria, yeasts and filamentous fungi. They may be aerobic or anaerobic and can be divided into three categories (examples shown in Table 3.2).

1 Heterotrophic methylotrophs, which can also grow heterotrophically on polycarbon compounds.

2 Obligate methylotrophs, unable to grow on polycarbon substrates or CO_2.

3 Autotrophic methylotrophs or C_1-utilizers that oxidize reduced C_1 compounds to CO_2, before fixing it via the Calvin cycle, as described above.

Heterotrophic and obligate methylotrophic bacteria operate one of two assimilation pathways: the **ribulose monophosphate pathway (Quayle cycle)**, which has two variants and may be an evolutionary antecedent of the Calvin cycle, or the **serine pathway**. Yeasts operate the **xylulose monophosphate pathway (dihydroxyacetone cycle)**. Key enzymes of each pathway are shown in Fig. 3.17. In all three assimilation pathways the reduced C_1 compounds are assimilated at the level of formaldehyde and there is the net generation of a three-carbon

Table 3.2 Examples of methylotrophic microorganisms

Heterotrophic methylotrophs
Bacteria
Arthrobacter P1 (RMP)
Methylobacterium organophylum (S)
Hyphomicrobium species (S)
Bacillus PM6 (RMP)
Yeasts
Candida boidinii (XMP)
Pichia angusta (*Hansenula polymorpha*) (XMP)
Filamentous fungi
Gliocladium deliquescens
Paecilomyces variotii
Trichoderma lignorum
Obligate methylotrophic bacteria
Methylococcus species (RMP)
Methylomonas species (RMP)
Methylocystis species (S)
Methylophilus methylotrophus (RMP)
Autotrophic methylotrophic bacteria (C_1-utilizers)
Paracoccus denitrificans (C)
Pseudomonas carboxydovorans (C)

C_1 assimilation pathway in parenthesis: C, Calvin cycle; RMP, ribulose monophosphate pathway; S, serine pathway; XMP, xylulose monophosphate pathway.

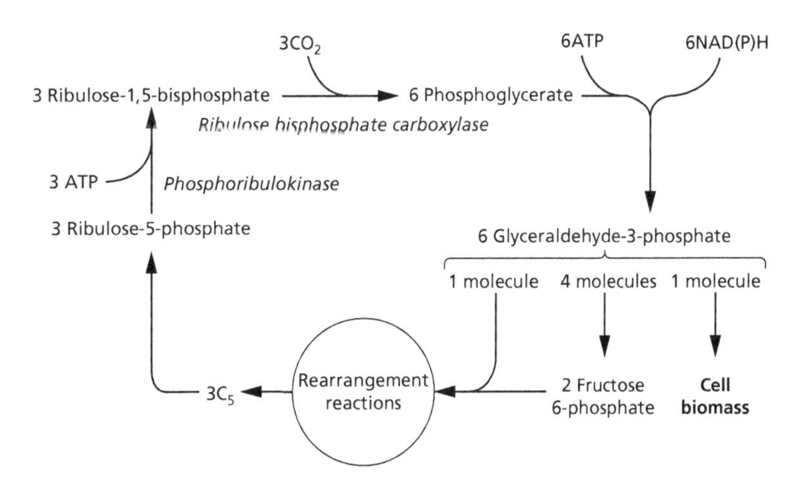

Fig. 3.16 The Calvin cycle.

Ribulose monophosphate pathway

Hexulose phosphate synthase *Hexulose phosphate isomerase*

Formaldehyde + ribulose 5-phosphate → hexulose 6-phosphate → fructose 6-phosphate

Serine pathway

Serine transhydroxymethylase

Formaldehyde + glycine → serine

Xylulose monophosphate pathway

Dihydroxyacetone synthase
(a special transketolase)

Formaldehyde + xylulose 5-phosphate → glyceraldehyde 3-phosphate + dihydroxyacetone

Calvin cycle

Ribulose bisphosphate carboxylase

CO_2 + H_2O + ribulose 1,5-bisphosphate → 2 phosphoglycerate

Fig. 3.17 Characteristic enzymes of C_1 assimilation in methylotrophs.

compound from the fixation of three formaldehyde molecules, or three CO_2 molecules in the case of the Calvin cycle. These C_3 compounds are then available to enter biosynthetic pathways (Fig. 3.18).

The methylotrophs have functional separation of carbon assimilation and dissimilation (energy generation). Most organisms generate energy by the dissimilation of a portion of the formaldehyde produced, via a linear pathway, to form CO_2 (Fig. 3.18), although, in some bacteria, this dissimilation occurs in the cyclic phosphogluconate pathway.

Metabolic regulation

Microorganisms naturally have tight control over their metabolism, but this often needs to be overcome when they are used in industrial fermentations, e.g. in attempting to over-produce specific metabolites. Consequently, the study of metabolic regulation is of special importance in industrial microbiology. The knowledge obtained allows the metabolic pathways of industrial microorganisms to be specifically manipulated to suit the requirements of the industrial process. This may be achieved through initial strain selection, strain development and the optimization of fermentation conditions (see Chapter 4).

Most metabolic pathways are controlled by a combination of several different systems organized to conserve energy and raw materials, while maintaining cell components at the necessary optimum concentrations. Regulation of catabolism and anabolism are somewhat different. Both may be regulated by their end-products and the concentration of ATP, adenosine diphosphate (ADP), adenosine monophosphate (AMP) and NAD^+, but in anabolism end-product regulation is of greater importance (see below). Anabolic and catabolic pathways may also be compartmentalized within separate cellular locations and they often use different cofactors. For example, catabolism mostly produces NADH, but it is NADPH that is normally used in biosynthesis. The regulation is imposed at two levels, the modification of enzyme activity and the control of enzyme synthesis.

Modification of enzyme activity

This involves the stimulation or inhibition of specific enzymes to rapidly change the flow of carbon within a pathway.

PHYSICAL SEPARATION OF ENZYME AND SUBSTRATE

The simplest mechanism for the regulation of enzyme activity merely involves controlling the concentration of available substrate and coenzymes. The most obvious mechanism is the control of substrate entry into a cell across the cytoplasmic membrane. This is mediated by specific transport systems that control the uptake of

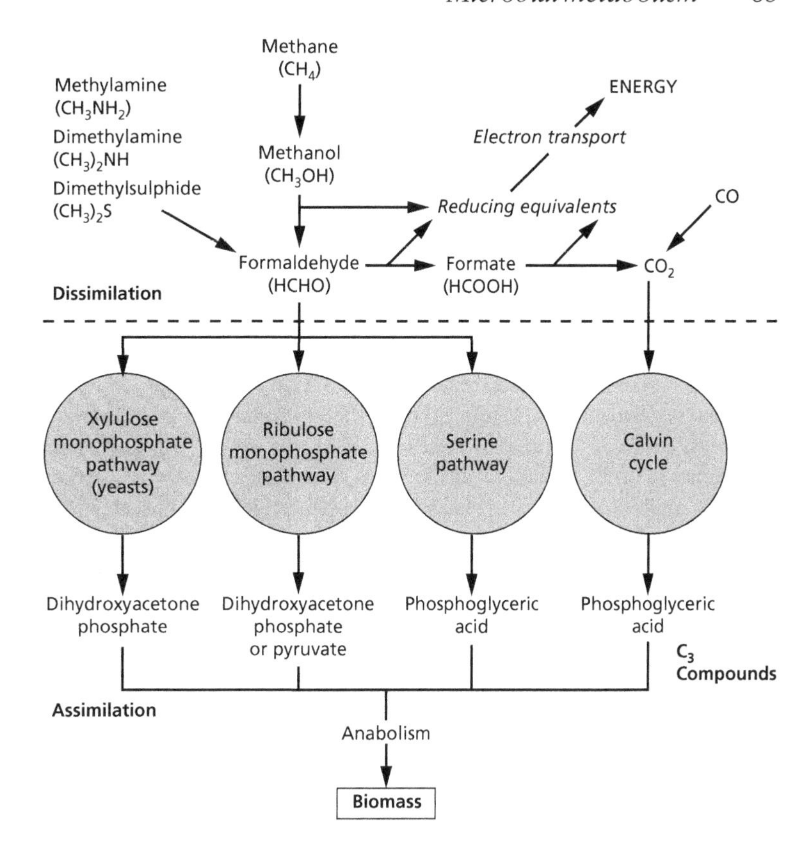

Fig. 3.18 A simplified diagram of dissimilation and assimilation pathways in methylotrophs.

many nutrients and may act as a simple means of regulation for some intracellular enzymes.

In eukaryotic cells, which have numerous membrane-bound organelles, compartmentalization can be used to control the transport of metabolites and coenzymes between different regions of the cell. Compartmentalization also allows the simultaneous, but separate, operation and regulation of similar pathways. Prokaryotes lack organelles, but development of concentration gradients of metabolites and coenzymes, between different regions of the cytoplasmic matrix, may play a similar regulatory role.

REGULATORY ENZYMES

Many metabolic pathways contain **regulatory enzymes** whose activity can be adjusted to provide a mechanism for fine tuning the flux of carbon through a pathway. Some of these enzymes are **allosteric**, usually composed of several subunits, and do not follow Michaelis–Menten kinetics. Their activity can be altered by a small molecule called an effector or modulator. This effector molecule binds reversibly through non-covalent forces to the regulatory site on the enzyme, causing the enzyme to change shape. As a result, the catalytic site is altered. Positive effectors bring about an increase in enzyme activity and negative effectors cause enzyme inhibition. In addition, some regulatory enzymes are switched on and off by their reversible covalent modification, such as phosphorylation and dephosphorylation.

Every metabolic pathway has at least one **pacemaker enzyme** that catalyses the slowest or rate-limiting reaction. This is usually the first step or first unique step of the pathway and it is often irreversible. It controls the overall rate of flux through the pathway and is normally inhibited by the product of the last step in the pathway, which is called feedback inhibition or end-product inhibition. If levels of the end-products become too high, the pathway is slowed via this mechanism. Pathways with branches are further controlled by having similar regulatory enzymes at each arm of the branch point, allowing separate control of the branches. End-products of each branch often inhibit both the enzyme at the start of their respective branch and the pacemaker enzyme at the beginning of the main pathway. In some branched pathways, several isoenzymes (isozymes) may control the first step, rather than having a single initial pacemaker enzyme (Fig. 3.19a). This provides a more subtle means of control for the flow of carbon, as each isozyme can be controlled by a different end-product. Alternatively, when there is just one enzyme, it may be regulated by all end-products of the branches. In certain cases, all end-products must be present at elevated levels to completely inhibit the enzyme, a concerted effect (Fig. 3.19b), or they may have a cumulative effect (Fig. 3.19c). In some branched pathways control is sequential. First, the intermediate at the branch point accumulates, due to end-product inhibition of the enzymes responsible for its onward metabolism, whereupon it inhibits the enzyme at the beginning of the pathway (Fig. 3.19d).

Control of enzyme synthesis

The concentration of any enzyme is determined by its rate of synthesis, rate of dilution (where growth proceeds when synthesis of the enzyme ceases), and the rate of inactivation and degradation by proteases. However, it is enzyme synthesis that generally plays the most important role in regulation, albeit providing a slower means of regulating flux through metabolic pathways than does modification of enzyme activity. It may be considered as a coarse tuning mechanism, as opposed to the fine tuning provided by regulatory enzymes, and is more difficult to reverse.

Some enzymes are constantly synthesized and are **constitutive** components of the cell, whereas others are synthesized only when required. Even the constitutive enzymes are produced at different concentrations, as some genes are more highly expressed than others. The expression is controlled by the strength of their gene promoter, which lies upstream from the structural gene for the enzyme. Strong promoters are very useful in genetic engineering. Their attachment to heterologous genes that are cloned into a host organism usually ensures a high level of gene expression.

Many catabolic enzymes are regulated by **induction**, as it is efficient for them to be synthesized only when their substrates are present, whereas synthesis of anabolic enzymes is mostly controlled through **repression**. When a cell accumulates excess end-product of an anabolic pathway, such as an amino acid, or if a supply of that compound is encountered in the environment, it responds by slowing or stopping the synthesis of enzymes in the biosynthetic pathway for this product. A more immediate effect, as discussed above, is that the regulatory enzymes of this pathway are inhibited, thus preventing needless further carbon flux through the pathway.

The translation stage of protein synthesis does not appear to be a major control point in prokaryotes, but may be of greater significance in eukaryotic organisms. Control of enzyme synthesis in prokaryotes is normally at

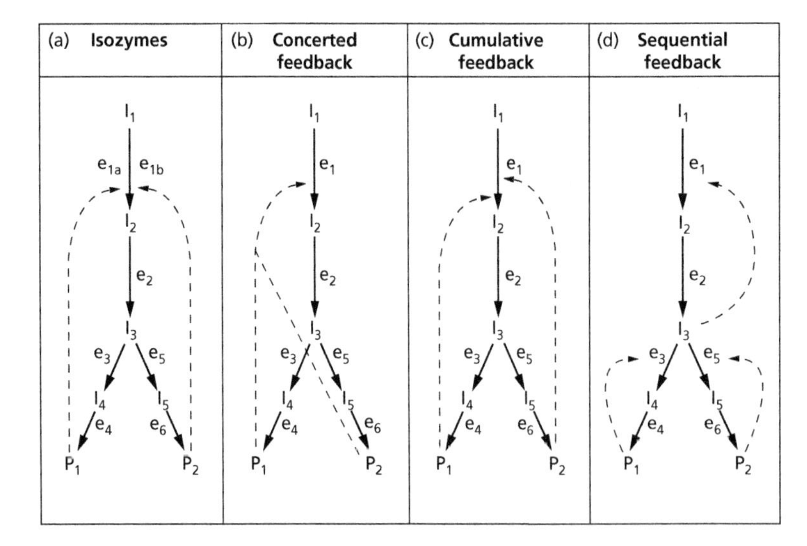

Fig. 3.19 Examples of feedback control of branched pathways (e = enzymes; I = intermediates; P = end-products and the broken lines are the paths of inhibition).

the level of transcription (mRNA synthesis). Studies on the induction of β-galactosidase in *E. coli* have provided a considerable amount of information regarding the mechanisms involved. This enzyme catalyses the hydrolysis of lactose to glucose and galactose, and is not actively synthesized if lactose is not present. In the absence of lactose, a repressor protein, produced under the direction of regulatory genes, binds to the operator region of the DNA, upstream from the structural gene for β-galactosidase, thereby blocking the synthesis of mRNA by RNA polymerase. However, when the inducer, lactose (specifically allolactose), is available, it binds reversibly to repressor molecules, which changes their shape to inactivate them. The change of conformation prevents them from binding to the operator region of the DNA and allows transcription of the gene to proceed. For enzyme synthesis controlled by repression, the opposite occurs. Repressors change to active forms in the presence of a corepressor, enabling them to bind to the operator region thereby inhibiting transcription. The corepressor is the metabolite whose synthesis is controlled by the enzyme and it effectively regulates its own production. For example, in tryptophan biosynthesis, excess tryptophan acts as a corepressor which activates the repressor protein.

When a metabolite is catabolized or synthesized by a series of enzymes, it is usually advantageous for all enzymes of the pathway to be switched on or off simultaneously. In bacteria this coordinated action is more easily achieved as the genes for these enzymes are often located together as a cluster on the chromosome called an **operon**. For example, the lactose (lac) operon consists of three structural genes for β-galactosidase, β-galactoside permease and β-galactoside transacylase, whose synthesis is controlled simultaneously by both the lac repressor protein and the catabolite activator protein (see below). In eukaryotes the situation is much more complex, as genes for related activities tend to be scattered, rather than clustered together as an operon.

Enzyme synthesis may also be controlled by a mechanism called **attenuation**, where regulation of gene expression is via termination of transcription, rather than control of initiation. This involves termination sites within the operon that result in truncated mRNA. Tryptophan biosynthesis, which involves an operon of five enzymes, is coordinated through both repression, as mentioned above, and attenuation.

Mechanisms of general regulation

REGULATION OF RESPIRATION AND FERMENTATION

Aerobic respiration and fermentation can be regulated by environmental conditions, which include the availability of oxygen and sugars. Under aerobic conditions, many organisms exhibit a slower rate of sugar catabolism via glycolysis than when maintained under anaerobic conditions. This is because fewer carbon units have to be metabolized to obtain the same quantity of ATP, as aerobic respiration and associated oxidative phosphorylation is considerably more efficient than fermentation. This suppression of glycolysis by oxygen is a regulatory phenomenon referred to as the **Pasteur effect**, which is apparent only at low sugar concentrations (less than 5 mmol/L for some yeasts). The flux through glycolysis appears to be primarily controlled by the allosteric enzyme phosphofructokinase, which is regulated by the relative concentrations of ATP and AMP. When ATP concentrations fall relative to AMP, enzyme activity increases.

Several yeasts exhibit a further phenomenon, the **Crabtree effect**, where at high sugar concentrations, even in the presence of oxygen, fermentation overrides respiration. Consequently, NADH generated via the EMP pathway is mainly oxidized by fermentative means rather than by mitochondrial respiration. This may be explained by the saturation of the limited respiratory capacity of Crabtree-positive organisms.

CARBON CATABOLITE REPRESSION

In many microorganisms the enzymes for glucose catabolism are constitutive, whereas additional enzymes must usually be produced for the catabolism of any other carbon and energy source. If glucose is present in the medium, along with other sugars, it is always the first to be used. The result is sequential utilization that produces a diauxic (biphasic) growth pattern (Fig. 3.20). This is because the production of enzymes for the uptake and utilization of the other sugars, and several other enzymes, is repressed by glucose or an initial product of its metabolism. In some cases, it is due to glucose-induced inactivation of key enzymes; for others, repression results from rapid glucose uptake, which causes the intracellular level of cyclic adenosine monophosphate (cAMP) to fall. This important compound is involved in the positive control of many catabolic enzymes, e.g. the

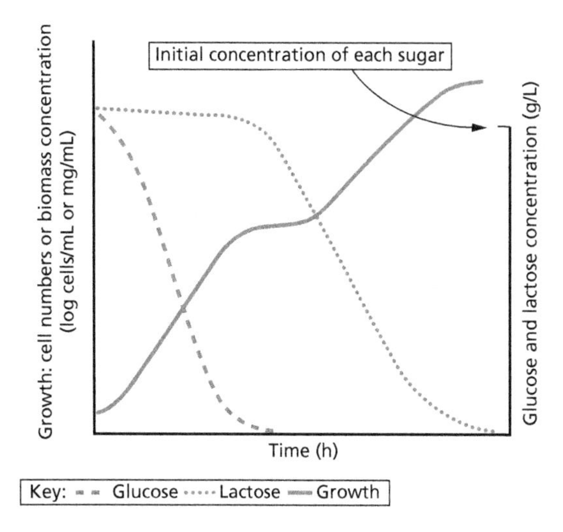

Fig. 3.20 Diauxic growth of a microorganism on a mixture of glucose and lactose.

lac operon. cAMP is the activator of a protein called catabolite activator protein (CAP), which in the active state binds to a site within the promoter region of the gene. Without it, RNA polymerase is unable to bind and start gene transcription. Therefore, in the absence of cAMP only inactive CAP molecules are present and transcription does not take place. Levels of cAMP rise only when the rate of glucose transport into the cell falls below a certain level. Although glucose has been the most extensively studied, this phenomenon is not restricted to glucose, as other carbon substrates have been shown to act in a similar repressive manner.

Carbon catabolite regulation of primary metabolism provides a selective advantage for the organism, as it ensures the economical use of metabolic machinery. It also plays a role in controlling secondary metabolism. For example, glucose has a repressive effect on the production of many secondary metabolites, including antibiotics (β-lactams and tetracylines) and some alkaloids. Consequently, if glucose is used as a substrate for their industrial production, it has to be fed at low rates in order to overcome this repressive effect.

ADENYLATE ENERGY CHARGE

Adenylates (ATP, ADP and AMP) both individually and collectively have important control functions within the cell. The overall adenylate energy charge ratio (EC) of a cell is determined by the following formula:

$$EC = \frac{ATP + \frac{1}{2}ADP}{AMP + ADP + ATP}$$

If all adenylate is in the form of ATP, the EC value is 1.0 ('fully charged'). Values for actively growing cells are normally in the range of 0.8–0.95. Enzymes involved in ATP synthesis tend to be inhibited at high EC levels and, conversely, those anabolic enzymes that consume ATP are often stimulated if the EC is high. The EC may also play a role in the regulation of secondary metabolism.

Secondary metabolism and its control

Secondary metabolism is performed by many filamentous fungi, but is rather less widespread amongst bacteria. Microorganisms that produce secondary metabolites exhibit two phases during batch culture, the **trophophase** and **idiophase** (Fig. 2.1). Trophophase is the growth phase of a culture and idiophase is the following period when the secondary metabolites are formed. Production frequently begins when active growth has ceased, often as the growth-limiting substrate concentration approaches the K_s value for the culture (see Chapter 2). The success of any further biosynthesis in the idiophase is dependent on the preceding trophophase. Secondary metabolism during the idiophase utilizes primary metabolites to produce species-specific and chemically diverse end-products that are not essential for growth of the microorganism. Their production involves metabolic pathways which are not used during the growth phase, at least not during exponential growth (Table 3.3). These pathways are often not as well characterized as those for primary metabolism.

Although secondary metabolism can be easily demonstrated in batch culture, it can be studied in continuous systems, as it occurs at low dilution (growth) rates. Here, secondary metabolism can be demonstrated to operate at the same time as, and not after, primary metabolism.

Much energy can be expended in the production of secondary metabolites. However, many of these products appear to have no recognizable metabolic function for the producing organism, yet they are not merely end-products of metabolism, such as fermentation end-products. Those that perform specific roles for their producer organism include sideramines (ferrichromes and ferrioxamines) and ionophore antibiotics (macrotetralides). Both groups are involved in the uptake of

Table 3.3 Some metabolic pathways involved in the production of secondary metabolites by fungi

Secondary metabolite	Pathways
Ergot alkaloids	Mixed biosynthetic origin (amino acid–isoprenoid)
Griseofulvin	Derived from 7 acetate units via **polyketide pathways**
Gibberellic acid	Derived from mevalonate via **isoprenoid pathways**
Gallic acid	Derived from **shikimate pathway** intermediates
Kojic acid	Derived directly from glucose
β-lactams (e.g. penicillin)	Derived from the amino acids L-α-aminoadipic acid D-cysteine and L-valine

Table 3.4 Some examples of industrially important secondary metabolites and their producer microorganisms

Agricultural aids	Gibberellins (plant hormones)—*Fusarium moniliforme*
Analgesics	Ergot alkaloids—*Claviceps purpurea*
Antibiotics	Cephalosporin (a β-lactam)—*Acremonium chrysogenum* Erythromycin (a macrolide)—*Saccharapolyopora erythraea* Penicillin (a β-lactam)—*Penicillium chrysogenum* Streptomycin (an aminoglycoside)—*Streptomyces griseus*
Anticancer drugs	Anthracycline antibiotics (e.g. adriamycin)—*Streptomyces peucetius* Actinomycin D—*Streptomyces antibioticus* Bestatin—*Streptomyces olivoreticuli*
Hallucinogens	Lysergic acid—*Claviceps paspali*
Immunosuppressants	Cyclosporin—*Tolypocladium inflatum*
Pigments	Carotenoids—*Blakeslea trispora*
Plastics	Kojic acid—*Aspergillus oryzae*
Toxins	Diphtheria toxin—*Corynebacterium diphtheriae* Tetanus toxin—*Clostridium tetani* Mycotoxins, e.g. aflatoxins—*Aspergillus flavus*
Ulcer treatment	Pepstatin (pepsin inhibitor)—*Streptomyces testaceus*

cations. Some compounds may inhibit competing organisms and others such as gramicidins are associated with the promotion of spore formation. Fortuitously, many secondary metabolites exhibit properties that have proved very useful and have become major industrial microbial products (Table 3.4).

The synthesis of secondary metabolites is usually tightly regulated by the cell. Some regulatory mechanisms are common to both primary and secondary metabolism, including the cell's adenylate EC and carbon catabolite repression. Carbon sources that support high growth rates tend to be repressive; for example, glucose suppresses the production of several antibiotics including penicillin and chloramphenicol.

Additional control of secondary metabolism is provided by the following.

1 **Non-ribosomal peptide synthesis.** Normally, protein synthesis involves transcription of the genetic code of DNA to mRNA, which is then translated to form a polypeptide by a ribosome. However, several small peptides, such as some antibiotics, are formed via an alternate mechanism, independent of mRNA and ribosomes. The peptide is synthesized through the action of peptide synthetases, sometimes referred to as antibiotic-committed enzymes, produced at the end of exponential growth. These enzymes, rather than mRNA, act as the template for peptide synthesis. For example, gramicidin S, produced by *Bacillus brevis*, is composed of two

pentapeptide dimers, and each dimer is formed from four activated amino acids (amino-acyl derivatives). The other amino acid, L-phenylalanine, is activated by gramicidin synthase II and converted to the D-form. This is then transferred on to gramicidin synthase I and the other activated amino acids are linked in turn to form a pentapeptide. Two pentapeptide dimers then combine to form the active cyclic gramicidin S molecule.

Similar peptide synthetases are involved in the synthesis of the antifungal compounds echinocandin (a cyclic lipopeptide produced by *Aspergillus nidulans*) and cyclosporin (an undecapeptide from *Tolypocladium inflatum*). Cyclosporin is exploited as an immunosuppressant that is extensively used in the treatment of autoimmune diseases and in transplant surgery. The synthetase involved in its formation is a very large single polypeptide with a molecular mass of 1.7 million Daltons.

2 **Autoregulation.** This may involve small molecules, such as 'factor A' (2 (*S*)-isocapryloyl-3-(*S*)-hydroxymethyl-γ-butyrolactone), which is produced by *Streptomyces* species. They appear to function as hormone-like autoregulatory compounds that stimulate the formation of secondary metabolites. Factor A controls streptomycin biosynthesis, streptomycin resistance and sporulation, and is effective at concentrations as low as 1×10^{-9} mol/L.

3 **End-product regulation.** As in primary metabolism, end-products may play a role in the feedback inhibition of certain secondary metabolic pathways. For example, the antibiotic chloramphenicol inhibits aryl amine synthatase, an enzyme involved in chloramphenicol biosynthesis.

4 **Inducible effects.** Production of several antibiotics is stimulated by the addition of inducer compounds to the fermentation medium. Many inducers are primary metabolites; for example, the amino acid methionine induces cephalosporin production in *Acremonium* (*Cephalosporium*) *chrysogenum*. The alkaloid ergoline, which is produced by *Claviceps purpurea*, is stimulated by the addition of L-tryptophan or tryptophan analogues to the culture media during the exponential phase of growth.

5 **Nitrogen and phosphate regulation.** Readily metabolizable sources of nitrogen, e.g. ammonia, are often seen to inhibit the production of secondary metabolites such as penicillin. Also, formation of candicidin, streptomycin and tetracycline is inhibited by phosphate concentrations in excess of 10 mmol/L. Usually a maximum level of 1 mmol/L phosphate is necessary to ensure that inhibition does not occur. This phenomenon may be associated with changes in EC, which has been shown to be directly influenced by the rate of phosphate flux into some cells. A high rate of flux increases ATP formation and raises the overall EC. In certain cases, the ATP may act as a corepressor in the synthesis of key enzymes involved in the biosynthesis of secondary metabolites.

In order to obtain high yields of secondary metabolites from industrial fermentations, environmental conditions that elicit these regulatory mechanisms, particularly repression and feedback inhibition, must be avoided. If potentially suppressive nutrients are used, they must be provided at low subsuppressive rates. In addition, overproducing mutant strains are often developed for these processes.

Further reading

Papers and reviews

Maden, B. E. (1995) No soup for starters? Autotrophy and the origins of metabolism. *Trends in Biochemical Science* **20**, 337–341.

Marians, K. J. (1992) Prokaryotic DNA replication. *Annual Review of Biochemistry* **61**, 673–719.

Mitchell, P. (1961) Coupling of phosphorylation and hydrogen transfer by a chemiosmotic type of mechanism. *Nature* **191**, 144–148.

Russell, J. B. & Cook, G. M. (1995) Energetics of bacterial growth: balance of anabolic and catabolic reactions. *Microbiology Reviews* **59**, 48–62.

Sebald, M. & Hauser, D. (1994) Pasteur, oxygen and the anaerobe revisited. *Anaerobe* **1**, 11–16.

Shively, J. M., van Keulen, G. & Meijer, W. G. (1998) Something from almost nothing: carbon dioxide fixation in chemoautotrophs. *Annual Review of Microbiology* **52**, 191–230.

Tijian, R. (1995) Molecular machines that control genes. *Scientific American* **258**, 56–63.

Books

Berry, D. R. (ed.) (1988) *Physiology of Industrial Fungi*. Blackwell Scientific Publications, Oxford.

Caldwell, D. R. (1995) *Microbial Physiology and Metabolism*. William C. Brown, Oxford.

Danson, M. J., Hough, D. W. & Lunt, G. G. (1992) *The Archaebacteria: Biochemistry and Biotechnology*. Portland Press, London.

Dawes, I. W. & Sutherland, I. W. (1992) *Microbial Physiology*, 2nd edition. Blackwell Scientific Publications, Oxford.

Dickinson, J. R. & Schweizer, M. (eds) (1998) *The Metabolism and Molecular Physiology of* Saccharomyces cerevisiae. Taylor & Francis, London.

Griffin, D. H. (1994) *Fungal Physiology*, 2nd edition. Wiley, New York.

Koch, A. L. (1995) *Bacterial Growth and Form*. Chapman & Hall, New York.

Large, P. J. & Bamforth, C. W. (1988) *Methylotrophy and Biotechnology.* Longman, Harlow.

Moat, A. G. & Foster, J. W. (1995) *Microbial Physiology*, 3rd edition. Wiley, New York.

Walker, G. M. (1998) *Yeast Physiology and Biotechnology.* Wiley, Chichester.

Part 2 Bioprocessing

4 Industrial microorganisms

Microoorganisms are used extensively to provide a vast range of products and services (Table 4.1). They have proved to be particularly useful because of the ease of their mass cultivation, speed of growth, use of cheap substrates (which in many cases are wastes) and the diversity of potential products. Their ability to readily undergo genetic manipulation has also opened up almost limitless further possibilities for new products and services from the fermentation industries.

Traditional fermentations were originally performed (and still are in some cases) by a mixture of wild microorganisms emanating from the raw materials or the local environment, e.g. some food and alcoholic beverage fermentations. Initial attempts to improve the microorganisms involved occurred little more than 120 years ago, when they were first isolated from these processes as pure cultures from which the most useful strains were then selected. Those fermentation processes developed during the first 80 years of the 20th century have mostly used monocultures. The specific microorganisms employed were often isolated from the natural environment, which involved the random screening of a large number of isolates. Alternatively, suitable microorganisms were acquired from culture collections (see p. 78). Most of these microorganisms, irrespective of their origins, were subsequently modified by conventional strain improvement strategies, using mutagenesis or breeding programmes, to improve their properties for industrial use. Several processes developed in the last 20 years have involved recombinant microorganisms and genetic engineering technology has increasingly been used to improve established industrial strains.

In most cases, regulatory considerations are of major importance when choosing microorganisms for industrial use. Fermentation industries often prefer to use established GRAS (generally regarded as safe) microorganisms (Table 4.2), particularly for the manufacture of food products and ingredients. This is because requirements for process and product approval using a 'new' microorganism are more stringent and associated costs are much higher. Where pathogens and some genetically manipulated microorganisms (GMMs) are used as the producer organism, additional safety measures must be taken. Special containment facilities are employed and it may be possible to use modified ('crippled') strains that cannot exist outside the fermenter environment (see Chapter 8).

Isolation of suitable microorganisms from the environment

Strategies that are adopted for the isolation of a suitable industrial microorganism from the environment can be divided into two types, **'shotgun'** and **objective** approaches. In the shotgun approach, samples of free living microorganisms, biofilms or other microbial communities are collected from animal and plant material, soil, sewage, water and waste streams, and particularly from unusual man-made and natural habitats. These isolates are then screened for desirable traits. The alternative is to take a more objective approach by sampling from specific sites where organisms with the desired characteristics are considered to be likely components of the natural microflora. For example, when attempting to isolate an organism that can degrade or detoxify a specific target compound, sites may be sampled that are known to be contaminated by this material. These environmental conditions may select for microorganisms able to metabolize this compound.

Once the samples have been collected a major problem is deciding on the growth media and cultivation conditions that should be used to isolate the target microorganisms. An initial step is often to kill or repress the proliferation of common organisms and encourage the growth of rare ones. **Enrichment cultures** may then be performed in batch culture, or often more suitably in continuous systems. This encourages the growth of those organisms with the desired traits and increases the quantity of these target organisms, prior to isolation

Table 4.1 Examples of industrial fermentation products and their producer microorganisms

	Bacteria	Yeasts and filamentous fungi
Traditional products		
Bread, beer, wine and spirits		Mainly *Saccharomyces cerevisiae*
Cheeses, other dairy products	Lactic acid bacteria	
Ripening of blue and Camembert-type cheeses		*Penicillium* species
Fermented meats and vegetables	Mostly lactic acid bacteria	
Mushrooms		*Agaricus bisporus, Lentinula edodes*
Soy sauce		*Aspergillus oryzae* *Zygosaccharomyces rouxii*
Sufu (soya bean curd)		*Mucor* species
Vinegar	*Acetobacter* species	
Agricultural products		
Gibberellins		*Fusarium moniliforme*
Fungicides		*Coniothyrium minitans*
Insecticides	*Bacillus thuringiensis*	
Silage	Lactic acid bacteria	
Amino acids		
L-Glutamine	*Corynebacterium glutamicum*	
L-Lysine	*Brevibacterium lactofermentum*	
L-Tryptophan	*Klebsiella aerogenes*	
Enzymes		
Carbohydrases		
α-amylase	*Bacillus subtilis*	
β-amylase		*Aspergillus niger*
amyloglucosidase		*Aspergillus niger*
glucose isomerase	*Streptomyces olivaceus*	
invertase		*Kluyveromyces* species
lactase (β-galactosidase)		*Kluyveromyces lactis*
Cellulases		*Trichoderma viride*
Lipases		*Candida cylindraceae*
Pectinases		*Aspergillus wentii*
Proteases		
subtilisin (alkaline)	*Bacillus licheniformis*	
neutral		*Aspergillus oryzae*
microbial rennet (acid)		*Rhizomucor miehei*
Fuels and chemical feedstocks		
Acetone	*Clostridium* species	
Butanol	*Clostridium acetobutylicum*	
Ethanol	*Zymomonas mobilis*	*Saccharomyces cerevisiae*
Glycerol		*Zygosaccharomyces rouxii*
Methane	Methanogenic archaeans	
Nucleotides		
5′-Inosine monophosphate	*Bacillus subtilis*	
5′-Guanosine monophosphate	*Brevibacterium ammoniagenes*	
Organic acids		
Acetic	*Acetobacter xylinum*	
Citric		*Aspergillus niger* *Yarrowia lipolytica*
Fumaric		*Rhizopus* species
Gluconic	*Acetobacter suboxydans*	
Itaconic		*Aspergillus itaconicus*
Kojic		*Aspergillus flavus*
Lactic	*Lactobacillus delbrueckii*	

Continued

Table 4.1 *continued*

	Bacteria	Yeasts and filamentous fungi
Pharmaceuticals and related compounds		
Alkaloids		
ergotamine		*Claviceps purpurea*
ergometrine		*Claviceps fusiformis*
D-lysergic acid		*Claviceps paspali*
Antibiotics		
Aminoglycosides		
streptomycin	*Streptomyces griseus*	
β-Lactams		
penicillins		*Penicillium chrysogenum*
cephalosporins		*Acremonium chrysogenum*
clavulanic acid	*Streptomyces clavuligerus*	
Lantibiotics		
nisin	*Lactococcus lactis*	
Macrolides		
erythromycin	*Saccharapolyopora erythraea*	
Peptides		
bacitracin	*Bacillus licheniformis*	
gramicidin	*Bacillus brevis*	
Tetracyclines		
chlortetracycline	*Streptomyces aureofasciens*	
Hormones		
Human growth hormone	Recombinant *Escherichia coli*	Recombinant *Saccharomyces cerevisiae*
Insulin	Recombinant *Escherichia coli*	Recombinant *Saccharomyces cerevisiae*
Immunosuppressants		
Cyclosporin		*Trichoderma polysporum*
Interferon	Recombinant *Escherichia coli*	Recombinant *Saccharomyces cerevisiae*
Steroids	*Arthrobacter* species	*Rhizopus* species
Vaccines	*Bacillus anthracis*	
	Clostridium tetani	
	Recombinant *Escherichia coli*	
	Salmonella typhi	
Vitamins		
B_{12} (cyanocobalamin)	*Pseudomonas denitrificans*	
β-Carotene (provitamin A)		*Blakeslea trispora*
Ascorbic acid (vitamin C)	*Acetobacter suboxydans*	
Riboflavin	Recombinant *Bacillus subtilis*	*Ashbya gossypii*
Polymers		
Alginates	*Azotobacter vinelandii*	
Cellulose	*Acetobacter xylinum*	
Dextran	*Leuconostoc mesenteroides*	
Gellan	*Sphingomonas paucimobilis*	
Polyhydroxybutyrate	*Ralstonia eutropha*	
Pullulan		*Aureobasidium pullulans*
Scleroglucan		*Sclerotium rolfsii*
Xanthan	*Xanthomonas campestris*	
Single cell protein	*Methylococcus capsulatus*	*Candida utilis*
	Methylophilus methylotrophus	*Fusarium venenatum*
		Kluyveromyces marxianus
		Paecilomyces variotii
		Saccharomyces cerevisiae

Table 4.2 Examples of microorganisms classified as GRAS (generally regarded as safe)

Bacteria
Bacillus subtilis
Lactobacillus bulgaricus
Lactococcus lactis
Leuconostoc oenos
Yeasts
Candida utilis
Kluyveromyces marxianus
Kluyveromyces lactis
Saccharomyces cerevisiae
Filamentous fungi
Aspergillus niger
Aspergillus oryzae
Mucor javanicus (Mucor circinelloides f. *circinelloides)*
Penicillium roqueforti

Note: Normally, these microorganisms require no further testing if used under acceptable cultivation conditions.

and screening. However, this mode of selection is suitable only for cases where the desired trait provides a competitive advantage for the organisms.

Subsequent isolation as pure cultures on solid growth media involves choosing or developing the appropriate selective media and growth conditions. Once isolated as pure cultures, each must be screened for the desired property; production of a specific enzyme, inhibitory compound, etc. However, at this stage the level of activity or concentration of the target product *per se* is not of major concern, as strain development can normally be employed to vastly improve performance. Selected isolates must also be screened for other important features, such as stability and, where necessary, non-toxicity.

These isolation and screening procedures are more easily applied to the search for a single microorganism. However, it is much more difficult to isolate consortia which together have the ability/characteristic that is sought and whose composition may vary with time. Such groups can be more efficient, particularly where the ability to degrade a complex recalcitrant compound is involved.

Culture collections

Microbial culture collections provide a rich source of microorganisms that are of past, present and potential future interest. There are almost 500 culture collections around the world; most of these are small, specialized collections that supply cultures or other related services only by special agreement. Others, notably national collections, publish catalogues listing the organisms held and provide extensive services for industrial and academic organizations (Table 4.3). In the UK for example, the National Culture Collection (UKNCC) is made up of several collections. They are housed in separate institutions and tend to specialize in bacteria, yeasts, filamentous fungi or algae of either industrial or medical importance; whereas in the USA there is a main centralized collection, the American Type Culture Collection (ATCC), which holds all types of microorganisms.

The prime functions of a culture collection are to maintain the existing collection, to continue to collect new strains and to provide pure, authenticated culture samples of each organism. Problems of culture maintenance have been aided by the development and use of cryopreservation and freeze-drying (lyophilization) techniques, along with miniaturized storage methods. One convenient method involves adsorption of cells to glass beads (2 mm diameter) that may be placed in frozen storage, from which individual beads may be removed without thawing the whole sample.

Use of microorganisms selected from a culture collection obviously provides significant cost savings compared with environmental isolation and has the advantage that some characterization of the microorganism will have already been performed. However, the disadvantage is that competitors have access to the same microorganism.

Industrial strains and strain improvement

Irrespective of the origins of an industrial microorganism, it should ideally exhibit:

1 genetic stability;
2 efficient production of the target product, whose route of biosynthesis should preferably be well characterized;
3 limited or no need for vitamins and additional growth factors;
4 utilization of a wide range of low-cost and readily available carbon sources;
5 amenability to genetic manipulation;
6 safety, non-pathogenicity and should not produce toxic agents, unless this is the target product;
7 ready harvesting from the fermentation;
8 ready breakage, if the target product is intracellular; and

Table 4.3 Examples of some important culture collections useful to industrial microbiologists*

Culture collection	Type of microorganisms held
American Type Culture Collection (ATCC) Manassa, Virginia, USA	All
Centraalbureau voor Schimmelcultures (CBS) Baarn, The Netherlands	Filamentous fungi and yeasts
Collection Nationale de Cultures de Microorganismes (CNCM) Paris, France	All
Deutsche Sammlung von Mikroorganismen und Zellkulture (DSMZ) Braunscheig, Germany	All
UK microbial culture collections	
Culture Collection of Algae & Protozoa (Marine) (CCAP), Dunstaffnage Marine Laboratory, Oban	Algae and protozoa (marine)
Culture Collection of Algae & Protozoa (Freshwater) (CCAP), Institute for Freshwater Ecology, Ambleside	Algae and protozoa (freshwater)
European Collection of Animal Cell Cultures (ECACC), Centre for Applied Microbiological Research (CAMR), Porton Down	Animal cell cultures
CABI Bioscience UK Centre, Egham (formerly International Mycological Institute)	Filamentous fungi
National Collection of Food Bacteria (NCFB), Aberdeen	Food bacteria
National Collection of Industrial & Marine Bacteria (NCIMB), Aberdeen	Bacteria (general, industrial and marine)
National Collection of Pathogenic Fungi (NCPF), Public Health Laboratory, Bristol	Pathogenic fungi
National Collection of Plant Pathogenic Bacteria (NCPPB), Central Science Laboratory, Sand Hutton, York	Plant pathogenic bacteria
National Collection of Type Cultures (NCTC), Central Public Health Laboratory, Colindale	Medical microorganisms
National Collection of Wood Rotting Fungi (NCWRF), Building Research Establishment, Watford	Wood-rotting fungi
National Collection of Yeast Cultures (NCYC), Institute for Food Research, Norwich	Yeasts (other than known pathogens)

* For a comprehensive list see World Data Centre for Microorganisms.

9 production of limited byproducts to ease subsequent purification problems.

Other features that may be exploited are thermophilic or halophilic properties, which may be useful in a fermentation environment. Also, particularly for cells grown in suspension, they should grow well in conventional bioreactors to avoid the necessity to develop alternative systems. Consequently, they should not be shear sensitive, or generate excessive foam, nor be prone to attachment to surfaces.

Strain improvement

Further strain improvement is a vital part of process development in most fermentation industries. It provides a means by which production costs can be reduced through increases in productivity or reduction of manufacturing costs. Examples of some targets for strain improvement are given in Table 4.4.

In many cases strain improvement has been accomplished using natural methods of genetic recombination, which bring together genetic elements from two different genomes into one unit to form new genotypes. An alternative strategy is via mutagenesis. Those recombinants and mutants are then subjected to screening and selection to obtain strains whose characteristics are more specifically suited to the industrial fermentation process. However, such strains are unlikely to survive

Table 4.4 Examples of targets for strain improvement

Targets
Rapid growth
Genetic stability
Non-toxicity to humans
Large cell size, for easy removal from the culture fluid
Ability to use cheaper substrates
Modification of submerged morphology
Elimination of the production of compounds that may interfere with downstream processing
Catabolite derepression
Phosphate deregulation
Permeability alterations to improve product export rates
Metabolite resistance
Production of
additional enzymes
compounds to inhibit contaminant microorganisms
heterologous proteins that may also be engineered with downstream processing 'aids', e.g. polyarginine tails (see Chapter 7, p. 122)

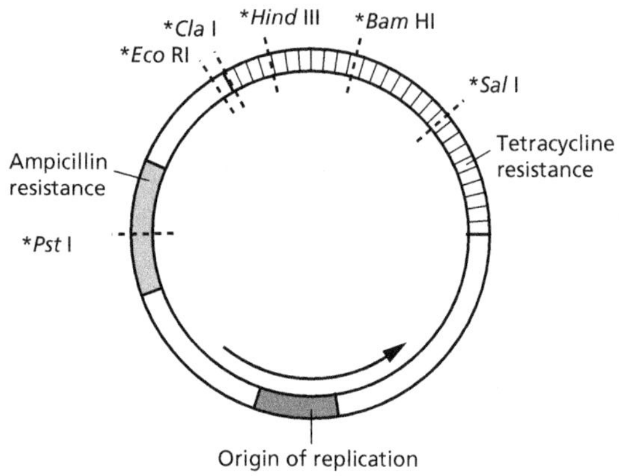

Fig. 4.1 Plasmid pBR322, which contains 4361 base pairs and is a typical example of a cloning vector (*restriction sites).

well in nature, as they often have altered regulatory controls that create metabolic imbalances. Also, they must then be maintained on specific media that select for, and help retain, the special characteristic(s).

NATURAL RECOMBINATION

Bacterial DNA is usually in the form of a single chromosome and plasmids; the latter are autonomous self-replicating accessory pieces of DNA (Fig. 4.1). Each plasmid carries up to a few hundred additional genes and there may be as many as 1000 copies of a plasmid per cell. They contain supplemental genetic information coding for traits not found in the bacterium's chromosomal DNA. Unlike most eukaryotic organisms, bacteria have no form of sexual reproduction. However, they are able to exchange some genetic material via the processes of conjugation, transduction and transformation.

Conjugation involves cell-to-cell contact, where the donor contacts the recipient with a filamentous protein structure called a sex pilus, which draws the two cells close together. The donor copies all or a part of its plasmid or chromosomal DNA and passes it through the pilus to the recipient. In **transduction**, a bacterial virus (bacteriophage) acts as a vector in transferring genes between bacteria. The bacteriophage attaches to a bacterial cell and injects its DNA into the host to become incorporated into the host chromosome. During bacteriophage replication the phage may acquire pieces of the adjacent host DNA. If the phages go on to enter new hosts, they are able to integrate their original DNA, and the genes picked up from their previous host, into the new host's chromosome. Bacteriophages, like plasmids, may also acquire transposons, which are pieces of DNA that can 'jump' from one piece of DNA to another, e.g. from a plasmid to a chromosome and vice versa. The bacteriophages can carry transposons on to new host bacterial cells, where they are able to 'jump' onto a plasmid or the host chromosomal DNA. The third process, **transformation**, involves cellular uptake of a naked piece of DNA from the surrounding medium, which then becomes incorporated into the cell. In natural environments this is a totally random process, the DNA fragments available for uptake being derived from cells that have lysed. The DNA fragments can be relatively large and may contain several genes. However, they are capable of entering and thus transforming only so-called **'competent'** cells, which are in a specific physiological state rendering them permeable to DNA.

In eukaryotes, genetic recombination naturally occurs during sexual reproduction. New genotypes result from the combination of parental chromosomes and as a consequence of crossing-over events during meiosis. The latter involves breakage of sections of chromosomal DNA and the exchange of these segments between homologous chromosomes to form new combinations. Some industrially important fungi, including *Penicillium* and *Aspergillus*, do not have a true sexual phase. However, a **parasexual cycle** has provided a route by which new strains can be produced. This is promoted when two genetically different haploid strains are grown together, allowing fusion of their hyphae. These events result in the formation of a heterokaryon, composed of mycelium containing nuclei derived from each strain. Direct formation of heterokaryons can now be performed *in vitro* by fusing protoplasts, which are cells that have had their walls removed. Also, certain eukaryotes, including some yeasts and filamentous fungi, possess autonomous plasmids, such as the 2 μm plasmid of *Saccharomyces cerevisiae*, which have proved useful as vectors in genetic engineering (see below).

MUTAGENESIS: A CONVENTIONAL TOOL FOR STRAIN IMPROVEMENT

Mutations result from a physical change to the DNA of a cell, such as deletion, insertion, duplication, inversion and translocation of a piece of DNA, or a change in the number of copies of an entire gene or chromosome. Subjection of microorganisms to repeated rounds of mutagenesis, followed by suitable selection and screening of the survivors, has been a very effective tool in improving many industrial microorganisms. As mutants can arise naturally or be induced, they are considered to be the product of natural events. Consequently, there are fewer problems in gaining approval from the regulatory authorities than when recombinant DNA technology is used to develop an industrial microorganism.

Spontaneous mutation rates are low; in most bacterial genes for example, the rate is approximately 10^{-10} per generation per gene. The mutation rate can be greatly increased by using mutagens, which are of two types. Physical mutagens include ultraviolet, γ and X radiation; and chemical mutagens are compounds such as ethane methane sulphonate (EMS), nitroso methyl guanidine (NTG), nitrous acid and acridine mustards. Mutants are formed when the mutagens induce modifications of the base sequences of DNA that result in base-pair substitutions, frame-shift mutations or large deletions that go unrepaired. Mutagenesis can also be induced using transposons delivered by a suitable vector. They produce insertion mutants whose normal nucleotide sequence is interrupted by the transposon sequence.

Mutagenesis methods generally have rather limited use as they primarily achieve either loss of an undesirable characteristic or increasing production of a product, due to impairment of a control mechanism. These traditional methods have been successfully employed in removing the yellow colour of early penicillin preparations caused by chrysogenin, a yellow pigment produced by *Penicillium chrysogenum*. Mutagenesis programmes have also been highly effective in increasing the yield of penicillin in industrial strains of the same organism. Other notable examples of impairment of control processes, resulting in greater product yields, are seen in several microorganisms used for amino acid production.

More recently, methods have been developed to enhance both the overall mutability and mutation rate of specific genes, in order to obtain the maximum frequency of desired mutant types. This **directed mutagenesis** obviously requires a knowledge of the genes that control the target product and often a genetic map of the organism. In addition, *in vitro* mutagenesis is now used in combination with genetic engineering (see below) to modify isolated genes or parts of genes.

GENETIC ENGINEERING OF MICROORGANISMS

Over the last 20 years the development of recombinant DNA technology and methods of cell fusion, such as hybridoma formation for monoclonal antibody production (Chapter 17), have had a major impact on industrial microbiology. In contrast to natural recombination processes, modern recombinant DNA technology provides almost unlimited opportunities for the production of novel combinations of genes. These methods are also highly specific and well controlled, and a vast range of genetic information is available from almost any living and even extinct organisms. Recombinant DNA technology has allowed specific gene sequences to be transferred from one organism to another and allows additional methods to be introduced into strain improvement schemes. This can be used to increase the product yield by removing metabolic bottlenecks in pathways and by amplifying or modifying specific metabolic steps.

Overall, genetic engineering procedures allow totally new properties to be added to the capabilities of industrial microorganisms. Microorganisms may be manipulated to synthesize and often excrete enhanced ranges of enzymes, which may facilitate the production of novel compounds or allow the utilization of cheaper complex substrates. As there is no restriction to the origins of the genes that microorganisms express, the production of plant and animal proteins is made possible. Valuable products already produced include human growth hormone, insulin and interferons (see Chapter 11). Nevertheless, these methods have not totally replaced traditional mutatagenesis methods and the two approaches should be viewed as complementary strategies for strain improvement.

Strategies for the genetic engineering of bacteria

Genetic engineering involves manipulation of DNA outside the cell. It necessitates the initial isolation and recovery of the gene(s) of interest from the donor organism's genome. Isolated DNA sequences may then be modified and the regulation of their expression altered, before insertion into host organisms via a suitable easily manipulated vector system. The first step requires total DNA extraction from the donor organism, which is then cut into smaller sequences using a specific **restriction endonuclease.** Many of these restriction enzymes, found in various species of bacteria, make a staggered cut through a double-stranded DNA molecule at a specific sequence or **palindrome** (Fig. 4.2). As a result, the ends of cut molecules have complementary single-stranded sequences. The small sections of DNA (restriction fragments) can then be joined or spliced into vector DNA molecules that have been cut with the same restriction enzyme. Splicing is performed by an enzyme, DNA ligase, and creates a synthetic DNA molecule. Plasmids and bacteriophages have been the most useful cloning vectors. They play an important role as delivery systems to introduce the recombinant molecules into host cells via transformation or transduction. Once inside they are capable of autonomous replication, which maintains the recombinant DNA within the host cell.

Introduction of recombinant plasmids into bacterial cells can be achieved following calcium chloride treatment, which renders the cell membranes more permeable to DNA. After introduction the plasmids replicate autonomously. In some cases, numerous copies are produced within the host cell to increase the amount of the recombinant DNA per cell. Plasmids can be designed to contain selectable genetic markers, such as antibiotic resistance, vitamin requirement, etc. These markers may be used to select only those host cells that have incorporated the plasmid during transformation, e.g. the 4.3 kb plasmid pBR322, carrying ampicillin and tetracycline resistance markers (Fig. 4.1).

Bacteriophages are particularly useful cloning vectors as up to half of their genome can be removed and replaced with foreign DNA. This is achieved *in vitro* using restriction enzymes in a similar manner to plasmid manipulation. Suitable DNA fragments are then packaged into phage particles, which are able to infect a selected host.

The mixture of restriction fragments, originating from a whole DNA extract, once packaged within phages or plasmids, is used to transform or transduce host cells. This generates a **DNA library** consisting of individual clones that contain different recombinant DNA molecules, representing all DNA sequences/genes of the donor genome. Once the library has been established, the clones are allowed to form colonies on solid selective media. At this stage, the specific clone containing the recombinant DNA molecule of interest can be identified. If the foreign gene is successfully expressed in the host bacterium and a heterologous protein is made, detection can be achieved by use of a specific antibody reaction with the protein. Alternatively, if the recombinant protein is an enzyme that is not normally produced by the host, the enzyme activity can be detected.

Limitations of prokaryotic hosts

The genetic engineering procedures briefly described above, and the example of a cloning strategy outlined in

Enzyme	Source	Restriction site
Eco RI	*Escherichia coli* RY13	5'-G↓AATTC-3' 3'-CTTAA↑G-5'
Bam HI	*Bacillus amyloliquefaciens* H	5'-G↓GATCC-3' 3'-CCTAG↑G-5'
Pst I	*Providencia stuartii* 164	5'-CTGCA↓G-3' 3'-G↑ACGTC-5'

Fig. 4.2 Examples of restriction endonucleases and their restriction sites.

Fig. 4.3, are rather simplistic. In practice the situation may be rather more complex. Often, the whole purpose of cloning a gene is to obtain large quantities of its product. If *Escherichia coli* is used as the host and the gene introduced is not from an *E. coli* strain, i.e. a heterologous gene, it may not be expressed. This problem can be overcome using expression vectors in which the foreign gene is inserted in a configuration that puts it under regulatory controls recognized by the host. Additionally, to maximize production of the foreign protein, the expression vector must replicate to a high copy number and be stable. The foreign gene should ideally be linked to a strong promoter that has a high affinity for RNA polymerase and the transcribed mRNA should be efficiently translated. Sometimes, it may be advantageous for the expression of the cloned gene to be manipulated by placing it under the control of a regulatory switch, in order that production of the recombinant protein does not occur until required.

Generally, Gram-negative bacteria are able to express genes from Gram-positive bacteria. However, the converse is not always as readily achievable. Additional problems also arise if the objective is to clone and express genes from a eukaryotic organism in a bacterium such as *E. coli*. Here the differences between prokaryotic and eukaryotic gene expression must be taken into

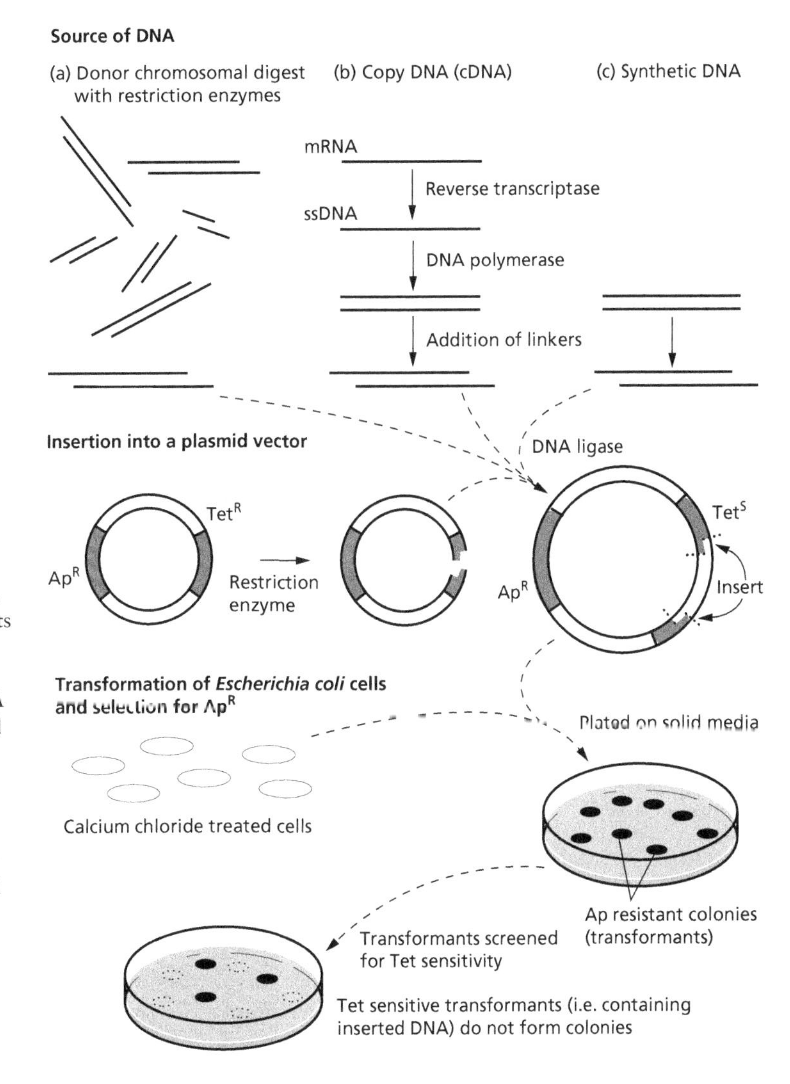

Fig. 4.3 Outline of a strategy for cloning DNA in *E. coli* (after Brown *et al.* (1987)). The source DNA is usually DNA fragments from bacteria, or, for eukaryotic sources, synthetic or cDNA (derived via reverse transcription of mRNA) is used. The DNA fragment, from whatever source, is ligated into a plasmid vector containing suitable markers; in this case, resistance to ampicillin (Ap^R) and tetracycline (Tet^R). Ligation is into the tetracycline resistance gene thus inactivating it (Tet^S). The vector is then used to transform the host cells. All cells successfully transformed exhibit ampicillin resistance. However, of those, only cells containing foreign DNA are tetracycline sensitive. Those containing foreign DNA can then be screened for the desired gene or gene product.

account. Eukaryotic genes contain non-coding regions, introns, which would obviously cause problems if the gene was directly transferred to a prokaryotic system. However, introns are excised during normal RNA processing. Consequently, eukaryotic mRNA can be used to synthesize a gene that can function within a prokaryote. This requires the use of reverse transcriptase to generate a complementary DNA copy or cDNA from the RNA. Alternatively, if the gene nucleotide sequence or amino acid sequence of the product protein is known, a synthetic gene can be synthesized.

Secretion of recombinant proteins is often preferred for product stability and it may make downstream processing to recover the product less problematical. However, it presents a further challenge as some organisms excrete more efficiently than others. In Gram-negative bacteria, secretion is often into the periplasmic space rather than directly into the medium, as proteins cannot readily traverse their outer membrane. For some purposes, this may be beneficial as it often simplifies downstream processing. Where secretion into the medium is necessary, it may be achieved by using cell-wall-less bacteria (L-forms). Secretion of proteins is further complicated by the fact that they must be synthesized with an extra amino acid sequence at their N-terminal end. This signal sequence of 20–25 amino acids aids the passage of the protein across the cell membrane and is removed during secretion.

Other general problems in the expression of heterologous proteins, which may also be encountered with some eukaryotic hosts (see below) include:

1 instability of certain gene products within the host microorganism, leading to their rapid degradation by proteases;

2 incorrect folding of the polypeptide that generates an inactive molecule, which may accumulate to form inclusion bodies within the cell; and

3 difficulties in achieving post-translational modification of proteins, such as cleavage, glycosylation or amidation.

Eukaryotic hosts

A preferred alternative strategy for the expression of heterologous eukaryotic genes is often the employment of a suitable eukaryotic host that will naturally perform any necessary post-translational protein modification and secretion. Initially, *Saccharomyces cerevisiae* was a popular choice because it is safe and a vast amount of information has been accumulated about its genetics, physiology and performance in industrial fermentation processes. Also, it has a relatively rapid growth rate and readily undergoes genetic manipulation, but unlike higher eukaryotes (animal and plant cells) this yeast is easily and cheaply grown on an industrial scale. Several heterologous eukaryotic proteins have been successfully mass produced by *S. cerevisiae*. However, product yields are relatively low at 1–5% of total protein and some proteins are retained within the periplasm. Other yeasts may be better hosts, particularly the methylotrophs, *Pichia angusta* (formerly *Hansenula polymorpha*) and *Pichia pastoris*, which have a number of advantages over *S. cerevisiae*. They have strong inducible promoters and are capable of generating post-translational protein modification similar to those performed by human cells. Downstream processing is also less problematical as they do not secrete many of their own proteins into the medium. *P. angusta* has often been preferred for industrial applications due to greater versatility; examples of established heterologous products from this yeast include hepatitis B vaccine, the feed additive enzyme phytase and the antithrombotic hirudin.

Developments in the field of recombinant DNA technology are progressing at a rapid rate and gene cloning in animal and plant cells is now relatively routine. For example, vectors based on viruses, such as bovine papilloma virus, retroviruses and simian virus 40 are used to stably transform mammalian cells. Transformation of dicotyledonous plant cells can be performed using plasmids derived from the soil bacteria *Agrobacterium tumefaciens* and *A. rhizogenes*. *In vivo*, *A. tumefaciens* induces tumours that involves its Ti plasmid, and *A. rhizogenes* induces formation of hairy roots, which is mediated by the Ri plasmid.

Individual proteins can now be engineered by changing a few component amino acids in order to modify their properties, and there are opportunities to 'build-in' features that aid downstream processing (see Chapter 7). In addition, the complete sequencing of many microbial genomes, including several food grade organisms, is stimulating further advances in metabolic engineering. Relatively minor metabolic engineering has already been implemented to improve the production of existing metabolites, allow the production of new metabolites, impart new catabolic activities and improve fermentation performance. However, whole metabolic networks within a microorganism may now be restructured and such extensive metabolic engineering has major implications within industrial microbiology.

Strain stability

A key factor in the development of new strains is their stability. An important aspect of this is the means of preservation and storage of stock cultures so that their carefully selected attributes are not lost. This may involve storage in liquid nitrogen or lyophilization. Strains transformed by plasmids must be maintained under continual selection to ensure that plasmid stability is retained. Instability may result from deletion and rearrangements of recombinant plasmids, which is referred to as **structural instability**, or complete loss of a plasmid, termed **segregational stability**. Some of these problems can be overcome by careful construction of the plasmid and the placement of essential genes within it. Segregational instability can also be overcome by constructing so-called suicidal strains that require specific markers on the plasmid for survival. Consequently, plasmid-free cells die and do not accumulate in the culture. These strains are constructed with a lethal marker in the chromosome and a repressor of this marker is located on the plasmid. Cells express the repressor as long as they possess the plasmid, but if it is lost the cells express the lethal gene. However, integration of a gene(s) into the chromosome is normally the best solution, as it overcomes many of these instability problems.

Further reading

Papers and reviews

Baneyx, F. (1999) Recombinant protein expression in *Escherichia coli*. *Current Opinion in Biotechnology* 10, 411–421.

Braun, P., Gerritse, G., van Dijl, J.-M. & Quax, W. J. (1999) Improving protein secretion by engineering components of the bacterial translocation machinery. *Current Opinion in Biotechnology* 10, 376–381.

Cameron, D. C. & Chaplen, F. W. R. (1997) Developments in metabolic engineering. *Current Opinion in Biotechnology* 8, 175–180.

Hochney, R. C. (1994) Recent developments in heterologous protein production in *Escherichia coli*. *Trends in Biotechnology* 12, 456–463.

Horikoshi, K. (1995) Discovering novel bacteria, with an eye to biotechnological applications. *Current Opinion in Biotechnology* 6, 292–297.

Nakayama, K. (1981) Sources of industrial microorganisms. In: *Biotechnology*, Volume 1 (eds H.-J. Rehm & G. Reed). VCH, Weinheim, pp. 355–410.

Nielsen, J. (1998) The role of metabolic engineering in the production of secondary metabolites. *Current Opinion in Biotechnology* 1, 330–336.

Ogawa, J. & Shimizu, S. (1999) Microbial enzymes: new industrial applications from traditional screening methods. *Trends in Biotechnology* 17, 13–19.

Roessner, C. A. & Scott, A. I. (1996) Genetically engineered synthesis of natural products: from Alkaloids to Corrins. *Annual Review of Microbiology* 50, 467–490.

Stephanopoulos, G. (1994) Metabolic engineering. *Current Opinion in Biotechnology* 5, 196–200.

Sudbery, P. E. (1996) The expression of recombinant proteins in yeasts. *Current Opinion in Biotechnology* 7, 517–524.

Books

Anke, T. (1997) *Fungal Biotechnology*. Chapman & Hall, London.

Brown, C. M., Campbell, I. & Priest, F. G. (1987) *Introduction to Biotechnology*. Blackwell Scientific Publications, Oxford.

Dale, J. W. (1998) *Molecular Genetics of Bacteria*, 3rd edition. Wiley, Chichester.

Demain, A.L., Davies, J. E. & Atlas, R. M. (1999) *Manual of Industrial Microbiology and Biotechnology*. American Society for Microbiology, Washington, D.C.

Drlica, K. (1997) *Understanding DNA and Gene Cloning*, 3rd edition. Wiley, Chichester.

Glazer, A. N. & Nikaido, H. (1995) *Microbial Biotechnology*. W. H. Freeman, New York.

Glick, B. R. & Pasternak, J. J. (1998) *Molecular Biotechnology: Principles and Applications of Recombinant DNA*, 2nd edition. ASM Press, Washington, D.C.

Murooka, Y. & Imanaka, T. (eds) (1994) *Recombinant Microbes for Industrial and Agricultural Applications*. Marcel Dekker, New York.

Old, R. W. & Primrose, S. B. (1994) *Principles of Gene Manipulation*, 5th edition. Blackwell Scientific Publications, Oxford.

Peters, P. (1993) *Biotechnology: a Guide to Genetic Engineering*. William C. Brown, Dubuque, IA.

Snyder, L. & Champness, W. (1997) *Molecular Genetics of Bacteria*. ASM Press, Washington.

Walker, G. M. (1998) *Yeast Physiology and Biotechnology*. Wiley, Chichester.

5 Fermentation media

Media formulation

Most fermentations require liquid media, often referred to as broth, although some solid-substrate fermentations are operated. Fermentation media must satisfy all the nutritional requirements of the microorganism and fulfil the technical objectives of the process. The nutrients should be formulated to promote the synthesis of the target product, either cell biomass or a specific metabolite. In most industrial fermentation processes there are several stages where media are required. They may include several inoculum (starter culture) propagation steps, pilot-scale fermentations and the main production fermentation. The technical objectives of inoculum propagation and the main fermentation are often very different, which may be reflected in differences in their media formulations. Where biomass or primary metabolites are the target product, the objective is to provide a production medium that allows optimal growth of the microorganism. For secondary metabolite production, such as antibiotics, their biosynthesis is not growth related. Consequently, for this purpose, media are designed to provide an initial period of cell growth, followed by conditions optimized for secondary metabolite production. At this point the supply of one or more nutrients (carbon, phosphorus or nitrogen source) may be limited and rapid growth ceases.

Most fermentations, except those involving solid substrates, require large quantities of water in which the medium is formulated. General media requirements include a carbon source, which in virtually all industrial fermentations provides both energy and carbon units for biosynthesis, and sources of nitrogen, phosphorus and sulphur. Other minor and trace elements must also be supplied, and some microorganisms require added vitamins, such as biotin and riboflavin. Aerobic fermentations are dependent on a continuous input of molecular oxygen, and even some anaerobic fermentations require initial aeration of media, e.g. beer fermentations. Usually, media incorporate buffers, or the pH is controlled by acid and alkali additions, and antifoam agents may be required. For some processes, precursor, inducer or inhibitor compounds must be introduced at certain stages of the fermentation.

The initial step in media formulation is the examination of the overall process based on the stoichiometry for growth and product formation. This primarily involves consideration of the input of the carbon and nitrogen sources, minerals and oxygen, and their conversion to cell biomass, metabolic products, carbon dioxide, water and heat. From this information it should be possible to calculate the minimum quantities of each element required to produce a certain quantity of biomass or metabolite. Typically, the main elemental formula of microbial cells is approximately $C_4H_7O_2N$, which on a basis of dry weight is 48% C, 7% H, 32% O and 14% N. The elemental composition of baker's yeast, for example, is $C_{3.72}H_{6.11}O_{1.95}N_{0.61}S_{0.017}P_{0.035}$ and $K_{0.056}$. Elemental composition varies slightly with growth rate, but the range is relatively small compared with interspecies differences, particularly between bacteria and fungi. Ideally, a knowledge of the complete elemental composition of the specific industrial microorganism allows further media refinement. This ensures that no element is limiting, unless this is desired for a specific purpose.

Once the elemental requirements of a microorganism have been established, suitable nutrient sources can be incorporated into the media to fulfil these demands. However, it is important to be aware of potential problems that can arise when using certain compounds. For example, those that are rapidly metabolized may repress product formation. To overcome this, intermittent or continuous addition of fresh medium may be carried out to maintain a relatively low concentration that is not repressive. Certain media nutrients or environmental conditions may affect not only the physiology and biochemistry, but also the morphology of the microorganism. In some yeasts the single cells may develop into pseudo-mycelium or flocculate, and filamentous fungi may form pellets. This may or may not be desirable, as such morphological changes can influence product yield and other fermentation properties.

The media adopted also depend on the scale of the fermentation. For small-scale laboratory fermentations pure chemicals are often used in well-defined media. However, this is not possible for most industrial-scale fermentation processes, simply due to cost, as media components may account for up to 60–80% of process expenditure. Industrial-scale fermentations primarily use cost-effective complex substrates, where many carbon and nitrogen sources are almost undefinable. Most are derived from natural plant and animal materials, often byproducts of other industries, with varied and variable composition. The effects of such batch-to-batch variations must be determined. Small-scale trials are usually performed with each new batch of substrate, particularly to examine the impact on product yield and product recovery.

The main factors that affect the final choice of individual raw materials are as follows.

1 Cost and availability: ideally, materials should be inexpensive, and of consistent quality and year round availability.
2 Ease of handling in solid or liquid forms, along with associated transport and storage costs, e.g. requirements for temperature control.
3 Sterilization requirements and any potential denaturation problems.
4 Formulation, mixing, complexing and viscosity characteristics that may influence agitation, aeration and foaming during fermentation and downstream processing stages.
5 The concentration of target product attained, its rate of formation and yield per gram of substrate utilized.
6 The levels and range of impurities, and the potential for generating further undesired products during the process.
7 Overall health and safety implications.

The final composition of industrial media is not solely the concern of the fermentation stage. Crude substrates provide initial cost savings, but their higher levels of impurities may necessitate more costly and complex recovery and purification steps downstream, and possibly increased waste treatment costs. Also, the physical and chemical properties of the formulated medium can influence the sterilization operations employed. A medium that is easily sterilized with minimum thermal damage is vitally important. Thermal damage not only reduces the level of specific ingredients, but can also produce potentially inhibitory byproducts that may also interfere with downstream processing. Other media characteristics can affect product recovery and purification, and the ease with which the cells are separated from the spent medium (see Chapter 7).

Carbon sources

A carbon source is required for all biosynthesis leading to reproduction, product formation and cell maintenance. In most fermentations it also serves as the energy source. Carbon requirements may be determined from the biomass yield coefficient (Y), an index of the efficiency of conversion of a substrate into cellular material.

$$Y_{\text{carbon}}(\text{g/g}) = \frac{\text{biomass produced (g)}}{\text{carbon substrate utilized (g)}}$$

For commercial fermentations the determination of yield coefficients for all other nutrients is usually essential. Each may be determined by conducting a series of batch culture experiments where the specific substrate is the only growth-limiting media component and all other nutrients are in excess (see Chapter 2). By varying the initial concentration of the growth-limiting substrate and then plotting total growth against substrate concentration for each batch, the growth yield (Y) can be estimated. However, the value obtained relates to a specific set of operating conditions; varying pH, temperature, etc., can alter the yield coefficient. Various organisms may exhibit different yield coefficients for the same substrate (Table 5.1), due primarily to the pathway by which the compound is metabolized. Differences can also be seen within an individual. For example, *Saccharomyces cerevisiae* grown on glucose has biomass yield coefficients of 0.56 and 0.12 g/g under aerobic and anaerobic conditions, respectively.

As most carbon substrates also serve as energy sources, the organism's efficiency of both adenosine triphosphate (ATP) generation and its utilization are obviously additional key factors. Often, it is very useful, although rather difficult, to estimate how much ATP is required for growth. However, estimates of Y_{ATP} (yield of cells per mole of ATP generated during growth) can be calculated if the metabolism of the organism has been fully elucidated (also see Chapter 14, Single cell protein production).

Carbohydrates are traditional carbon and energy sources for microbial fermentations, although other sources may be used, such as alcohols, alkanes and

Table 5.1 Growth yield (Y_{carbon}) on minimal medium plus various carbon and energy sources

	$Y_{glucose}$	$Y_{ethanol}$	$Y_{methanol}$	Y_{octane}
Aerobic growth				
Aspergillus nidulans	0.61			
Candida utilis	0.51	0.68		
Escherichia coli	0.52			
Pichia angusta			0.36	
Penicillium chrysogenum	0.43			
Pseudomonas aeruginosa	0.43			
Pseudomonas species			0.54	1.07
Saccharomyces cerevisiae	0.56	0.63		
Anaerobic growth				
Moorella thermacetica	0.11			
Escherichia coli	0.13			
Klebsiella pneumoniae	0.12			
Saccharomyces cerevisiae	0.12			

organic acids. Animal fats and plant oils may also be incorporated into some media, often as supplements to the main carbon source.

MOLASSES

Pure glucose and sucrose are rarely used for industrial-scale fermentations, primarily due to cost. Molasses, a byproduct of cane and beet sugar production, is a cheaper and more usual source of sucrose. This material is the residue remaining after most of the sucrose has been crystallized from the plant extract. It is a dark coloured viscous syrup containing 50–60% (w/v) carbohydrates, primarily sucrose, with 2% (w/v) nitrogenous substances, along with some vitamins and minerals. Overall composition varies depending upon the plant source, the location of the crop, the climatic conditions under which it was grown and the factory where it was processed. The carbohydrate concentration may be reduced during storage by contaminating microorganisms. A similar product, hydrol molasses, can also be used. This byproduct of maize starch processing primarily contains glucose.

MALT EXTRACT

Aqueous extracts of malted barley can be concentrated to form syrups that are particularly useful carbon sources for the cultivation of filamentous fungi, yeasts and actinomycetes. Extract preparation is essentially the same as for malt wort production in beer brewing (Chapter 12). The composition of malt extracts varies to some extent, but they usually contain approximately 90% carbohydrate, on a dry weight basis. This comprises 20% hexoses (glucose and small amounts of fructose), 55% disaccharides (mainly maltose and traces of sucrose), along with 10% maltotriose, a trisaccharide. In addition, these products contain a range of branched and unbranched dextrins (15–20%), which may or may not be metabolized, depending upon the microorganism. Malt extracts also contain some vitamins and approximately 5% nitrogenous substances, proteins, peptides and amino acids.

Sterilization of media containing malt extract must be carefully controlled to prevent over-heating. The constituent reducing sugars and amino acids are prone to generating Maillard reaction products when heated at low pH. These are brown condensation products resulting from the reaction of amino groups of amines, amino acids and proteins with the carbonyl groups of reducing sugars, ketones and aldehydes. Not only does this cause colour change, but it also results in loss of fermentable materials and some reaction products may inhibit microbial growth.

STARCH AND DEXTRINS

These polysaccharides are not as readily utilized as monosaccharides and disaccharides, but can be directly metabolized by amylase-producing microorganisms, particularly filamentous fungi. Their extracellular enzymes hydrolyse the substrate to a mixture of glucose, maltose or maltotriose to produce a sugar spectrum similar to that found in many malt extracts.

Maize starch is most widely used, but it may also be obtained from other cereal and root crops. To allow use in a wider range of fermentations, the starch is usually converted into sugar syrup, containing mostly glucose. It is first gelatinized and then hydrolysed by dilute acids or amylolytic enzymes, often microbial glucoamylases that operate at elevated temperatures.

SULPHITE WASTE LIQUOR

Sugar containing wastes derived from the paper pulping industry are primarily used for the cultivation of yeasts. Waste liquors from coniferous trees contain 2–3% (w/v) sugar, which is a mixture of hexoses (80%) and pentoses (20%). Hexoses include glucose, mannose and galactose, whereas the pentose sugars are mostly xylose and arabinose. Those liquors derived from deciduous trees contain mainly pentoses. Usually the liquor requires processing before use as it contains sulphur dioxide. The low pH is adjusted with calcium hydroxide or calcium carbonate, and these liquors are supplemented with sources of nitrogen and phosphorus.

CELLULOSE

Cellulose is predominantly found as lignocellulose in plant cell walls, which is composed of three polymers: cellulose, hemicellulose and lignin (see Chapter 10, Bioconversion of lignocellulose). Lignocellulose is available from agricultural, forestry, industrial and domestic wastes. Relatively few microorganisms can utilize it directly, as it is difficult to hydrolyse. The cellulose component is in part crystalline, encrusted with lignin, and provides little surface area for enzyme attack. At present it is mainly used in solid-substrate fermentations to produce various mushrooms (see Chapter 14). However, it is potentially a very valuable renewable source of fermentable sugars once hydrolysed, particularly in the bioconversion to ethanol for fuel use (see Chapter 10).

WHEY

Whey is an aqueous byproduct of the dairy industry. The annual worldwide production is over 80 million tonnes, containing over 1 million tonnes of lactose and 0.2 million tonnes of milk protein. This material is expensive to store and transport. Therefore, lactose concentrates are often prepared for later fermentation by evaporation of the whey, following removal of milk proteins for use as food supplements. Lactose is generally less useful as a fermentation feedstock than sucrose, as it is metabolized by fewer organisms. *S. cerevisiae,* for example, does not ferment lactose. This disaccharide was formerly used extensively in penicillin fermentations and it is still employed for producing ethanol, single cell protein, lactic acid, xanthan gum, vitamin B_{12} and gibberellic acid.

ALKANES AND ALCOHOLS

n-Alkanes of chain length C_{10}–C_{20} are readily metabolized by certain microorganisms. Mixtures, rather than a single compound, are usually most suitable for microbial fermentations. However, their industrial use is dependent upon the prevailing price of petroleum. Methane is utilized as a carbon source by a few microorganisms, but its conversion product methanol is often preferred for industrial fermentations as it presents fewer technical problems. High purity methanol is readily obtained and it is completely miscible with water. Methanol has a high per cent carbon content and is relatively cheap, although only a limited number of organisms will metabolize it. Also, unlike many other carbon sources, only low concentrations, 0.1–1% (v/v), are tolerated by microorganisms, higher levels being toxic. During fermentations on methanol, the oxygen demand and heat of fermentation are high, but this is even more problematic when growing on alkanes. Several companies used methanol in microbial protein production in the 1970s and early 1980s, but these processes are currently uneconomic (see Chapter 14, Microbial biomass).

Ethanol is less toxic than methanol and is used as a sole or cosubstrate by many microorganisms, but it is too expensive for general use as a carbon source. However, its biotransformation to acetic acid by acetic acid bacteria remains a major fermentation process (see Chapter 12, Vinegar manufacture).

FATS AND OILS

Hard animal fats that are mostly composed of glycerides of palmitic and stearic acids are rarely used in fermentations. However, plant oils (primarily from cotton seed, linseed, maize, olive, palm, rape seed and soya) and occasionally fish oil, may be used as the primary or supplementary carbon source, especially in antibiotic production. Plant oils are mostly composed of oleic and linoleic acids, but linseed and soya oil also have a substantial amount of linolenic acid. The oils contain more energy per unit weight than carbohydrates. In addition, the carbohydrates occupy a greater volume, because they are usually prepared as aqueous solutions of concentrations no greater than 50% (w/w). Consequently, oils can be particularly useful in fed-batch operations, as less spare capacity is needed to accommodate further additions of the carbon source.

Nitrogen sources

Most industrial microbes can utilize both inorganic and organic nitrogen sources. Inorganic nitrogen may be supplied as ammonium salts, often ammonium sulphate and diammonium hydrogen phosphate, or ammonia. Ammonia can also be used to adjust the pH of the fermentation. Organic nitrogen sources include amino acids, proteins and urea. Nitrogen is often supplied in crude forms that are essentially byproducts of other industries, such as corn steep liquor, yeast extracts, peptones and soya meal. Purified amino acids are used only in special situations, usually as precursors for specific products.

CORN STEEP LIQUOR

Corn steep liquor is a byproduct of starch extraction from maize and its first use in fermentations was for penicillin production in the 1940s. The exact composition of the liquor varies depending on the quality of the maize and the processing conditions. Concentrated extracts generally contain about 4% (w/v) nitrogen, including a wide range of amino acids, along with vitamins and minerals. Any residual sugars are usually converted to lactic acid (9–20%, w/v) by contaminating bacteria. Corn steep liquor can sometimes be replaced by similar liquors, such as those derived from potato starch production.

YEAST EXTRACTS

Yeast extracts may be produced from waste baker's and brewer's yeast, or other strains of *S. cerevisiae*. Alternate sources are *Kluyveromyces marxianus* (formerly classified as *K. fragilis*) grown on whey and *Candida utilis* cultivated using ethanol, or wastes from wood and paper processing. Those extracts used in the formulation of fermentation media are normally salt-free concentrates of soluble components of hydrolysed yeast cells. Yeast extracts with sodium chloride concentrations greater than 0.05% (w/v) cannot be used in fermentation processes due to potential corrosion problems.

Yeast cell hydrolysis is often achieved by autolysis, using the cell's endogenous enzymes, usually without the need for additional hydrolytic enzymes. Autolysis can be initiated by temperature or osmotic shock, causing cells to die but without inactivating their enzymes. Temperature and pH are controlled throughout to ensure an optimal and standardized autolysis process. Temperature control is particularly important to prevent loss of vitamins. Autolysis is performed at 50–55°C for several hours before the temperature is raised to 75°C to inactivate the enzymes. Finally, the cells are disrupted by plasmolysis or mechanical disruption. Cell wall materials and other debris are removed by filtration or centrifugation and the resultant extract is rapidly concentrated. Extracts are available as liquids containing 50–65% solids, viscous pastes or dry powders. They contain amino acids, peptides, water-soluble vitamins (Table 5.2) and some glucose, derived from the yeast storage carbohydrates (trehalose and glycogen).

PEPTONES

Peptones are usually too expensive for large-scale industrial fermentations. They are prepared by acid or enzyme hydrolysis of high protein materials: meat, casein, gelatin, keratin, peanuts, soy meal, cotton seeds, etc. Their amino acid compositions vary depending upon the original protein source. For example, gelatin-derived peptones are rich in proline and hydroxyproline, but are almost devoid of sulphur-containing amino acids; whereas keratin peptone is rich in both proline and cystine, but lacks lysine. Peptones from plant

Table 5.2 Protein and vitamin composition of yeast extract

Total proteins, peptides & amino acids (%, w/v)	73–75
free amino acids	35–40
peptides less than 600 Da	10–15
material above 600 Da	20–30
Vitamins (μg/g)	
thiamin	30
riboflavin	120
niacin	700
pyridoxine	20
folic acid	30
calcium pantothenate	300
biotin	2.5

Note: mineral content varies with the processing steps used.

sources invariably contain relatively large quantities of carbohydrates.

SOYA BEAN MEAL

Residues remaining after soya beans have been processed to extract the bulk of their oil are composed of 50% protein, 8% non-protein nitrogenous compounds, 30% carbohydrates and 1% oil. This residual soya meal is often used in antibiotic fermentations because the components are only slowly metabolized, thereby eliminating the possibility of repression of product formation.

Water

All fermentation processes, except solid-substrate fermentations, require vast quantities of water. In many cases it also provides trace mineral elements. Not only is water a major component of all media, but it is important for ancillary equipment and cleaning. A reliable source of large quantities of clean water, of consistent composition, is therefore essential. Before use, removal of suspended solids, colloids and microorganisms is usually required. When the water supply is 'hard', it is treated to remove salts such as calcium carbonate. Iron and chlorine may also require removal. For some fermentations, notably plant and animal cell culture, the water must be highly purified.

Water is becoming increasingly expensive, necessitating its recycle/reusage wherever possible. This minimizes water costs and reduces the volume requiring waste-water treatment.

Minerals

Normally, sufficient quantities of cobalt, copper, iron, manganese, molybdenum, and zinc are present in the water supplies, and as impurities in other media ingredients. For example, corn steep liquor contains a wide range of minerals that will usually satisfy the minor and trace mineral needs. Occasionally, levels of calcium, magnesium, phosphorus, potassium, sulphur and chloride ions are too low to fulfil requirements and these may be added as specific salts.

Vitamins and growth factors

Many bacteria can synthesize all necessary vitamins from basic elements. For other bacteria, filamentous fungi and yeasts, they must be added as supplements to the fermentation medium. Most natural carbon and nitrogen sources also contain at least some of the required vitamins as minor contaminants. Other necessary growth factors, amino acids, nucleotides, fatty acids and sterols, are added either in pure form or, for economic reasons, as less expensive plant and animal extracts.

Precursors

Some fermentations must be supplemented with specific precursors, notably for secondary metabolite production. When required, they are often added in controlled quantities and in a relatively pure form. Examples include phenylacetic acid or phenylacetamide added as side-chain precursors in penicillin production. D-threonine is used as a precursor in L-isoleucine production by *Serratia marsescens*, and anthranillic acid additions are made to fermentations of the yeast *Hansenula anomala* during L-tryptophan production.

Inducers and elicitors

If product formation is dependent upon the presence of a specific inducer compound or a structural analogue, it must be incorporated into the culture medium or added at a specific point during the fermentation. In plant cell culture the production of secondary metabolites, such as flavonoids and terpenoids, can be triggered by adding elicitors. These may be isolated from various microorganisms, particularly plant pathogens (see Chapter 17).

Inducers are often necessary in fermentations of genetically modified microorganisms (GMMs). This is because the growth of GMMs can be impaired when the cloned genes are 'switched on', due to the very high levels of their transcription and translation. Consequently, inducible systems for the cloned genes are incorporated that allow initial maximization of growth to establish high biomass density, whereupon the cloned gene can then be 'switched on' by the addition of the specific chemical inducer.

Inhibitors

Inhibitors are used to redirect metabolism towards the target product and reduce formation of other metabolic intermediates; others halt a pathway at a certain point to prevent further metabolism of the target product. An example of an inhibitor specifically employed to redirect metabolism is sodium bisulphite, which is used in the production of glycerol by *S. cerevisiae* (see Chapter 10).

Some GMMs contain plasmids bearing an antibiotic resistance gene, as well as the heterologous gene(s). The incorporation of this antibiotic into the medium used for the production of the heterologous product selectively inhibits any plasmid-free cells that may arise.

Cell permeability modifiers

These compounds increase cell permeability by modifying cell walls and/or membranes, promoting the release of intracellular products into the fermentation medium. Compounds used for this purpose include penicillins and surfactants. They are frequently added to amino acid fermentations, including processes for producing L-glutamic acid using members of the genera *Corynebacterium* and *Brevibacterium* (see Chapter 10).

Oxygen

Depending on the amount of oxygen required by the organism, it may be supplied in the form of air containing about 21% (v/v) oxygen, or occasionally as pure oxygen when requirements are particularly high. The organism's oxygen requirements may vary widely depending upon the carbon source. For most fermentations the air or oxygen supply is filter sterilized prior to being injected into the fermenter.

Antifoams

Antifoams are necessary to reduce foam formation during fermentation. Foaming is largely due to media proteins that become attached to the air–broth interface where they denature to form a stable foam. If uncontrolled the foam may block air filters, resulting in the loss of aseptic conditions; the fermenter becomes contaminated and microorganisms are released into the environment. Of possibly most importance is the need to allow 'freeboard' in fermenters to provide space for the foam generated. If foaming is minimized, then throughputs can be increased.

There are three possible approaches to controlling foam production: modification of medium composition, use of mechanical foam breakers and addition of chemical antifoams. Chemical antifoams are surface-active agents which reduce the surface tension that binds the foam together. The ideal antifoam should have the following properties:

1 readily and rapidly dispersed with rapid action;
2 high activity at low concentrations;
3 prolonged action;
4 non-toxic to fermentation microorganisms, humans or animals;
5 low cost;
6 thermostability; and
7 compatibility with other media components and the process, i.e. having no effect on oxygen transfer rates or downstream processing operations.

Natural antifoams include plant oils (e.g. from soya, sunflower and rapeseed), deodorized fish oil, mineral oils and tallow. The synthetic antifoams are mostly silicon oils, poly alcohols and alkylated glycols. Some of these may adversely affect downstream processing steps, especially membrane filtration.

Animal cell culture media

Animal cell culture media are normally based on complex basal media, such as Eagle's cell culture medium, which contains glucose, mineral salts, vitamins and amino acids. For mammalian cells a serum is usually added, such as fetal calf serum, calf serum, newborn calf serum or horse serum. Sera provide a source of essential growth factors, including initiation and attachment factors, and binding proteins. They also supply hormones, trace elements and protease inhibitors.

The highly complex composition of sera makes substitution with lower cost ingredients very difficult. Sterilization of formulated animal culture media and media constituents is also more problematic as many components are thermolabile, requiring filter sterilization. Normally, sera constitute 5–10% (v/v) of the medium, but attempts have been made to reduce and ultimately eliminate its use. This is necessary due to its high cost and the fact that it is a potential source of prions and viruses. In some circumstances levels have now been lowered to 1–2% (v/v) and some cell lines have been developed that grow in serum-free media.

Plant cell culture media

In contrast to most animal cell culture media, those used for plant cell culture are usually chemically defined. They contain an organic carbon source (as most plant cells are grown heterotrophically), a nitrogen source, mineral salts and growth hormones. Sucrose is frequently incorporated as the carbon source, particularly for secondary metabolite production, but glucose, fructose, maltose and even lactose have been used. Nitrate is the usual nitrogen source, often supplemented with ammonium salts. However, some species may require organic nitrogen, normally in the form of amino acids. The combination and concentration of plant hormones provided depend upon the specific fermentation. Auxins are usually supplied, along with cytokinins to promote cell division. A two-phase culture has often proved to be useful in increasing productivity, particularly for producing secondary metabolites such as shikonin (see Chapter 17). The first phase uses a medium optimized for growth, the second promotes product formation.

Culture maintenance media

These media are used for the storage and subculturing of key industrial strains. They are designed to retain good cell viability and minimize the possible development of genetic variation. In particular, they must reduce the production of toxic metabolites that can have strain-destabilizing effects. If strains are naturally unstable, they should be maintained on media selective for the specific characteristic that must be retained.

Further reading

Papers and reviews

Blanch, H. W. & Yee, L. (1993) Defined media optimization for growth of recombinant *Escherichia coli* X90. *Biotechnology and Bioengineering* **41**, 221–230.

Kennedy, M. & Krouse, D. (1999) Strategies for improving fermentation medium performance: a review. *Journal of Industrial Microbiology and Biotechnology* **23**, 456–475.

Russell, J. B. & Cook, G. M. (1995) Energetics of bacterial growth: Balance of anabolic and catabolic reactions. *Microbiological Reviews* **59**, 48–62.

Vardar-Sukan, F. (1992) Foaming and its control. In: *Recent Advances in Biotechnology* (eds F. Vardar-Sukan & S. Suha-Sukan), pp. 113–146. Kluwer, Dordrecht.

Books

Atkinson, B. & Mavituna, F. (1991) *Biochemical Engineering and Biotechnology Handbook*, 2nd edition. Macmillan Press, Basingstoke.

Blanch, H. W. & Clark, D. S. (1997) *Biochemical Engineering*. Marcel Dekker, New York.

Demain, A. L., Davies, J. E. & Atlas, R. M. (1999) *Manual of Industrial Microbiology and Biotechnology*. American Society for Microbiology, Washington D.C.

Isaac, S. & Jennings, D. (1995) *Microbial Culture*. BIOS Scientific Publishers, Oxford.

Stanbury, P. F., Whitaker, A. & Hall, S. J. (1995) *Principles of Fermentation Technology*, 2nd edition. Butterworth-Heinemann (Pergamon), Oxford.

6 Fermentation systems

Microbiologists use the term fermentation in two different contexts. First, in metabolism, fermentation refers to energy-generating processes where organic compounds act as both electron donor and acceptor (see Chapter 3). Second, in the context of industrial microbiology, the term also refers to the growth of large quantities of cells under aerobic or anaerobic conditions, within a vessel referred to as a fermenter or bioreactor. Apart from their use for cell cultivation, and for live vegetative cell and spore biotransformations, similar vessels are used in processes involving cell-free and immobilized enzyme transformations. However, here we will be considering fermenters for microbial, plant and animal cell culture. Although we will be primarily examining conventional fermenters, it must be remembered that, particularly with the advent of recombinant DNA technology, alternate systems for producing specific cell products are now available. Monoclonal antibodies are already extracted from the ascitic fluid of rodents (see Chapter 17), vaccines can be produced in sheep's milk or in fruit (e.g. biosynthesis of malaria vaccine in bananas) and various other recombinant products may be manufactured in agricultural plants.

Although solid-substrate fermentations are operated (see p. 105), most fermentations use liquid media, often referred to as broth, under aerobic or anaerobic conditions. Some, like beer and wine fermentations, are non-stirred, non-aerated and are not operated aseptically (see Chapter 12), whereas many others are stirred, aerated and aseptic (Table 6.1). Fermentations are also broadly classified according to the organization of the biological phase (Table 6.2), whether it is in suspension or in the form of a supported film. For **suspended growth**, the cells are freely dispersed in the growth medium and interact as individual or flocculated units. These systems may appear simple, but in reality nutrient and oxygen gradients may develop. In **supported growth**, the cells develop as a biofilm, normally on an inert support material and result in the formation of a complex interacting community of cells. Supported growth systems can be subdivided into fixed film processes, where the medium flows over the static support material; and fluidized/expanded systems, where particles of support material are suspended in a liquid medium. Also, some microorganisms can be grown as unattached surface films at a liquid–air interface.

Fermenter design and construction

The main function of a fermenter is to provide a suitable environment in which an organism can efficiently produce a target product that may be cell biomass, a metabolite or bioconversion product. Most are designed to maintain high biomass concentrations, which are essential for many fermentation processes, whereas control strategies largely depend on the particular process and its specific objectives. The performance of any fermenter depends on many factors, but the key physical and chemical parameters that must be controlled are agitation rate, oxygen transfer, pH, temperature and foam production.

Laboratory fermentations may be performed in bottles or conical flasks that can be shaken to provide aeration where necessary. These vessels are normally plugged with cotton wool or a styrofoam bung to prevent microbial contamination, but this can lead to evaporation losses and restricted exchange of gases. Consequently, even on a laboratory scale, vessels specifically constructed for fermentations are usually preferred. For industrial processes, fermenters with capacities up to several hundred thousand litres are used. These are mostly purpose-built and designed for a specific process, although some flexibility may be necessary in certain instances. Their design, quality of construction, mode of operation and the level of sophistication largely depend upon the production organism, the optimal operating conditions required for target product formation, product value and the scale of production. Overriding considerations are reliability and the need to minimize capital investment and running costs.

Table 6.1 Examples of aseptic and non-aseptic fermentations

Aseptic		Non-aseptic	
Aerobic	Anaerobic	Aerobic	Anaerobic
Animal and plant cell cultures	Acetone	Acetification of ethanol in vinegar production*	Alcoholic beverages; beer, wine, etc.*
Alkaloids	Butanol	Ripening of some cheeses	Primary dairy fermentations*
Amino acids	Ethanol	Mushroom production	Silage production
Most antibiotics	Glycerol	Aerobic waste-water treatment	Anaerobic waste-water treatment
Most biomass (SCP) production	Lactic acid		
Most enzymes	Some toxins		
Most organic acids			
rDNA proteins			
Steroid biotransformations			
Some toxins			
Most vaccines			
Most vitamins			
Xanthan gum			

* Usually a clean operation often referred to as 'commercially sterile'.

Table 6.2 Classification of industrial fermentations according to the organization of the biological phase

Suspended mode		Supported mode	
Individual cells	Flocs and aggregates	Fixed film	Films on fluidized supports
Cylindroconical vessels *Saccharomyces cerevisiae* (beer)	Activated sludge reactor mixed culture (waste-water treatment)	Trickling film generator acetic acid bacteria (vinegar)	Fluidized bed reactor mixed culture (waste-water treatment)
Airlift fermenter *Methylophilus methylotrophus* (biomass)	Stirred tank reactor *Aspergillus niger* (citric acid production)	Trickle filters mixed culture (waste-water treatment)	Fluidized bed reactor animal cells (monoclonal antibodies)
Stirred tank reactor *Bacillus subtilis* (enzymes)	Stirred tank reactor *Penicillium chrysogenum* (penicillin)	Hollow fibre fermenter animal cells (monoclonal antibodies)	

Large volume, low value products that include many of the traditional fermentation products, such as alcoholic beverages, are usually produced using relatively simple fermenters and may not operate under aseptic conditions. There are often fewer risks when operating such non-aseptic fermentations at extreme pH or high temperatures, or where protected substrates are used. These are substrates that few microorganisms will utilize. Nevertheless, strict adherence to good manufacturing practices reduces the risk of microbial contamination of pure culture (axenic) fermentations. Other fermentations do not involve pure culture inoculum and actively encourage the development of indigenous microorganisms, e.g. some food fermentations and waste-water treatment. Conversely, fermentations producing high value, relatively low volume products, especially pharmaceuticals, invariably demand more elaborate systems and operate under strict aseptic conditions.

Traditionally, fermenters have been open cylindrical or rectangular vessels made from wood or stone. Some of these are still used, particularly for certain food and beverage fermentations. However, most fermentations are now performed in closed vessels designed to exclude microbial contamination. These fermenters must withstand repeated sterilization and cleaning, and should be constructed from non-toxic, corrosion-resistant materials. Small fermentation vessels of a few litres capacity are constructed from glass and/or stainless steel. Pilot-scale and many production vessels are normally made of stainless steel with polished internal surfaces, whereas very large fermenters are often constructed from mild steel lined with glass or plastic, in order to reduce the cost. If aseptic operation is required, all associated pipelines transporting air, inoculum and nutrients for the fermentation need to be sterilizable, usually by steam. Normally, fermenters up to 1000 L capacity have an external jacket, and larger vessels have internal coils. Both provide a mechanism for vessel sterilization and temperature control during the fermentation.

As previously stated, fermentations can be carried out under non-aseptic conditions where the risk of contamination is not a major concern. However, many fermenters must be designed for prolonged aseptic operation (Table 6.1). The design rules for an aseptic bioreactor demand that there is no direct contact between the sterile and non-sterile sections to eliminate microbial contamination. Any connections to the fermenter should be suitable for steam treatment to kill any resident microorganisms and systems must be designed to allow aseptic inoculation, sampling and harvesting. Every individual part should be easily maintained, cleaned and independently steam sterilizable, particularly valves. Most vessel cleaning operations are now automated using spray jets, which are located within the vessels. They efficiently disperse cleaning fluids and this cleaning mechanism is referred to as cleaning-in-place (CIP).

Associated pipework must also be designed to reduce the risk of microbial contamination. There should be no horizontal pipes or unnecessary joints and dead stagnant spaces where material can accumulate, otherwise this may lead to ineffective sterilization. Overlapping joints are unacceptable and flanged connections should be avoided as vibration and thermal expansion can result in loosening of the joints to allow ingress of microbial contaminants. Butt-welded joints with polished inner surfaces are preferred.

Other features that must be incorporated are pressure gauges and safety pressure valves, which are required during sterilization and operation. The safety valves prevent excess pressurization, thus reducing potential safety risks. They are usually in the form of a metal foil disc held in a holder set into the wall of the fermenter. These discs burst at a specified pressure and present a much lower contamination risk than spring-loaded valves. Pumps should be avoided if aseptic operation is required, as they can be a major source of contamination. Centrifugal pumps may be used, but their seals are potential routes for contamination. These pumps generate high shear forces and are not suitable for pumping suspensions of shear-sensitive cells. Other pumps used include magnetically coupled, jet and peristaltic pumps. Alternate methods of liquid transfer are gravity feeding or vessel pressurization. In fermentations operating at high temperatures or containing volatile compounds, a sterilizable condenser may be required to prevent evaporation loss. For safety reasons, it is particularly important to contain any aerosols generated within the fermenter by filter-sterilizing the exhaust gases. Also, fermenters are often operated under positive pressure to prevent entry of contaminants. However, this may not be applicable if pathogens or certain recombinant DNA microorganisms are being used (see Chapter 8).

Control of chemical and physical conditions

The basic concept of a bioreactor is to separate, by use of boundaries, the internal fermentation environment from the external environment. Therefore, anything entering or leaving the fermentation can be monitored and this introduces the basic notion of 'energy and mass balances'. However, some parameters in a system cannot be balanced. These are **intensive properties** (temperature, concentration, pressure and specific heat) whose properties are independent of the size of the system, whereas **extrinsive properties** (mass, volume, entropy and energy) can be balanced. For example, if 10 g of water at 30°C is added to 35 g of water at 30°C, the resulting water has a temperature (intensive property) of 30°C not 60°C, but the mass of water (extrinsive property) is additive at 45 g.

The key extrinsive parameters are **mass** and **energy**, consequently the number of atoms of carbon, nitrogen, oxygen, etc., and the energy present in the system at the start of the operation, and any further input during the fermentation, must all be accounted for at the end of the process. If the inventory for mass and energy in the system at the start and finish balance, then the system is understood. This provides a powerful analytical tool, especially when combined with the determination of thermodynamic properties (heat transfer, density, rheology and temperature) and rate equations (biomass production, nutrient utilization and waste product formation). Together this information can be used to build mathematical and computer models that may be used to monitor and control future fermentations.

Agitation

Agitation of suspended cell fermentations is performed in order to mix the three phases within a fermenter. The liquid phase contains dissolved nutrients and metabolites, the gaseous phase is predominantly oxygen and carbon dioxide, and the solid phase is made up of the cells and any solid substrates that may be present. Mixing should produce homogeneous conditions and promote nutrient, gas and heat transfer. Heat transfer is necessary during both sterilization and for temperature maintenance during operation. Efficient mixing is particularly important for oxygen transfer in aerobic fermentations, as microorganisms can take up oxygen only from the liquid phase. Transfer into liquid from the gaseous phase is enhanced by agitation. It prolongs retention of air bubbles in suspension, reduces bubble size to increase the surface area for oxygen transfer, prevents bubble coalescence and decreases the film thickness at the gas-liquid interface (see p. 99, Aeration). However, maintenance of suitable shear conditions during the fermentation is very important, because certain agitation systems develop high shear that may damage shear-sensitive cells. Conversely, low shear systems can lead to cell flocculation or unwanted growth on surfaces, such as on the vessel walls, stirrer and electrodes.

Fermenter agitation requires a substantial input of energy and there are three principal mechanisms that may be used.

1 Stirred tank reactors (STRs) have mechanically moving agitators or impellers within a baffled cylindrical vessel (Fig. 6.1). Baffles are usually flat vertical plates whose width is about one-tenth of the vessel diameter. Normally, 4–6 baffle plates are fitted to the inside vessel walls to aid mixing and mass transfer by increasing turbulence, preventing vortex formation and eliminating 'dead spaces'.

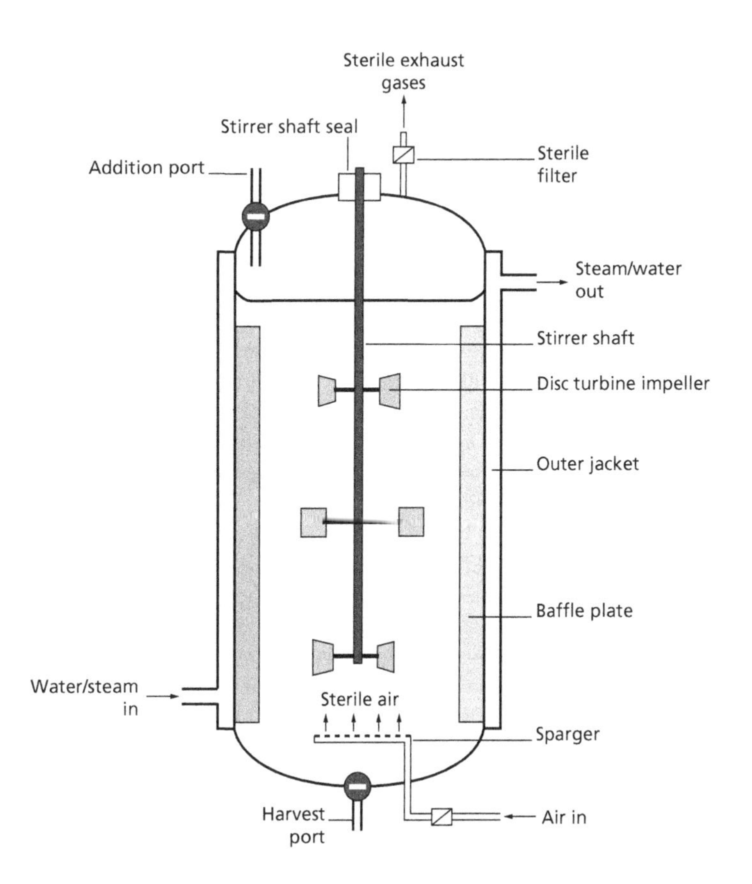

Fig. 6.1 Diagram of a stirred tank reactor.

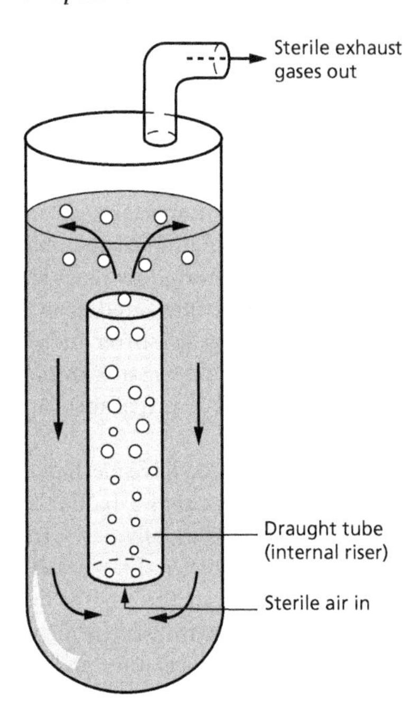

Fig. 6.2 A diagram illustrating the principle of an airlift fermenter.

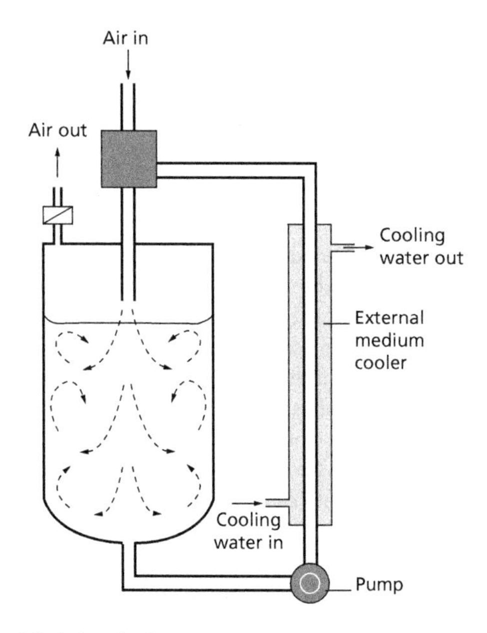

Fig. 6.3 A deep-jet fermenter.

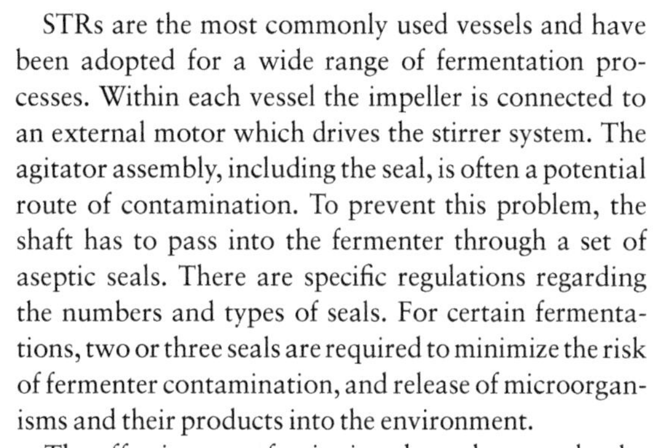

STRs are the most commonly used vessels and have been adopted for a wide range of fermentation processes. Within each vessel the impeller is connected to an external motor which drives the stirrer system. The agitator assembly, including the seal, is often a potential route of contamination. To prevent this problem, the shaft has to pass into the fermenter through a set of aseptic seals. There are specific regulations regarding the numbers and types of seals. For certain fermentations, two or three seals are required to minimize the risk of fermenter contamination, and release of microorganisms and their products into the environment.

The effectiveness of agitation depends upon the design of the impeller blades, speed of agitation and the depth of liquid. Most STRs have height–diameter aspect ratios of 3 : 1 or 4 : 1. STRs must create high turbulence to maintain transfer rates, but this also generates considerable shear force that is detrimental to certain cells. For instance, many animal and plant cells are shear-sensitive and excessive stirring may result in cell disruption. In these cases STRs may be unsuitable without modification, and airlift or supported biofilm reactors may be preferred.

2 **Pneumatic systems**, such as **airlift fermenters**, have no moving parts and use the expansion of compressed gas to bring about the mixing (Fig. 6.2). These systems have lower energy requirements and create less shear than STRs. Liquid movement is initiated by the injection of compressed air at the bottom of the internal or external riser column and the air bubbles expand in the riser causing the upward movement of liquid and initiating its cycling within the fermenter. Even large fermenters do not require internal cooling coils as a jacket can normally provide sufficient heat transfer, due to the rapid movement of fluid within the vessel.

3 **Hydrodynamic mechanisms** use liquid kinetic energy to mix the fermenter contents, which is achieved by using an external liquid pump for external circulation and reinjection, e.g. **deep-jet fermenters** (Fig. 6.3).

The mixing of nutrients and gaseous exchange within any fermenter is complex. It is influenced by medium density and rheology, size and geometry of the vessel, and the amount of power used in the system. STRs have agitators with multiple impellers to give a well-mixed homogeneous environment. Nevertheless, in reality, non-uniform conditions normally prevail in vessels greater than 500 L capacity. The flow patterns of fluid motion in these stirred tanks can be of two types, laminar and turbulent flow, as a function of the **Reynolds number** (Re) of the impeller.

$$\text{Re} = \frac{\text{inertial forces}}{\text{viscous forces}}$$

Laminar and turbulent flow are characterized by high and low Re, respectively. The Re value that marks the transition between the two regimes depends on the geometry of the impeller and vessel.

The Re is an example of a natural variable or dimensionless number, whose magnitude, unlike substantial variables (length, mass, volume, temperature, etc.), does not require units. This is because the dimensions of the numerator exactly cancel those of the denominator, such as ratios of substantial variables, e.g. the aspect ratio of a cylinder (length divided by diameter). Dimensionless analysis is used extensively for solving problems in chemical/biochemical engineering and is particularly useful here, where such complex hydrodynamics are associated with physical transfer processes operating within a fermenter.

In liquid culture the rheological behaviour, fluid flow properties, are vitally important as they have a major impact on mixing and mass transfer, particularly for oxygen transfer in aerobic fermentations. When fluids are stirred they may behave as follows.

1 **Newtonian** fluids, which obey Newton's law of viscosity. Their viscosity does not vary with shear or agitation rate.

$$\text{Fluid viscosity or fluid resistance to flow } (\eta) = \frac{\text{shear stress or force per unit area}}{\text{shear rate or velocity gradient}}$$

2 **Non-Newtonian** fluids, whose viscosity varies with shear or agitation rates. For example, **pseudoplastic** fluids exhibit decreasing apparent viscosity with increasing shear or agitation rate, whereas in **dilatant** solutions the opposite occurs. For **Bingham-plastic** behaviour, flow does not occur unless a stress is first imposed.

3 **Viscoelastic** fluids, which do not observe normal liquid-state properties when stirred, such as some polymers.

Many bacterial and yeast fermentations exhibit Newtonian fluid characteristics. Mycelial cultures, and those involving polymeric substrates and products, particularly polysaccharide gels (e.g. xanthan), often display non-Newtonian properties that inhibit high flow dynamics within the fermenter. However, few fermentations behave as viscoelastic solutions.

Heat transfer

In fermenter design, efficient heat transfer is important in controlling the temperature during sterilization operations and maintaining the required operating temperature throughout the fermentation run. Heat generated in the fermentation is primarily due to metabolic activity of microorganisms and mechanical agitation processes. For most fermentations this heat needs to be dissipated by cooling. Conversely, for fermentations that operate above ambient temperature, such as those involving thermophilic organisms, there needs to be an input of heat.

Heat transfer is primarily achieved using an outer jacket surrounding the internal phase or via internal coils. No direct contact exists between the cooling/heating system and the fermentation medium. The heat is conducted through the vessel wall, coils and baffles. These systems are also used to sterilize the vessel and contents before inoculation, by the injection of pressurized steam (see p. 104, Sterilization). Automatic temperature control during the fermentation is accomplished by injecting either cold or hot water into the outer jacket and/or internal coils. In some circumstances alternative cooling media may be used, e.g. glycol.

Mass transfer

Transfer of nutrients from the aqueous phase into the microbial cells during fermentation is relatively straightforward as the nutrients are normally provided in excess. However, oxygen transfer in aerobic fermentations is rather more complex.

AERATION

Some fermentations operate anaerobically, but the majority are aerobic and require the provision of large quantities of normally sterile air or oxygen that must be dispersed throughout the fermenter. Aeration is also useful for purging unwanted volatile metabolic products from the medium. Compressed air entering a fermenter is usually stripped of moisture and any oil vapours that may originate from the compressor. To prevent the risk of contamination, gases introduced into the fermenter should be passed through a sterile filter. A similar filter on the air exhaust system avoids environmental contamination. Sterile filtered air or oxygen normally enters the fermenter through a sparger system, and air flow rates for large fermenters rarely exceed 0.5–1.0 volumes of air per volume of medium per minute (vvm). To promote aeration in stirred tanks, the sparger is usually located directly below the agitator.

Sparger structure can affect the overall transfer of oxygen into the medium, as it influences the size of the gas bubbles produced. Small bubbles are desirable because the smaller the bubble, the larger the surface area to volume ratio, which provides greater oxygen transfer. However, spargers with small pores that are effective in producing small air bubbles are more prone to blockage and require a higher energy input.

Oxygen is only sparingly soluble in aqueous solution and the solubility decreases as the temperature rises. This adds to the other difficulties, particularly those caused by the large volume of the vessel, wherein there will be regions where mixing is less efficient. When high biomass concentrations are used to increase productivity it also creates an enormous demand for oxygen. Consequently, the operation of aerobic processes is generally more demanding, as it is difficult to prevent oxygen from becoming a rate-limiting factor.

Oxygen transfer is complex, as it involves a phase change from its gaseous phase to the liquid phase, and is influenced by the following factors.

1 the prevailing physical conditions; temperature, pressure and surface area of air/oxygen bubbles;
2 the chemical composition of the medium;
3 the volume of gas introduced per unit reactor volume per unit time;
4 the type of sparger system used to introduce air into the fermenter;
5 the speed of agitation; or
6 a combination of these factors.

During aerobic fermentations molecular oxygen must be maintained at optimal concentrations to ensure maximum productivity. The two steps associated with an oxygen mass balance are the rate at which oxygen can be delivered to the biological system (oxygen transfer rate, OTR) and the rate at which it is utilized by the microorganisms (critical oxygen demand). If the rate of oxygen utilization is greater than the OTR, anaerobic conditions will develop, which may limit growth and productivity. However, attempts can be made to raise the OTR by elevating the pressure, enriching the inlet air with oxygen, and increasing both agitation and airflow rates.

The OTR is determined by the driving force, the oxygen gradient, and the resistance to oxygen transfer.

$$\text{rate of oxygen transfer} = \frac{\text{driving force}}{\text{resistance}} = \frac{\text{oxygen gradient } (C^* - C_L)}{K_L a} \quad 6.1$$

where C^* = saturated dissolved oxygen concentration (mmol/dm^3); C_L = oxygen concentration at time, t (mmol/dm^3); K_L = mass transfer coefficient (cm/h), i.e. the sum of reciprocals for the residencies of oxygen transfer from the gaseous to liquid phase; and a = gas–liquid interface area per liquid volume (cm^2/cm^3). It should be noted that both K_L and a are difficult to measure individually and are usually linked together as K_La, the volumetric mass transfer coefficient (per hour).

The driving force for oxygenation is the oxygen gradient ($C^* - C_L$). Consequently, the rate of oxygenation is faster at low dissolved oxygen concentrations, compared with higher concentrations. However, the overall K_La is not affected.

In order for oxygen to transfer from the gaseous phase to an individual cell or site of reaction, it must pass through several points of resistance (Fig. 6.4):

1 resistance within the gas film to the phase boundary;
2 penetration of the phase boundary between the gas bubble and bulk liquid;
3 transfer from the phase boundary to the bulk liquid;
4 movement within the liquid;
5 transfer to the surface of the cell;
6 entry into cell; and
7 transport to the site of reaction within the cell.

The rate-limiting step (controlling factor) in oxygen transfer is the movement of oxygen from the gaseous phase to the gas–liquid boundary layer, particularly for viscous media. Gaseous oxygen molecules move rapidly, due to their kinetic energy. However, to enter the liquid they have to cross this boundary layer at the surface of the bubble. This is composed of a thin layer of oxygen molecules that line the inside of the bubble and a thicker layer of water molecules coating the bubble surface. Diffusion across this boundary is particularly influenced by temperature, solutes and surfactants.

The rate of transfer of oxygen from an air bubble to the liquid phase is described by

$$\frac{dC_L}{dt} = K_La(C^* - C_L) \quad 6.2$$

Integration of equation 6.2 gives

$$C^* - C_L = e^{-K_Lat} \quad 6.3$$

and in terms of natural logarithms, for use in K_La determination below, this is

$$\ln(C^* - C_L) = -K_La * t \quad 6.4$$

Therefore, K_La is a measure of the aeration capacity of the fermenter and must be maintained above a

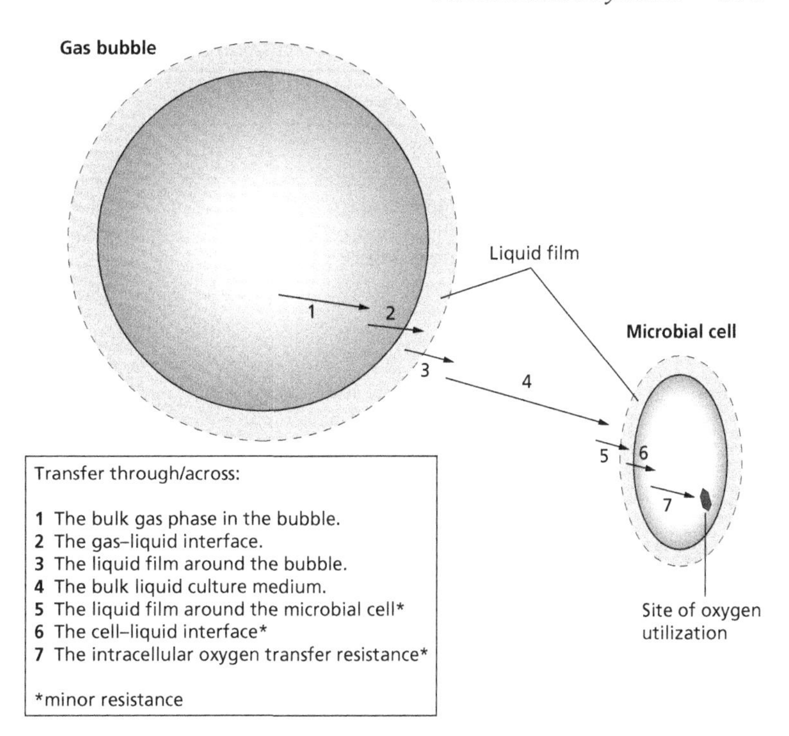

Fig. 6.4 Oxygen mass transfer steps from an air bubble to the site of utilization within a microbial cell (not to scale).

minimum critical level to supply the oxygen requirements of the organism.

Determination of K_La is relatively straightforward and is used to compare fermenters in both scale-up and scale-down. A dynamic method suitable for some vessels involves filling the fermenter under investigation with medium, and fixing specific rates of aeration and stirring. The percentage oxygen saturation can be monitored by using a rapid response polarographic or galvanic oxygen probe. When the dissolved oxygen reaches saturation and a steady state is achieved, the air supply is replaced with nitrogen. Once the oxygen in solution has fallen to a sufficiently low level, usually 10% of its saturation (equilibrium) value, air is reintroduced and the exponential reoxygenation profile is recorded. The K_La for these specific conditions is determined by a semilog plot of the reoxygenation profile, $\ln(C^* - C_L)$ against time, the slope of which is the negative of the mass transfer coefficient ($-K_La$) (see equation 6.4 and Fig. 6.5). This procedure can be repeated under different operating conditions, e.g. varying stirrer speed, air flow rate, etc.

Once in the liquid, the rate of oxygen acquisition by cells depends on the oxygen gradient between the oxygen in the bulk liquid and at the site of utilization. Movement in the bulk liquid is aided by good mixing. The rate of use by the biological system will be determined by the affinity and saturation characteristics of the terminal oxidase. As microorganisms exhibit different oxygen requirements, the level of aeration necessary will vary from fermentation to fermentation.

Fermenter control and monitoring

Fermentation systems must be efficiently controlled in order to optimize productivity and product yield, and ensure reproducibility. The key physical and chemical parameters involved largely depend on the bioreactor, its mode of operation and the microorganism being used. They are primarily aeration, mixing, temperature, pH and foam control. Control and maintenance at optimum levels inside the reactor is mediated by sensors (electrodes), along with compatible control systems and data logging (Table 6.3). Internal sensors that are in or above the fermentation medium (pH, oxygen, foam, redox, medium analysis and pressure probes) should be steam sterilizable and robust. Some sensors do not come into direct contact with any internal component of the bioreactor and do not need sterilization; for example, load cells, agitator shaft power and speed meters, and

Table 6.3 The sensors used in fermenter monitoring and control

Sensor	Measurement: Physical	Chemical	Biological
Electrodes	Temperature (thermistor, resistance thermometer, thermometer)	Dissolved oxygen Dissolved carbon dioxide Nutrients (biosensors, e.g. for glucose) pH Metal ions Foam level detection	Biosensors for biologically active products
Meters	Air flow rate in and out Agitation shaft power Speed of agitation, e.g. impeller tachometer	Acid/alkali addition	
Transducers	Pressure Liquid flow		
Mass spectra		Directly on-line or off-line nutrients and in flow and exhaust gases	Products
Spectrophotometers (determination on-line and off-line)			Biomass

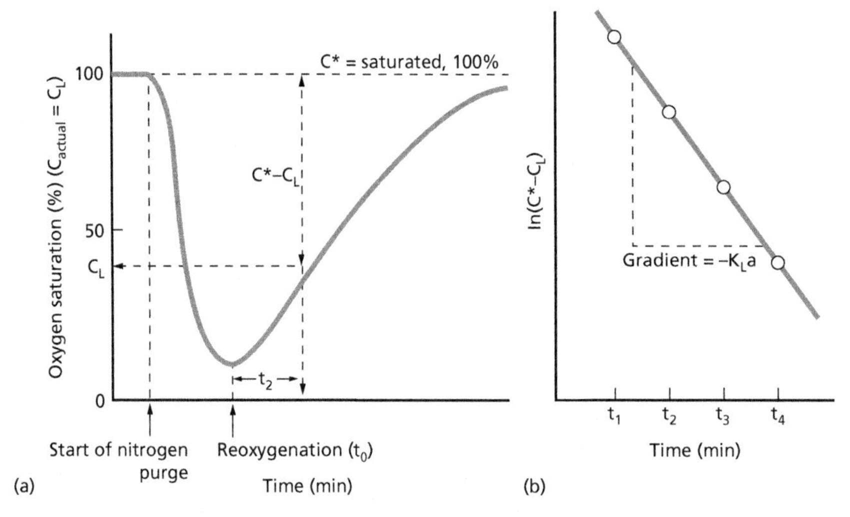

Fig. 6.5 Determination of the volumetric mass transfer coefficient (K_La). (a) reoxygenation curve; (b) plot of natural log (ln) of C^* (saturated dissolved oxygen concentration) – C_L (oxygen concentration at time, t) against time.

external sensors used to analyse samples regularly withdrawn from the bioreactor. Samples can be taken off-line for various analyses, such as cell counts and determination of DNA, RNA, lipids, specific proteins, carbohydrates and other key metabolites and substrates.

Control of pH is usually a major factor as many fermentations yield products that can alter the pH of the growth media. Fermentation media often contain buffering salts, usually phosphates, but their capacity to control pH can be exceeded and addition of acid or alkali may be required. The pH can be maintained at the desired value by their automatic addition in response to changes recorded by the pH electrode. Many fermentations produce acid and adjustment of the pH can be made with ammonium hydroxide, which may also act as a nitrogen source. Monitoring of temperature is usually via resistance thermometers and thermistors linked to automatic heating or cooling systems. Levels of dissolved O_2 and CO_2 are determined using O_2 and CO_2 electrodes.

Foam production in bioreactors is often a major problem, particularly in aerated fermentations. Formation of foam is due to the presence of surface-active agents, especially proteins, which produce stable foams. If not controlled this can lead to contamination and blockage of air filters. There are basically three methods used to control foam production: media modification, mechanical foam-breaking devices or the automatic addition of chemical antifoam agents (see Chapter 5).

The basic principle of control involves a sensory system linked to a control system and feedback loop (Fig. 6.6). Sensors are used to measure and record the events within the bioreactor. In conjunction with process control, they maintain the difference between the measured and desired values at a minimum level. Overall control can be manual or automated; newer systems have integral and derivative control systems. Data recorded from the sensors and control decisions are downloaded to a computer where appropriate calculations can be performed to determine production of biomass and product, overall oxygen and carbon dioxide transfer rates, nutrient utilization, power usage, etc. All of this information may be used to construct a mathematical/computer model of the process. This can be applied in future fermentations to compare the actual fermentation with the composite model. If a deviation is discovered, appropriate alarm and correctional systems are activated, giving greater control of the fermentation. However, any control system needs to be regularly calibrated when first installed and then regularly checked to conform to good manufacturing practices (GMP).

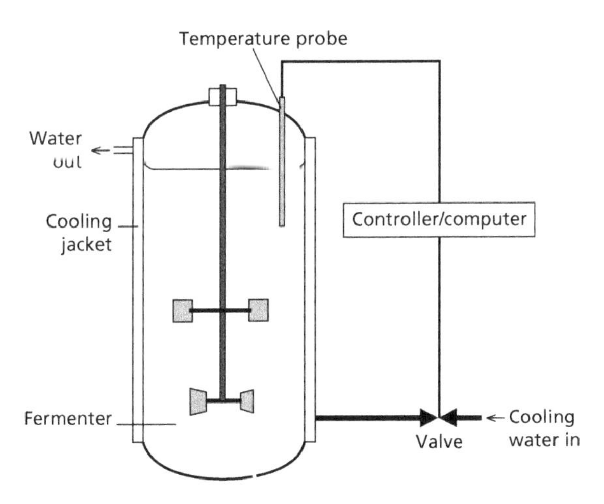

Fig. 6.6 A scheme for controlling fermenter temperature.

Operating modes

Industrial fermentations are operated as batch, fed-batch or continuous cultures (see Chapter 2, Microbial growth). Most are batch processes, which are closed systems where there are no additions following inoculation, apart from acid or alkali for pH control and input of air for aerobic fermentations. In batch fermentations there is a definite beginning and end to the process. A fermenter is loaded, sterilized and inoculated, and the organism is grown through a typical batch profile (see Chapter 2). The product is then harvested and the fermenter must be cleaned before restarting the cycle. This non-productive phase is referred to as 'down-time'. Examples of this mode of operation include the production of alcoholic beverages, most amino acids, enzymes, organic acids, etc. Variations include fed-batch systems that have been successfully used for producing baker's yeast and penicillin. For this mode of operation extra nutrients are added as the fermentation progresses, which increases the fermentation volume. Additions may be made continuously, intermittently or as a single supplementation, often when a batch culture approaches the end of the rapid growth phase. Fed-batch operation can extend the product formation phase and may overcome problems associated with the use of repressive, rapidly metabolized, substrates (see Chapter 14, Baker's yeast production). This method is also useful where a substrate causes viscosity problems or is toxic at high concentrations. Fed-batch with recycle of cells (biomass) can also be used for specific purposes, e.g. some ethanol fermentations and waste-water treatment processes.

The advantages of batch systems are that initial capital expenditure is lower and, if contamination occurs, it is relatively simple to terminate and restart a new fermentation cycle. As mentioned above, batch systems are successfully used in the production of many traditional fermentation products; and for producing secondary metabolites, such as antibiotics, where the cells are first grown beyond the rapid growth phase prior to the accumulation of these metabolites. However, batch fermentations are theoretically less effective for the production of biomass and primary (growth-associated) metabolic products. Only a small fraction of each batch fermentation cycle is productive, as there may be a considerable lag period and it is only in the later stages of the exponential phase that large quantities of the product are generated. Other disadvantages of batch systems are the batch-to-batch variability of the product; plus increased non-productive down-time, involving cleaning, sterilizing, refilling and poststerilization cooling. The increased frequency of sterilization may also cause greater stress on instruments and probes. In addition, the running costs are greater for preparing and maintaining stock cultures, and generally more personnel are required for operating batch processes.

Continuous culture is an open system where fresh medium is continuously added and culture is simultaneously removed at the same rate, resulting in a constant working volume. In continuous systems, cells grow exponentially for extended periods at a specified predetermined growth rate. Furthermore, the system has the property of reaching a steady state in which the concentration of limiting nutrient and the cell number do not vary with time (see Chapter 2). Consequently, in theory, such systems are more productive than batch systems. Continuous fermentations are particularly well suited for the production of biomass and growth-associated primary metabolites. Their reduced down-time and lower operating costs are also desirable attributes. However, they require higher initial capital expenditure and, to date, relatively few large-scale industrial examples have become established, other than for biomass, fuel/industrial ethanol and effluent treatment.

Problems associated with continuous culture processes, other than waste-water treatment, include the fact that throughout their 20–50 days or longer operation, sterility must be maintained and a continuous supply of media of constant composition provided. However, these difficulties can be overcome by GMP and good microbiological practices. Nevertheless, the operating conditions place strong selection pressure on the organism. Any genetic instability may lead to the generation of low-yielding mutants that may outgrow the original high-yielding strain.

Sterilization

The basic principles of sterilization have been discussed in Chapter 2. Here we will consider specific aspects relating to the fermentation process.

Air sterilization

To prevent contamination of either the fermentation by air-borne microorganisms or the environment by aerosols generated within the fermenter, both air input and air exhaust ports have air filters attached. These filters are designed to trap and contain microorganisms. Filters are made of glass fibre, mineral fibres, polytetrafluoroethylene (PTFE) or polyvinyl chloride (PVC), and must be steam sterilizable and easily changed. In some circumstances, particularly where pathogenic organisms are being grown, fermenter exhaust may also undergo dry heat sterilization (incineration) as an additional safety measure.

Media and vessel sterilization

For pilot-scale and industrial aseptic fermentations the fermenter can be sterilized empty. The vessel is then filled with sterile medium, prepared in a batch or continuous medium 'cooker' that may supply several fermentations. Alternatively, the fermenter is filled with formulated medium and the two are sterilized together in one operation. However, some industrial fermentations are not aseptic, but microbial contamination is still maintained at a minimum level by boiling or pasteurization of the media. Otherwise, a fast-growing contaminant could outgrow the industrial microorganism, or at least utilize some valuable nutrients. Microbial contaminants may also metabolize the target product, produce toxic compounds or secrete products that may block filters and interfere with downstream processing. If the contaminant is a bacteriophage it may lyse the culture, as can occur in fermentations involving lactic acid bacteria.

Small laboratory-scale fermenters of 1–5 L capacity are usually filled with medium and then sterilized in a steam autoclave. Here sterilization is normally performed using pressurized steam to attain a temperature of 121°C for 15 min. Care must be taken to avoid any pressure build-up inside the fermenter by venting without contaminating the contents. For pilot-scale and industrial fermenters more rigorous sterilization is necessary, involving increased sterilization time and/or higher temperature. Normally, the aim is to provide sterilization conditions that give an acceptable probability of contamination of 0.1% (1 in 1000). Consequently, if the original number of microbial cells (N_0) is known, the Del factor ($\nabla = \ln N_0/N_t$) can be calculated (see Chapter 2). The value of N_t, when a 0.1% risk is adopted, will be 10^{-3} cells. A sterilization profile of a typical fermenter is shown in Fig. 6.7. Destruction of cells occurs during heating (to 121°C in this case) and cooling (here from 121°C to the fermentation operating temperature). Therefore the overall Del factor may be represented as

$$\nabla_{total} = \nabla_{heating} + \nabla_{holding} + \nabla_{cooling}$$

Thus knowing both $\nabla_{heating}$ and $\nabla_{cooling}$ for a particular fermentation system, the necessary holding time needed to achieve the required overall Del factor can be calculated. Alternatively, the Richards approximation can be used, which employs only that part of the curve above 100°C (Fig. 6.7). This assumes that:

1 heating and cooling between 0°C and 100°C is unimportant for sterilization;
2 heating between 100°C and 121°C is at 1°C per minute, i.e. 20 min; and
3 cooling from 121°C to 100°C is at 1°C per minute, i.e. 20 min.

In practice, steam sterilization involves passing the steam under pressure into the vessel jacket and/or internal coils. Steam may also be injected directly into the head space above the fermentation medium. This aids sterilization, but can result in media volume changes. Steam sterilization is effective and cheaper than dry heat sterilization. However, certain media constituents may be heat labile and destroyed by excessive heat, e.g. glucose, some vitamins and components of animal cell culture media. Such heat-sensitive ingredients are often filter sterilized before use; alternatively, some can be heat sterilized with minimal degradation using a high temperature for a very short time, e.g. 140°C for 50 s. This is usually a continuous operation where the holding time is controlled by the flow rate through the sterilizer and the material is then rapidly cooled in a heat exchanger.

Solid-substrate fermentations

Solid-substrate fermentations have been used for producing various fermented foods in Asia for thousands of years, but this method is rarely used in Europe and North America. It involves the growth of microorganisms on solid, normally organic, materials in the absence or near absence of free water. The substrates used are often cereal grains, bran, legumes and lignocellulosic materials, such as straw, wood chippings, etc. Traditional processes are largely food fermentations producing oriental tempeh and sufu, cheeses (Chapter 12) and mushrooms (Chapter 14); along with compost and silage making (Chapter 15). In addition, enzymes, organic acids and ethanol are now produced by solid-substrate fermentations, particularly in areas where modern fermentation equipment is unavailable.

Solid-substrate fermentations lack the sophisticated control mechanisms that are usually associated with submerged fermentations. Their use is often hampered by lack of knowledge of the intrinsic kinetics of microbial growth under these operating conditions. Control of the environment within the bioreactors is also difficult to achieve, particularly the simultaneous maintenance of optimal temperature and moisture. However, in some instances, solid-substrate fermentations are the most suitable methods for the production of certain products (see Table 6.4). For example, most fungi do not form spores in submerged fermentations, but

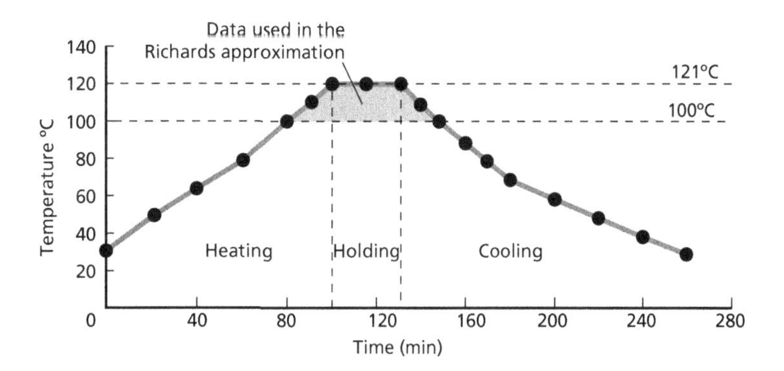

Fig. 6.7 Sterilization of a fermenter.

Table 6.4 Advantages and disadvantages of solid-substrate fermentations

Advantages	Disadvantages
Potentially provide superior productivity	Slower microbial growth
Low-cost media	Problems with heat build-up
Simple technology	Bacterial contamination can be problematic
Low capital costs	Difficulties often encountered on scale-up
Reduced energy requirements	Substrate moisture level difficult to control
Low waste-water output	
No problems with foaming	

sporulation is often accomplished in solid-substrate fermentations. This method is successfully employed in the production of *Coniothyrium minitans* spores for the biocontrol of the fungal plant pathogen, *Sclerotinia sclerotiorum*. Solid-substrate fermentations are normally multistep processes, involving:

1 pretreatment of a substrate that often requires mechanical, chemical or biological processing;
2 hydrolysis of primarily polymeric substrates, e.g. polysaccharides and proteins;
3 utilization of hydrolysis products; and
4 separation and purification of end-products.

The microorganisms associated with solid-substrate fermentations are those that tolerate relatively low water activity down to A_w values of around 0.7. They may be employed in the form of:

1 monocultures, as in mushroom production, e.g. *Agaricus bisporus*;
2 dual cultures, e.g. straw bioconversion using *Chaetomium cellulilyticum* and *Candida tropicalis*; and
3 mixed cultures, as used in composting and the preparation of silage, where the microorganisms may be indigenous or added mixed starter cultures (inoculants).

Environmental parameters that influence solid-substrate fermentations

WATER ACTIVITY, A_W (see Chapter 2)

Water is lost during fermentation through evaporation and metabolic activity. This is normally replaced by humidification or periodic additions of water. If moisture levels are too low, the substrate is less accessible, as it does not swell and microbial growth is reduced. However, if the moisture levels are too high there is a reduction in the porosity of the substrate, lowering the oxygen diffusion rates and generally decreasing gaseous exchange. Consequently, the rate of substrate degradation is reduced and there is also an increased risk of microbial contamination.

TEMPERATURE

Heat generation can be more problematic than in liquid fermentations and has a major influence on relative humidity within a fermentation. The temperature is largely controlled by aeration and/or agitation of the substrate.

AERATION

Most solid-substrate fermentations are aerobic, but the particular requirements for oxygen depend upon the microorganism(s) used and the specific process. Rates of aeration provided are also closely related to the need to dissipate heat, CO_2 and other volatile compounds that may be inhibitory. The kinetics of oxygen transfer in solid-substrate fermentations are poorly understood. However, the rate of oxygen transfer is greatly influenced by the size of the substrate particles, which determines the void space. Oxygen transfer within this void space is closely related to the moisture level, as the oxygen dissolves in the water film around the substrate particles. However, as mentioned above, if excess water fills the void spaces, it has a detrimental effect on oxygen transfer.

Bioreactors used for solid-substrate fermentations

Most solid-substrate fermentations are batch processes, although attempts are being made to develop semicontinuous and continuous systems. Some processes do not require bioreactors, they simply involve spreading the substrate onto a suitable floor. Those processes employing vessels exhibit considerable variations. A few anaerobic processes, such as silage production, require no mechanisms for agitation or aeration. However, the majority are aerobic fermentations, requiring aeration and occasional or continuous agitation. Bioreactors commonly used include the following (Fig. 6.8).

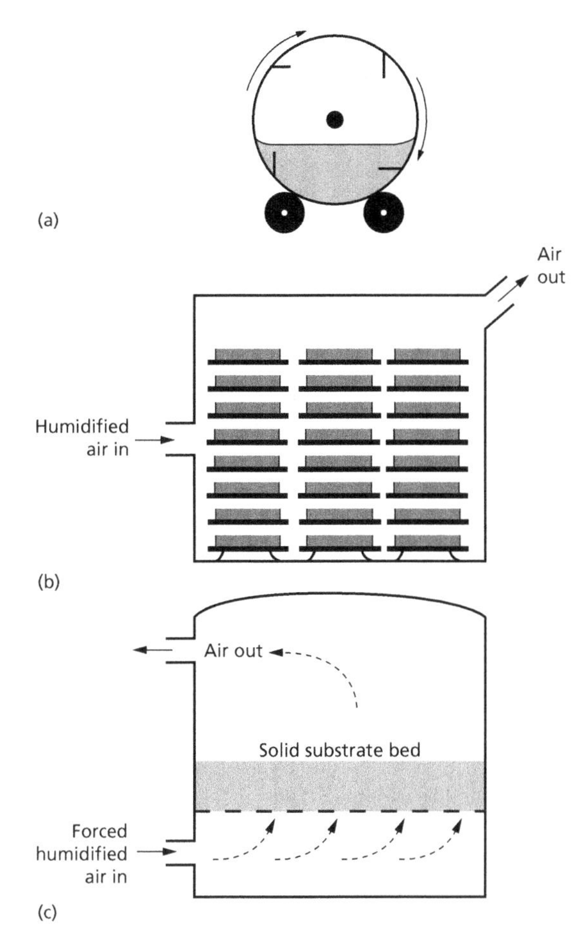

Fig. 6.8 Three types of solid-substrate fermenters: (a) rotating drum fermenter; (b) tray fermenter; (c) a bed system.

1 **Rotating drum fermenters**, comprising a cylindrical vessel of around 100 L capacity mounted on its side onto rollers that both support and rotate the vessel (Fig. 6.8a). These fermenters are used in enzyme and microbial biomass production. Their main disadvantage is that the drum is filled to only 30% capacity, otherwise mixing is inefficient.

2 **Tray fermenters**, which are used extensively for the production of fermented oriental foods and enzymes. Their substrates are spread onto each tray to a depth of only a few centimetres and then stacked in a chamber through which humidified air is circulated. These systems require numerous trays and large volume incubation chambers of up to 150 m^3 capacity (Fig. 6.8b).

3 **Bed systems**, as used in commercial koji production (see Chapter 12), consisting of a bed of substrate up to 1 m deep, through which humidified air is continuously forced from below (Fig. 6.8c).

4 **Column bioreactors**, consisting of a glass or plastic column, into which the solid substrate is loosely packed, surrounded by a jacket that provides a means of temperature control. These vessels are used to produce organic acids, ethanol and biomass.

5 **Fluidized bed reactors**, which provide continuous agitation with forced air to prevent adhesion and aggregation of substrate particles. These systems have been particularly useful for biomass production for animal feed.

Fermentation process development

Once a microorganism has been selected as the producer organism for a particular process, research is initially performed under laboratory-scale conditions using 1–10 L fermenters. This involves an examination of media formulation and 'feeding' strategies (batch, fed-batch, continuous, etc.), along with the selection of the most suitable type of fermentation system (stirred tank, airlift, packed bed, solid state, hollow fibre, etc.). Other factors that are considered include reactor configuration, and control of pH, dissolved oxygen, foam and temperature. Optimization of product yield in the laboratory is followed by process **scale-up**; first to pilot scale of 10–100 L and finally to industrial scale of 1000–100 000 L, or more, depending upon the specific process. However, during scale-up, decreased product yields are often experienced; the reason is that the conditions in the larger-scale fermenters are not identical to those experienced in the smaller-scale laboratory or pilot plant systems. Key factors that influence yield during the scale-up process are as follows.

1 Inoculum propagation procedures adopted, and the quality and quantity of inoculum used to start the fermentation.

2 Choice of medium; cheaper nutrient sources are often employed for large-scale operations due to cost constraints.

3 Industrial-scale sterilization protocols may result in greater degradation of heat-labile compounds, which affects the final quality of the medium.

4 Larger fermenters are often subject to the development of nutrient, temperature, pH and oxygen gradients, which were not experienced in smaller, well-mixed, fermenters.

5 Scale-up can also alter the generation of foam, shear forces and rate of removal of carbon dioxide.

Variations in any one of these factors may significantly change the operational conditions and consequently influence the product yield, compared with laboratory-scale experiments. Although desirable, it may not be possible to maintain all parameters at the same value at each stage of scale-up. However, it is possible to fix a key parameter and ensure that it remains unaltered throughout scale-up. Those parameters often chosen to remain firmly fixed are the fermenter height–diameter aspect ratio, power input per unit volume, oxygen transfer coefficient (K_La), dissolved oxygen level or impeller tip speed (to maintain similar shear). In **scale-down** studies, the opposite approach can be implemented, where the conditions in the large-scale fermenter are mimicked, as far as possible, in the small-scale systems.

Further reading

Papers and reviews

Cooney, C. L. (1983) Bioreactors: design and operation. *Science* **219**, 728–733.

Court, J. R. (1988) Computers in fermentation control: laboratory applications. *Progress in Industrial Microbiology* **25**, 1–45.

Daugulis, A. J. (1994) Integrated fermentation and recovery processes. *Current Opinion in Biotechnology* **5**, 192–195.

Daugaulis, A. J. (1997) Partitioning bioreactors. *Current Opinion in Biotechnology* **8**, 169–174.

Longobardi, G. P. (1994). Fed-batch vs. batch fermentation. *Bioprocess Engineering* **10**, 185–194.

Reisman, H. B. (1993) Problems in scale-up of biotechnology production processes. *CRC Critical Reviews in Biotechnology* **13**, 195–253.

Ritzka, A., Sosnitza, P., Ulber, R. & Scheper, T. (1997) Fermentation monitoring and process control. *Current Opinion in Biotechnology* **8**, 160–164.

Sharma, M. C. & Gurtu, A. K. (1993) Asepsis in bioreactors. *Advances in Applied Microbiology* **3**, 1–27.

van't Riet, K. (1983) Mass transfer in fermentation. *Trends in Biotechnology* **1**, 113–119.

Books

Asenjo, J. A. & Merchuk, J. C. (1995) *Bioreactor System Design*. Marcel Dekker, New York.

Atkinson, B. & Mavituna, F. (1991) *Biochemical Engineering and Biotechnology Handbook*, 2nd edition. Macmillan, Basingstoke.

Bastin, G. & Dochain, D. (1990) *On-line Estimation and Adaptive Control of Bioreactors*. Elsevier, Amsterdam.

Blanch, H. W. & Clark, D. S. (1997) *Biochemical Engineering*. Marcel Dekker, New York.

Bu'lock, J. & Kristiansen, B. (1987) *Basic Biotechnology*. Academic Press, London.

Demain, A.L., Davies, J. E. & Atlas, R. M. (1999) *Manual of Industrial Microbiology and Biotechnology*. American Society for Microbiology, Washington D.C.

Doran, P. M. (1995) *Bioprocess Engineering Principles*. Academic Press, London.

Jackson, A.T. (1990) *Process Engineering in Biotechnology*. Open University, Milton Keynes.

McNeil, B. & Harvey, L. M. (1990) *Fermentation: A Practical Approach*. IRL, Oxford.

Scragg, A.H. (1991) *Bioreactors in Biotechnology*. Ellis Horwood, New York.

Stanbury, P. F., Whitaker, A. & Hall, S. J. (1995) *Principles of Fermentation Technology*, 2nd edition. Butterworth-Heinemann (Pergamon), Oxford.

Vogel, H. C. & Todaro, C. L. (1997) *Fermentation and Biochemical Engineering Handbook—Principles, Process Design and Equipment*. Noyes Publications, Westwood, NJ.

7 Downstream processing

Industrial fermentations comprise both **upstream processing** (USP) and **downstream processing** (DSP) stages (Fig. 7.1). USP involves all factors and processes leading to, and including, the fermentation, and consists of three main areas. The first relates to aspects associated with the producer microorganism. They include the strategy for initially obtaining a suitable microorganism, industrial strain improvement to enhance productivity and yield, maintainance of strain purity, preparation of a suitable inoculum and the continuing development of selected strains to increase the economic efficiency of the process (see Chapter 4). The second aspect of USP involves fermentation media (see Chapter 5), especially the selection of suitable cost-effective carbon and energy sources, along with other essential nutrients. This media optimization is a vital aspect of process development to ensure maximization of yield and profit. The third component of USP relates to the fermentation (see Chapter 6), which is usually performed under rigorously controlled conditions, developed to optimize the growth of the organism or the production of a target microbial product.

DSP encompasses all processes following the fermentation. It has the primary aim of efficiently, reproducibly and safely recovering the target product to the required specifications (biological activity, purity, etc.), while maximizing recovery yield and minimizing costs. The target product may be recovered by processing the cells or the spent medium depending upon whether it is an intracellular or extracellular product. The level of purity that must be achieved is usually determined by the specific use of the product. Often, a product's purity will be defined by what is not present rather than what is. Purity of an enzyme, for example, is expressed as units of enzyme activity per unit of total protein. Not only is it important to reduce losses of product mass, but in many cases retention of the product's biological activity is vitally important.

Each stage in the overall recovery procedure is strongly dependent on the protocol of the preceding fermentation. Fermentation factors affecting DSP include the properties of microorganisms, particularly morphology, flocculation characteristics, size and cell wall rigidity. These factors have major influences on the filterability, sedimentation and homogenization efficiency. The presence of fermentation byproducts, media impurities and fermentation additives, such as antifoams, may interfere with DSP steps and accompanying product analysis. Consequently, a **holistic approach** is required when developing a new industrial purification strategy. The whole process, both upstream and downstream factors, needs to be considered. For example, the choice of fermentation substrate influences subsequent DSP. A cheap carbon and energy source containing many impurities may provide initial cost savings, but may necessitate increased DSP costs. Hence overall cost savings may be achieved with a more expensive but purer substrate. Also, adopting methods that use existing available equipment may be more cost-effective than introducing more efficient techniques necessitating investment in new facilities.

The physical and chemical properties of the product, along with its concentration and location, are obviously key factors as they determine the initial separation steps and overall purification strategy. It may be the whole cells themselves that are the target product or an intracellular product, possibly located within an organelle or in the form of inclusion bodies. Alternatively, the target product may have been secreted into the periplasmic space of the producer cells or the fermentation medium. Stability of the product also influences the requirement for any pretreatment necessary to prevent product inactivation and/or degradation.

DSP can be divided into a series of distinct **unit processes** linked together to achieve product purification (Fig. 7.2). Examples of the purification strategies for two products are shown in Fig. 7.3. Usually, the number of steps is kept to a minimum. This is not only because of cost, but because even though individual steps may obtain high yields, the overall losses

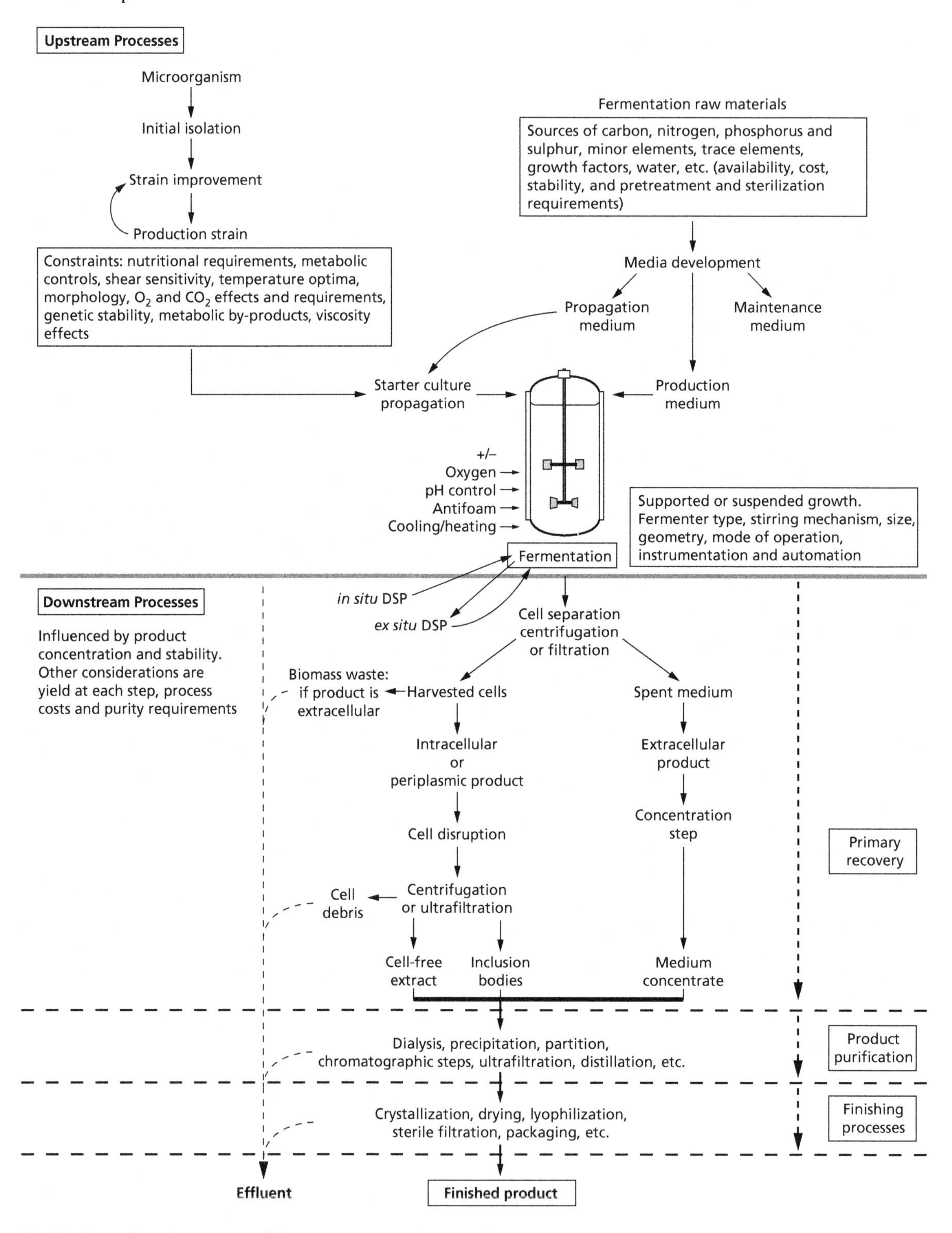

Fig. 7.1 An outline of upstream and downstream processing operations.

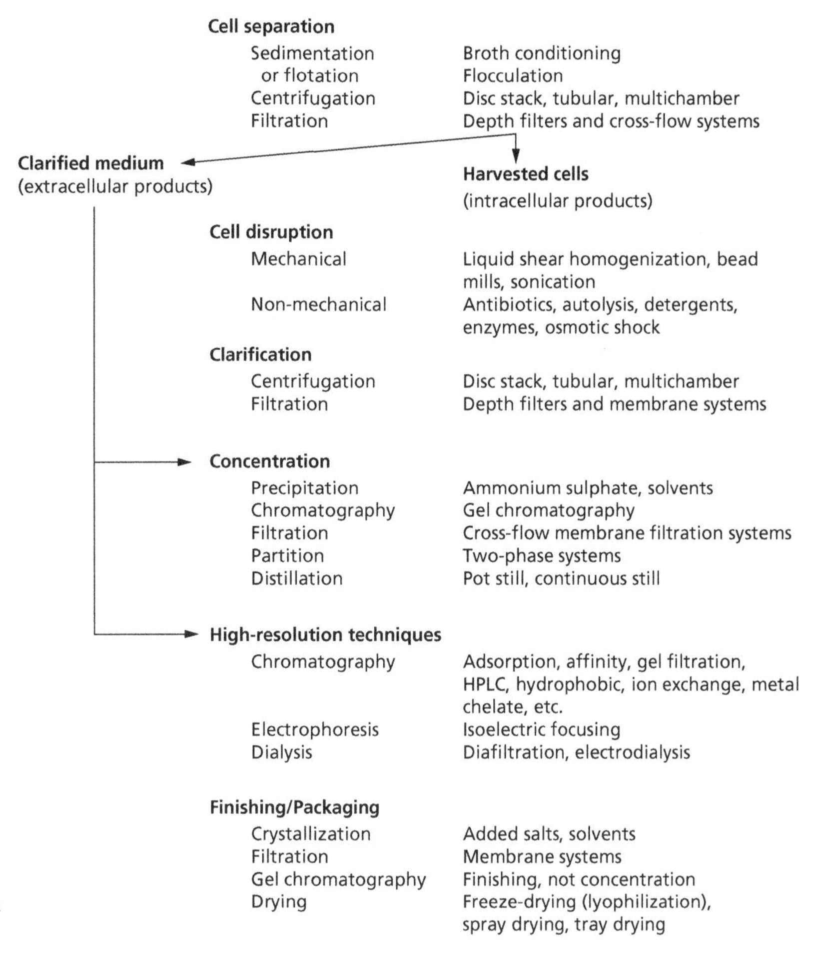

Fig. 7.2 Typical unit processes used in downstream processing.

of multistage purification processes may be prohibitive (Fig. 7.4). The specific unit steps chosen will be influenced by the economics of the process, the required purity of the product, the yield attainable at each step and safety aspects. In many cases, integration of fermentation and DSP is now preferred. Integration can often increase productivity, decrease the number of unit operations and reduce both the overall process time and costs.

Physical process integration may be achieved by placing separation units inside the fermenter or by directly linking the two systems together. Where product formation is coupled with growth, i.e. primary metabolites, higher productivity may be achieved by using integrated systems that maintain high cell densities through cell retention or recycling. Where products are inhibitory, a wide range of methods have been employed to partition bioreactors to allow the rapid *in situ* removal of products by extraction, adsorption or stripping, often increasing yield and productivity. Alternatively, the processing can be *ex situ*, where the product is removed outside the fermenter and the processed medium is returned to the fermentation. Such processes have been particularly successful for removing alcohols and solvents, but it is now also possible to extract some proteins. This mode of product removal can often improve raw material conversion, particularly where there are unfavourable thermodynamic equilibria, i.e. product removal 'pulls' the reaction in the direction of further product formation. Early product extraction can also enhance the yield for those products sensitive to prolonged exposure to the fermentation environment, where shear, proteolytic enzymes, oxidation, etc., may be destructive.

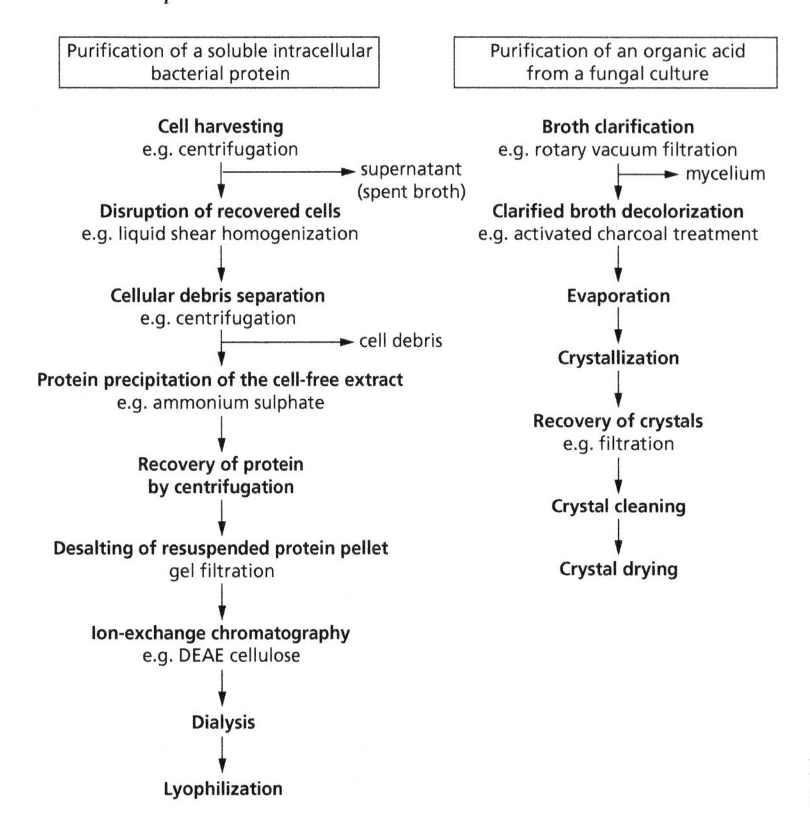

Fig. 7.3 Examples of unit downstream processing.

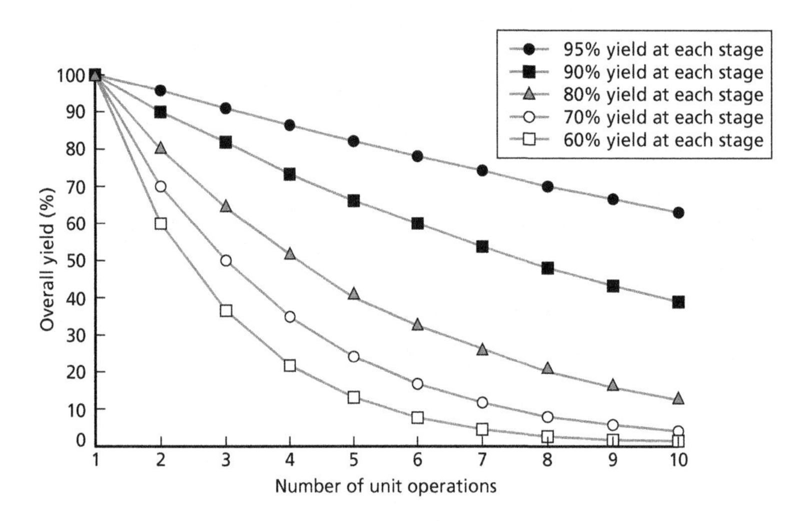

Fig. 7.4 The effect of yield in multistage purification.

Cell harvesting

The first step in the downstream processing of suspended cultures is a solid–liquid separation to remove the cells from the spent medium. Each fraction can then undergo further processing, depending on whether the product is intracellularly located, or has been secreted into the periplasmic space or the medium. Choice of solid–liquid separation method is influenced by the size and morphology of the microorganism (single cells, aggregates or mycelia), and the specific gravity, viscosity and rheology of the spent fermentation medium. These factors can also have major influences on the transfer of the liquid through pumps and pipes.

Broth conditioning

Broth conditioning techniques are mostly used in association with sedimentation and centrifugation for the separation of cells from liquid media. They alter or exploit some property of a microorganism, or other suspended material, such that it flocculates and usually precipitates. However, in certain cases it may be used to promote flotation. This uses the ability of some cells to adsorb to the gas–liquid interfaces of gas bubbles and float to the surface for collection, which occurs naturally in traditional ale and baker's yeast fermentations (see Chapter 12, Beer brewing). Certain floc precipitation methods are also used at the end of many traditional beer and wine fermentation processes, where the addition of finings (egg albumen, isinglass, etc.) may be employed to precipitate yeast cells. Major advantages of these techniques are their low cost and ability to separate microbial cells from large volumes of medium.

Some organisms naturally flocculate, which can be enhanced by chemical, physical and biological treatments. Such treatments can also be effective with cells that would not otherwise form flocs. Coagulation, the formation of small flocs from dispersed colloids, cells or other suspended material, can be promoted using coagulating agents (simple electrolytes, acids, bases, salts, multivalent ions and polyelectrolytes). Subsequent flocculation, the agglomeration of these smaller flocs into larger settleable particles, is often aided by inorganic salts (e.g. calcium chloride) or polyelectrolytes. These are high molecular weight, water soluble, anionic, cationic or non-ionic organic compounds, such as polyacrylamide and polystyrene sulphate.

Sedimentation

Sedimentation is extensively used for primary yeast separation in the production of alcoholic beverages, and in waste-water treatment. This low-cost technology is relatively slow and is suitable only for large flocs (greater than 100 μm diameter). The rate of particle sedimentation is a function of both size and density. Hence, the larger the particle and the greater its density, the faster the rate of sedimentation. The basis of this method of separation is sedimentation under gravity, which for a spherical particle can be represented by Stokes' Law:

$$V_g = \frac{d_p^2(p_s - p_l)g}{18\eta} \qquad 7.1$$

where V_g = rate of particle sedimentation (m/s); d_p = diameter of the particle (m); $p_s - p_l$ = difference in density between the particle and surrounding medium (kg/m^3); g = gravitational acceleration (m/s^2); and η = viscosity (Pascal seconds (Pa s)).

Therefore, for rapid sedimentation the difference in density between the particle and the medium needs to be large, and the medium viscosity must be low.

Centrifugation

If instead of simply using gravitational force to separate suspended particles, a centrifugal field is applied, the rate of solid–liquid separation is significantly increased and much smaller particles can be separated. Centrifugation may be used to separate particles as small as 0.1 μm diameter and is also suitable for some liquid–liquid separations. Its effectiveness, too, depends on particle size, density difference between the cells and the medium, and medium viscosity.

In a centrifuge, the terminal velocity of a particle is

$$V_c = \frac{d_p^2(p_s - p_l)\omega^2 r}{18\eta} \qquad 7.2$$

where V_c = centrifugal sedimentation rate or particle velocity (m/s); ω = angular velocity of the centrifuge (rad/s); and r = distance of the particle from the centre of rotation (m) (for η, $p_s - p_l$ and d_p, see equation 7.1).

Hence, the faster the operating speed (ω) and the greater the distance from the centre of rotation, the faster the sedimentation rate (V_c). Centrifuges can be compared using the relative centrifugal force (RCF) or **g** number (the ratio of the velocity in a centrifuge to the velocity under gravity = $\omega^2 r/g$). The choice of centrifuge depends on the particle size and density, and the viscosity of the medium. Higher-speed centrifuges are required for the separation of smaller microorganisms, such as bacteria, compared with yeasts. For example, relatively slow centrifugation effectively recovers residual yeast cells remaining in beer after the bulk has sedimented out. Conversely, an RCF of 20 000**g** may be required to recover suspended bacterial cells, cell debris and protein precipitates from liquid media.

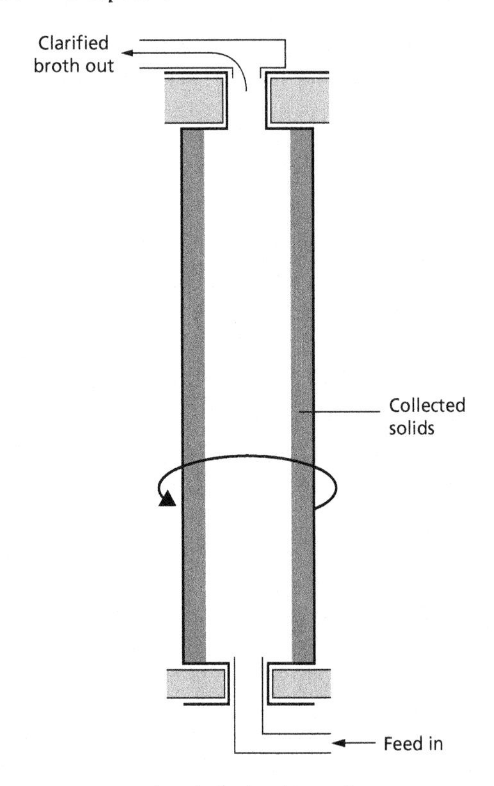

Fig. 7.5 Diagram of a tubular bowl centrifuge.

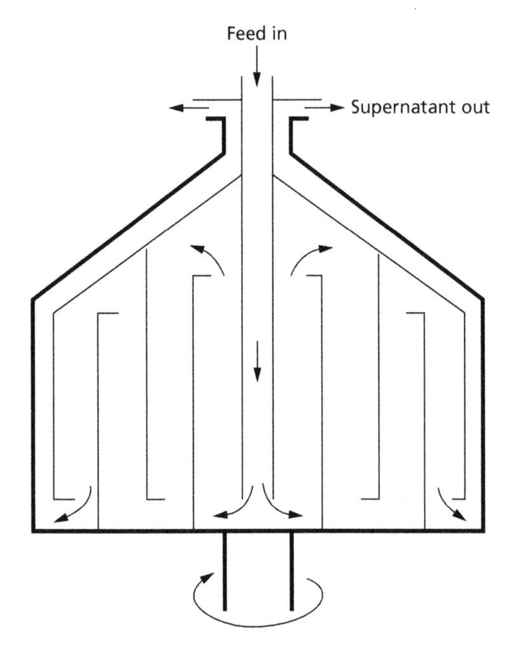

Fig. 7.6 Diagram of a multichamber bowl centrifuge.

Advantages of centrifugation include the availability of fully continuous systems that can rapidly process large volumes in small volume centrifuges. Centrifuges are steam sterilizable, allowing aseptic processing, and there are no consumable costs for membranes, chemicals or filter aids. However, the disadvantages of centrifugation are the high initial capital costs, the noise generated during operation and the cost of electricity. Also, physical rupture of cells may occur due to high shear and the temperature may not be closely controllable, which can affect temperature-sensitive products. Bioaerosol generation is a further major disadvantage, particularly when centrifuges are used for certain recombinant DNA organisms or pathogens. Under these circumstances the equipment must be contained (see Chapter 8).

INDUSTRIAL CENTRIFUGES

Centrifuges can be divided into small-scale laboratory units and larger pilot- and industrial-scale centrifuges. Laboratory batch centrifuges include, in ascending order of speed attainable: bench-top, high-speed and ultracentrifuges, capable of applying RCFs of 5000–500 000 *g*. Although industrial batch centrifuges are available, for most industrial purposes semicontinuous and continuous centrifuges are required to process the large volumes involved. However, the RCFs achieved are relatively low.

Four main types of industrial centrifuge are commonly used.

1 **Tubular centrifuges** usually produce the highest centrifugal force of 13 000–17 000 *g*. They have hollow tubular rotor bowls providing a long flow path for the suspension, which is pumped in at the bottom and flows up through the rotor (Fig. 7.5). Particulate material is thrown to the side of the bowl, and clarified liquid passes out at the top for continuous collection. As the particulate material accumulates on the inside of the bowl, the operating diameter becomes reduced. Consequently, there must be periodic removal of solids.

2 **Multichamber bowl centrifuges** consist of a bowl that is divided by vertically mounted interconnecting cylinders and are capable of operating at 5000–10 000 *g* (Fig. 7.6). The liquid feed passes from the centre through each chamber in turn, and the smaller particles collect in the outer chambers.

3 **Disc stack centrifuges** can operate at 5000–13 000 *g*. The centrifuge bowl contains a stack of conical discs whose close packing aids separation (Fig. 7.7). As liquid

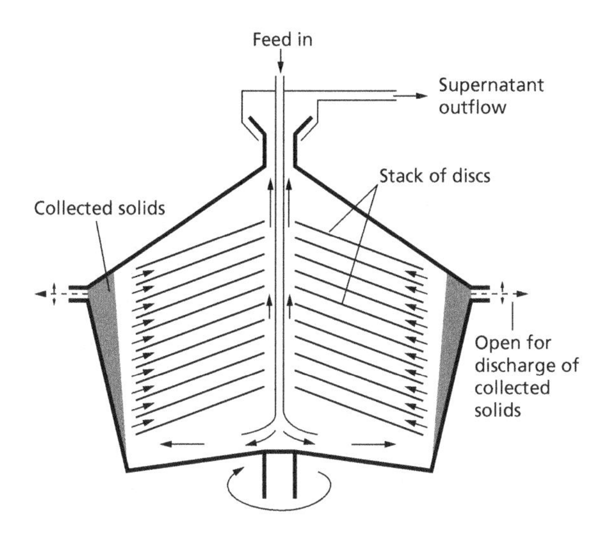

Fig. 7.7 Diagram of a disc stack centrifuge.

enters the centrifuge particulate material is thrown outwards, impinging on the underside of the cone discs. Particles then travel outwards to the bowl wall where they accumulate. These centrifuges usually have the facility to discharge the collected material periodically during operation.

4 Screw-decanter centrifuges operate continuously at 1500–5000g and are suitable for dewatering coarse solid materials at high solids concentrations. They are used in sewage systems for the separation of sludge, and for harvesting yeasts and fungal mycelium.

Filtration

Conventional filtration of liquids containing suspended solids involves depth filters composed of porous media (cloth, glass wool or cellulose) that retain the solids and allow the clarified liquid filtrate to pass through. As filtration proceeds collected solids accumulate above the filter medium, resistance to filtration increases and flow through the filter decreases. These techniques are generally useful for harvesting filamentous fungi, but are less effective for collecting bacteria. The two main types of conventional filtration commonly used in industry are as follows.

1 Plate and frame filters or **filter presses**, which are industrial batch filtration systems (Fig. 7.8). They are normally in the form of an alternating horizontal stack of porous plates and hollow frames. The stack is mounted in a support structure where it is held together with a

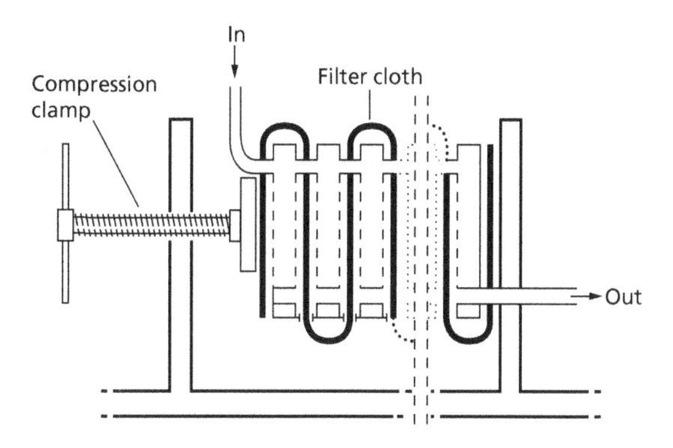

Fig. 7.8 A plate and frame filter (from Brown *et al.* (1987)).

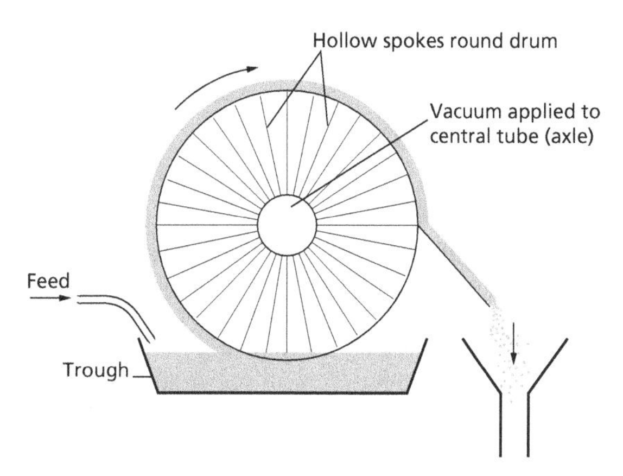

Fig. 7.9 A rotary vacuum filter showing the removal of filter cake using a knife.

hydraulic or screw ram. Filter cloths are held in place between the plates that contain flow channels for the feed and permeate streams. This essentially forms a series of cloth-lined chambers into which the cell suspension is forced under pressure. Following batch filtration the apparatus must be dismantled to remove the collected filter cake. These systems are used for harvesting microorganisms from fermentations, including the preparation of blocks of baker's yeast, the recovery of protein precipitates and the dewatering of sewage sludge. Similar horizontal and vertical pressure leaf filters are also available.

2 Rotary vacuum filters are simple continuous filtration systems that are used in several industrial processes, particularly for harvesting fungal mycelium during antibiotic manufacture, for baker's yeast production and in dewatering sludge during waste-water treatment. The device comprises a hollow perforated drum that supports the filter medium (Fig. 7.9). This drum slowly rotates in a continuously agitated tank containing the suspension to be filtered. Solids accumulate on the filter medium as liquid filtrate is drawn, under vacuum, through the filter medium into the hollow drum to a receiving vessel. As the drum rotates, collected solids held on the filter medium are removed by a knife that cuts/sloughs them off into a collection vessel. Filter media may be precoated with a filter aid, e.g. Kieselguhr (diatomaceous earth), which can be continuously replenished. The rate of filtration (flow of filtrate), for a constant-pressure (vacuum) and incompressible cake, is determined by the resistance of the cake and the filtration medium.

In terms of biosafety, neither filtration system is suitable for processing toxic products, pathogens or certain recombinant DNA microorganisms and their products.

MEMBRANE FILTRATION

Modern methods of filtration involve absolute filters rather than depth filters. These consist of supported membranes with specified pore sizes that can be divided into three main categories. They are, in decreasing order of pore size, **microfiltration**, **ultrafiltration** and **reverse osmosis** membranes. The suspension to be filtered is pumped across the membrane (cross-/tangential-flow) rather than at a right angle to it, as occurs with conventional filtration methods. This retards fouling of the membrane by particulate materials. Particles whose size is below the membrane 'cut-off' will pass through the membrane to become the **ultrafiltrate** or **permeate**, whereas the remainder is retained as the **retentate**. As filtration progresses, the flux across the membrane can slow due to membrane fouling. This may be caused by the accumulation of a layer of solute molecules on the surface of the membrane, referred to as concentration polarization. The presence of silicon antifoams may have a similar negative effect.

Microfiltration is used to separate particles of 10^{-2} μm to 10 μm, including removal of microbial cells from the fermentation medium. This method is relatively expensive due to the high cost of membranes, but it has several advantages compared with centrifugation. They include quiet operation, lower energy requirements, the product can be easily washed, good temperature control is possible, containment is readily achieved and no bioaerosols are produced. Consequently, it is suitable for handling pathogens and recombinant microorganisms.

Ultrafiltration is similar to microfiltration except that the membranes have smaller pore sizes, and are used to fractionate solutions according to molecular weight, normally within the range 2000–500 000 Da. The membranes have anisotropic structure, composed of a thin membrane with pores of specified diameter providing selectivity, lying on top of a thick, highly porous, support structure. A membrane manufactured with an exclusion size of 100 000 Da, for example, should produce a retentate of proteins and other molecules over 100 000 Da and an ultrafiltrate of all molecules below 100 000 Da. However, non-spherical proteins may exhibit different exclusion reactions to the membrane. Flat membranes are available, but for larger-scale operations hollow-fibre systems are usually preferred (Fig. 7.10). Several of these ultrafiltration units can be linked together to produce a sophisticated purification system. These methods are extensively employed for the purification of proteins, and for separating and concentrating

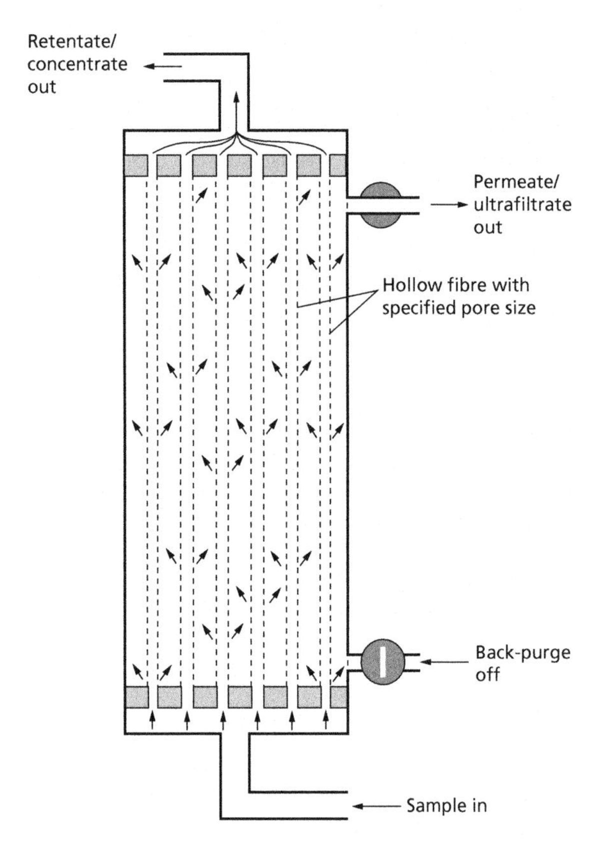

Fig. 7.10 A diagrammatic representation of a hollow-fibre ultrafiltration unit.

materials. Ultrafiltration is also effective in removing pyrogens (bacterial cell wall lipopolysaccharides), cell debris and viruses from media, and for whey processing.

Another variation on the ultrafiltration system is diafiltration, where water or other liquid is filtered to remove unwanted low molecular weight contaminants. This can be used as an alternative to gel filtration or dialysis for removing ammonium sulphate from a protein preparation precipitated by this salt (desalting, see p. 119), for changing a buffer or in water purification.

Reverse osmosis is used for dewatering or concentration steps and has been employed to desalinate sea water for drinking. In osmosis water will cross a semipermeable membrane if the concentration of osmotically active solutes, such as salt, is higher on the opposite side of the membrane. However, if pressure is applied to the 'salt side' then reverse osmosis will occur, and water will be driven across the membrane from the salt side. This reversal of osmosis requires a high pressure, e.g. a pressure of 30–40 bar is needed to dewater a 0.6 mmol/L salt solution (note: 1 bar = 100 kPa = 0.987 atm). Consequently, a strong metal casing is required to house this equipment. As the membranes have pore sizes of only 10^{-2} to 10^{-4} μm diameter, solute molecules can deposit on the surface, causing a large resistance to solvent flow. However, this fouling can be overcome by increasing the turbulence at the surface of the membrane.

Cell disruption

Some target products are intracellular, including many enzymes and recombinant proteins, several of which form inclusion bodies (see p. 122). Therefore, methods are required to disrupt the microorganisms and release these products. The breaching of the cell wall/envelope and cytoplasmic membrane can pose problems, particularly where cells possess strong cell walls. For example, a pressure of 650 bar is needed to disrupt yeast cells, although this may vary somewhat at different times during the growth cycle and depending upon the specific growth conditions.

General problems associated with cell disruption include the liberation of DNA, which can increase the viscosity of the suspension. This may also affect further processing, such as pumping the suspension on to the next unit process and flow through chromatography columns. A nucleic acid precipitation step or the addition of DNase can help to prevent this problem. If mechanical disruption is used then heat is invariably generated, which denatures proteins unless appropriate cooling measures are implemented. Products released from eukaryotic cells are often subject to degradation by hydrolytic enzymes (proteases, lipases, etc.) liberated from disrupted lysosomes. This damage can be reduced by the addition of enzyme inhibitors, cooling the cell extract and rapid processing. Alternatively, attempts may be made to produce mutant strains of the producer microorganism lacking the damaging enzymes.

Cell disruption can be achieved by both mechanical and non-mechanical methods. The disruption process is often quantified by monitoring changes in absorbance, particle size, total protein concentration or the activity of a specific intracellular enzyme released into the disrupted suspension.

Mechanical cell disruption methods

Several mechanical methods are available for the disruption of cells. Those based on **solid shear** involve extrusion of frozen cell preparations through a narrow orifice at high pressure. This approach has been used at the laboratory scale, but not for large-scale operations. Methods utilizing **liquid shear** are generally more effective. The French press (pressure cell) is often used in the laboratory and the high-pressure homogenizers, such as the Manton and Gaulin homogenizer (APV-type mill), are employed for pilot- and production-scale cell disruption. They may be used for bacterial and yeast cells, and fungal mycelium. In these devices the cell suspension is drawn through a check valve into a pump cylinder. At this point, it is forced under pressure (up to 1500 bar) through a very narrow annulus or discharge valve, over which the pressure drops to atmospheric. Cell disruption is primarily achieved by high liquid shear in the orifice and the sudden pressure drop upon discharge causes explosion of the cells.

The rate of protein release (efficiency of disruption) is independent of the cell concentration, but is a function of the pressure exerted, the number of cycles through the homogenizer and the temperature. Disruption of yeast cell preparations, for example, typically requires three passes through the pressure cell at 650 bar, whereas wild-type *Escherichia coli* generally needs 1100–1500 bar. During processing the temperature rises by about 2.2–2.4°C per 100 bar, i.e. by approximately 20°C over one pass at 800 bar. Consequently, precooling of the cell preparation is usually essential. The energy input necessary is approximately 0.35 kW per 100 bar and the throughput is up to 6000 L/h. A problem with this method of cell disruption is that all intracellular materials are released. As a result, the product of interest must be separated from a complex mixture of proteins, nucleic acids and cell wall fragments. Some fragments of cell debris are not readily separated, making the solution difficult to clarify. In addition, proteins may be denatured if the equipment is not sufficiently cooled and filamentous microorganisms may block the discharge valve. When used for certain categories of microorganisms, the homogenizers have to be contained to prevent the escape of aerosols.

On a small scale, manual grinding of cells with abrasives, usually alumina, glass beads, kieselguhr or silica, can be an effective means of disruption, but results may not be reproducible. In industry, **high-speed bead mills**, equipped with cooling jackets, are often used to agitate a cell suspension with small beads (0.5–0.9 µm diameter) of glass, zirconium oxide or titanium carbide. Cell breakage results from shear forces, grinding between beads and collisions with beads. The efficiency of cell breakage is a function of agitation speed, concentration of beads, bead density and diameter, broth density, flow rate and temperature. Cell concentration is also a major factor (optimum 30–60% dry weight), which is an important difference from the liquid shear homogenizers described above. Maximum throughput in these systems is about 2000 L/h.

Ultrasonic disruption of cells involves **cavitation**, microscopic bubbles or cavities generated by pressure waves. It is performed by ultrasonic vibrators that produce a high-frequency sound with a wave density of approximately 20 kilohertz/s. A transducer converts the waves into mechanical oscillations via a titanium probe immersed in the concentrated cell suspension. However, this technique also generates heat, which can denature thermolabile proteins. Rod-shaped bacteria are often easier to break than cocci, and Gram-negative organisms are more easily disrupted than Gram-positive cells. Sonication is effective on a small scale, but is not routinely used in large-scale operations, due to problems with the transmission of power and heat dissipation.

Cell disruption is a somewhat neglected area of bioprocessing, as there has been relatively little innovation and progress. Even the routinely used established mechanical methods were originally devised for other purposes. However, some newer disruption systems are being developed to give improved large-scale disruption, often with integral containment. They include a newly designed ball mill, the CoBall Mill; the Constant Systems high-pressure disrupter, which operates at up to 2700 bar; and two systems with no moving parts, the Microfluidics impingement jet system and the Glass-col nebulizer. The Parr Instruments cell disruption bomb is designed for disrupting mammalian cells. This is a relatively gentle method that works on the principle of nitrogen decompression and does not generate heat. Nitrogen is dissolved in cells under high pressure, and sudden pressure release then causes the cells to burst.

Non-mechanical cell disruption methods

An alternative to mechanical methods of cell disruption is to cause their permeabilization. This can be accomplished by autolysis, osmotic shock, rupture with ice crystals (freezing/thawing) or heat shock. Autolysis, for example, has been used for the production of yeast extract and other yeast products. It has the advantages of lower cost and uses the microbes' own enzymes, so that no foreign substances are introduced into the product. Osmotic shock is often useful for releasing products from the periplasmic space. This may be achieved by equilibrating the cells in 20% (w/v) buffered sucrose, then rapidly harvesting and resuspending in water at 4°C.

A wide range of other techniques have been developed for small-scale microbial disruption using various chemicals and enzymes. However, some of these can lead to problems with subsequent purification steps. Organic solvents, usually acetone, butanol, chloroform or methanol, have been used to liberate enzymes and other substances from microorganisms by creating channels through the cell membrane. Simple treatment with alkali or detergents, such as sodium lauryl sulphate or Triton X-100, can also be effective.

Several cell wall degrading enzymes have been successfully employed in cell disruption. For example, lysozyme, which hydrolyses β-1,4 glycosidic linkages within the peptidoglycan of bacterial cell walls, is useful for lysing Gram-positive organisms. Addition of ethylene diamine tetraacetic acid (EDTA) to chelate metal ions also improves the effectiveness of lysozyme and other treatments on Gram-negative bacteria. This is because EDTA has the ability to sequester the divalent cations that stabilize the structure of their outer membranes. Enzymic destruction of yeast cell walls can be achieved with snail gut enzymes that contain a mixture of β-glucanases. These enzyme preparations are also useful for producing living yeast spheroplasts or protoplasts.

The antibiotics penicillin and cycloserine may be used to lyse actively growing bacterial cells, often in combination with an osmotic shock. Other permeabilization techniques include the use of basic proteins such as protamine; the cationic polysaccharide chitosan is effective for yeast cells; and streptolysin permeabilizes mammalian cells.

Product recovery

Recovery of extracellular proteins is from the clarified medium, whereas disrupted cell preparations are used for both intracellular proteins and those held within the periplasmic space. In the case of some recombinant proteins expressed at high levels, they sometimes form inclusion bodies that are released by cell breakage (see p. 122).

Following cell disruption, soluble proteins are usually separated from cell debris by centrifugation. The resultant supernatant, containing the proteins, is then processed in a similar way to growth medium containing excreted proteins. This can involve several different types of methods, some of which have been previously described, such as ultrafiltration. An older, but nevertheless effective, method used at this stage is salting out of proteins, followed by the recovery of precipitated proteins by centrifugation. Precipitation is achieved by the addition of inorganic salts at high ionic strength, usually in the form of solid or saturated solutions of ammonium sulphate. Ammonium sulphate is popular because of its high solubility, low toxicity and low cost, but it can liberate ammonia at high pH values and is corrosive to metal surfaces, e.g. centrifuges. The solubility of the salt varies with temperature, so strict temperature control is required. Reduction of protein solubility can also be achieved by adding organic solvents, such as acetone, ethanol and isopropanol. This is performed at low temperature and reduces the dielectric constant of the medium, causing precipitation of the proteins.

Aqueous two-phase separation involves partitioning the protein between the two phases, depending upon its molecular weight and charge. Commonly used systems include dextran and polyethylene glycol (PEG), or PEG and potassium phosphate. The two phases can then be separated by centrifugation. This method is cheap, gentle and versatile, and can be scaled up for industrial applications, including the purification of enzymes, e.g. RNA polymerase from *E. coli*.

Many alkaloids, antibiotics, steroids and vitamins are recovered by liquid–liquid extraction methods using organic solvents. Antibiotics, for example, are extracted from culture media into solvents such as amyl acetate. Those solvents used should be non-toxic, selective, inexpensive, immiscible with broth and must have a high distribution coefficient for the product, i.e. the ratio of the product in the two phases. Efficient recovery of the solvent for reuse is an essential aspect of overall process economics.

Chromatography

Chromatographic techniques are usually employed for higher-value products. These methods, normally involving columns of chromatographic media (stationary phase), are used for desalting, concentration and purification of protein preparations. In choosing a chromatographic technique a number of considerations must to be taken into account. For protein products these factors include molecular weight, isoelectric point, hydrophobicity and biological affinity. Each of these properties can be exploited by specific chromatographic methods that may be scaled up to form an industrial unit process.

The order and choice of technique will depend on the particular product, but the following chromatographic parameters should be considered: **capacity**, **recovery** and **resolving power (selectivity)**. The capacity refers to the sample size that can be applied to the system in terms of protein concentration and volume, and the recovery is the yield of product at each stage. Yield values should be as high as possible otherwise the overall process will be uneconomic. Resolving power and selectivity relate to the ability to separate two components, one being the product and the other the impurity. This is particularly important at the final purification stage. Care must be taken not to allow a protein to denature during purification, so columns are usually run at 4°C. Changes of pH, excessive dilution and addition of certain chemicals can also affect the stability of the protein.

Adsorption chromatography separates according to the affinity of the protein, or other material, for the surfaces of the solid matrix. Alumina (Al_2O_3), hydroxyapatite ($Ca_{10}(PO_4)_6(OH)_2$) or silica (SiO_2) are used for purifying non-polar molecules, whereas polystyrene-based resins are effective matrices for polar molecules. This technique involves hydrogen bonding and/or van der Waals forces. The adsorbed proteins are usually eluted by increasing the ionic strength, often by using a gradient of increasing phosphate ion concentration.

Affinity chromatography is a particularly powerful and highly selective purification technique. It can often give up to several thousand-fold purification in a single step. However, this chromatographic method is expensive on an industrial scale. The technique involves specific chemical interactions between solute molecules, such as proteins, and an immobilized ligand (functional molecule). Ligands are covalently linked to the matrix material, e.g. agarose, via a spacer arm to avoid steric hindrance. Some ligands interact with a group of proteins, notably nicotinamide adenine dinucleotide, adenosine monophosphate, and Procion and Cibacron dyes; other ligands are highly specific, especially substrates, substrate analogues and antibodies. Since monoclonal antibodies have become more readily available, immunoaffinity chromatography methods have been developed for the purification of various antigens.

The loading capacity of affinity columns is large, as the volume of the sample is unimportant and high resolution can be attained. This is also a high-speed technique and elution is achieved using specific cofactors or substrates; alternatively, non-specific elution may be performed with salt or pH change. For industrial-scale operations, it is often necessary to sterilize the chromatographic media in order to comply with various regulatory requirements. This can be problematical as some ligands are sensitive to sterilizing agents.

Gel filtration chromatography essentially involves separation on the basis of molecular size (molecular sieving), although molecular shape can also influence separation performance. Consequently, it is particularly useful for desalting protein preparations. The stationary phase consists of porous beads composed of acrylic polymers, agarose, cellulose, cross-linked dextran, etc., which have a defined pore size. These support materials should be sterilizable, chemically inert, stable, highly porous and hydrophilic, containing some ionic groups. Mechanical rigidity is particularly important in order to maintain good flow rates. The initial choice of stationary phase material is also a key factor, as some may interact with the target product, e.g. carbohydrate-based matrices interact with glycoproteins.

Solute molecules below the exclusion size of the support material pass in and out of the beads. Molecules above the exclusion size pass only around the outside of the beads through the interstitial spaces and the apparent volume of the column is smaller for these larger excluded molecules. As a result, they flow faster down the column, separating from smaller molecules and eluting first. Smaller molecules able to enter the pores are then eluted in decreasing order of size.

Ion-exchange chromatography involves the selective adsorption of ions or electrically charged compounds onto ion-exchange resin particles by electrostatic forces. The matrix material is often based on cellulose substituted with various charged groups, either cations or anions. A commonly used example is the anion-exchange resin diethylaminoethyl (DEAE) cellulose. Proteins possess positive, negative or no charge depending on their isoelectric point (pI) and the pH of the surrounding buffer. If the pH of the buffer is below the pI, a protein has an overall positive charge, whereas a buffer at the pI results in the protein having no charge and pHs above its pI produce an overall negative charge. For example, if a protein has a pI of 4.2, at pH4.2 this protein is uncharged and will not bind to either positively or negatively charged resins. When the pH is raised to pH7.0 the protein is negatively charged and binds to positively charged resins. In conditions below pH4.0 the protein is positively charged and binds to negatively charged resins. If the protein is in an anionic state able to adsorb to DEAE cellulose, for example, any contaminants pass through the column. The product can then be desorbed as a purified fraction by altering the ionic strength of the buffer, often by using a gradient of increasing phosphate buffer concentration.

High-performance liquid chromatography (HPLC) was originally developed for the separation of organic molecules in non-aqueous solvents, but is now used for proteins in aqueous solution. This method uses densely packed columns containing very small rigid particles, 5–50 μm diameter, of silica or a cross-linked polymer. Consequently, high pressures are required. The method is fast and gives high resolution of solute molecules. Equipment for use in large-scale operations is now available.

Hydrophobic chromatography relies on hydrophobic interaction between hydrophobic regions or domains of a solute protein and hydrophobic functional groups of the support particles. These supports are often agarose substituted with octyl or phenyl groups. Elution from the column is usually achieved by altering the ionic strength, changing the pH or increasing the concentration of chaotropic ions, e.g. thiocyanate. This technique provides good resolution and, like ion-exchange chromatography, has a very high capacity as it is not limited by sample volume.

Metal chelate chromatography utilizes a matrix with attached metal ions, e.g. agarose containing calcium, copper or magnesium ions. The protein to be purified must have an affinity for this ion and binds to it by forming coordination complexes with groups such as the imidazole of histidine residues. Bound proteins are then eluted using solutions of free metal-binding ligands, e.g. amino acids.

Dialysis and electrodialysis

These membrane separation techniques are primarily used for the removal of low molecular weight solutes and inorganic ions from a solution. The membranes involved are size selective with specific molecular weight cut-offs. Low molecular weight solutes move across the membrane by osmosis from a region of high concentration to one of low concentration. Electrodialysis methods separate charged molecules from a solution by the application of a direct electrical current carried by mobile counter-ions. Membranes used contain ion-exchange groups and have a fixed charge; e.g. positively charged membranes allow the passage of anions and repel cations. They are essentially ion-exchange resins in sheet form and have also been used for the desalination of water.

Distillation

Distillation is used to recover fuel alcohol, acetone and other solvents from fermentation media, and for the preparation of potable spirits. Batch distillation in pot stills continues to be used for the production of some whiskies (see Chapter 12), but for most other purposes continuous distillation is the method of choice. With ethanol, for example, the continuous system produces a product with a maximum ethanol concentration of 96.5% (v/v). This azeotropic mixture is the highest concentration that can be achieved from aqueous ethanol, unless a dehydration step is introduced using a water entrainer such as benzene or cyclohexane.

Some continuous stills may be in the form of four or five separate columns, but the Coffey-type still comprises just two columns, the 'rectifier' and 'analyser', each containing a stack of 30–32 perforated plates (Fig. 7.11). Incoming fermentation broth is heated, as it passes down a coiled pipe within the rectifier column, by the ascending hot vapour produced by the analyser column. The now hot broth is released into a trough at the top of the analyser column and as it falls down the column it is heated by steam. Hot vapours generated are then conveyed from the top of the analyser column to the bottom of the rectifier column. As it passes upwards it is condensed on the coils carrying incoming broth. There is a temperature gradient in the rectifier column and each volatile compound condenses at its appropriate level, from where the fraction is collected.

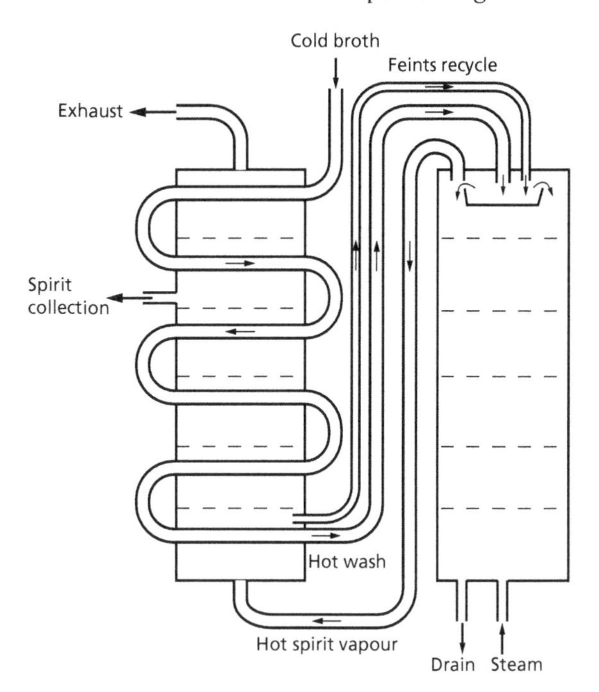

Fig. 7.11 A Coffey-type still (only six of the 30 column sections shown) (from Brown *et al.* (1987)).

Finishing steps

Crystallization

Product crystallization may be achieved by evaporation, low-temperature treatment or the addition of a chemical reactive with the solute. The product's solubility can be reduced by adding solvents, salts, polymers (e.g. non-ionic PEG) and polyelectrolytes, or by altering the pH.

Drying

Drying involves the transfer of heat to the wet material and removal of the moisture as water vapour. Usually, this must be performed in such a way as to retain the biological activity of the product. Parameters affecting drying are the physical properties of the solid–liquid system, intrinsic properties of the solute, conditions of the drying environment and heat transfer parameters. Heat transfer may be by direct contact, convection or radiation.

Rotary drum driers remove water by heat conduction. A thin film of solution is applied to the steam-heated surface of the drum, which is scraped with a knife to recover the dried product. In **vacuum tray driers** the material to be dried is placed on heated shelves within a chamber to which a vacuum is applied. This allows lower temperatures to be used due to the lower boiling point of water at reduced pressure. The method is suitable for small batches of expensive materials, such as some pharmaceuticals. **Spray drying** involves atomization and spraying of product solution into a heated chamber, and resultant dried particles are separated from gases using cyclones. **Pneumatic conveyor driers** use hot air that suspends and transports particles.

Freeze-drying (lyophilization) is often used where the final products are live cells, as in starter culture preparations, or for thermolabile products. This is especially useful for some enzymes, vaccines and other pharmaceuticals, where retention of biological activity is of major importance. In this method, frozen solutions of antibiotics, enzymes or microbial cell suspensions are prepared and the water is removed by sublimation under vacuum, directly from solid to vapour state. This method eliminates thermal and osmotic damage.

Inclusion bodies and the role of genetic engineering in downstream processing

As previously mentioned, many recombinant proteins are formed as inclusion bodies, which are insoluble inactive aggregates of over-expressed polypeptides. They arise due to the accumulation of partially folded nascent polypeptides that have exposed hydrophobic surfaces. Resulting hydrophobic interactions cause them to form insoluble aggregates. This is often fortuitous because it provides a concentrated form of the protein, as the aggregates may contain over 50% of target protein.

Once the producer cells are broken open and the large cell debris is removed, the inclusion bodies are easily recovered from the cell-free extracts by low-speed centrifugation at 5000–20 000 *g* for 15–60 min. The recovered protein must then be solubilized; however, at this point they are not soluble in ordinary aqueous buffers. Their solubilization requires concentrated solutions of chaotrophic agents (protein denaturants), such as 5–10 mmol/L urea and 4–8 mmol/L guanidine–HCl, or detergents. Where necessary, disulphide bonds are broken by the use of mercaptoethanol or dithiothreitol. Following dissolution the denaturants and reducing agents are removed by dialysis, diafiltration or gel filtration. This begins protein refolding (renaturation) to produce the functional conformation and restore biological activity. Refolding may be achieved by the slow removal of denaturants, along with additional treatments using specific buffers and various temperature regimes. For some recombinant proteins disuphide bonds must also be reformed, which is often accomplished by air oxidation or thiol–disulphide exchange reactions.

DSP can be aided by the introduction of specific aids at the upstream stages. Organisms can be modified to suppress the production of byproducts and enzymes that may interfere with DSP operations or degrade the target product. Recombinant protein products may be designed to be excreted from the cell, and cell breakage and release of intracellular products can also be assisted. The cell envelope of an organism may be modified so that it is permeable, or is at some point induced to become permeable to the product. For example, recombinant *E. coli* are often lysed by coexpressing lysozyme (under appropriate inducer control). Following lysozyme induction at the end of the fermentation, cell breakage is promoted by a freeze–thaw step.

Recombinant protein products may be engineered with a high affinity for certain separation matrices or joined to another molecule with desirable separation characteristics. A now well-established technique is the construction of fusion (hybrid or chimeric) proteins (Table 7.1). For example, the structural gene for the target recombinant peptide/protein can be fused/tagged with additional DNA encoding a natural or synthetic polypeptide at the 5′ or 3′ end of the gene. The tag can be a complete enzyme, such as β-galactosidase, chloramphenicolacetyltransferase or glutathione-*S*-transferase. These tags may be employed to protect short peptides that are often rapidly degraded, or aid in downstream processing and the specific assaying of the peptide/protein product. Tags may be used to allow affinity, ion-exchange, hydrophobic covalent or metal-chelate separation of the peptide/protein. Some methods are suitable for both secreted proteins and those forming inclusion bodies. An enzyme or chemical cleavage site can also be included in the linker between the target protein and its tag, in order to facilitate later removal of the tag. Enzyme cleavage sites include those for endopeptidases, enterokinase and thrombin, whereas chemical sites are those targeted by acid, cyanogen bromide or hydroxylamine.

Table 7.1 Example of fusion proteins to aid the purification of recombinant proteins

Purification tag	Matrix	Elution
Affinity		
chloramphenicol acetyltransferase	*p*-amino chloramphenicol–Sepharose	5 mmol/L chloramphenicol
glutathione-*S*-transferase	Glutathione–agarose	Excess reduced glutathione
protein A	IgG–Sepharose	0.5 mmol/L Acetic acid
Ion-exchange		
polyarginine	S–Sepharose	Sodium chloride gradient
Metal-chelate		
histidine–tryptophan dipeptide	Imindiacetic–Sepharose (Ni^{2+})	Low pH gradient

Tags coding for a specific dipeptide or a sequence of amino residues, such as arginine, phenylalanine or cysteine, may be added at the C-terminal end forming a 'tail' on the target protein. Polyarginine tails, for example, result in a particularly basic protein that unlike most other cellular proteins binds to cation-exchange resins. Once recovered from the resin, the polyarginine tails can be removed by passing the protein through a column of an immobilized exopeptidase, e.g. carboxpeptidase A. This mode of purification has been used for the preparation of urogastrone and interferon-γ. Alternatively, a gene for a protein that binds to IgG, such as staphylococcal protein A (SPA), can be fused to the target protein gene. The product with the SPA tag can then be purified by immunoaffinity chromatography using IgG as the ligand.

Further reading

Papers and reviews

Bernardez-Clark, E. D. (1998) Refolding of recombinant proteins. *Current Opinion in Biotechnology* 9, 157–163.

Christi, Y. & Moo-Young, M. (1986) Disruption of microbial cells for intracellular products. *Enzyme and Microbial Technology* 8, 194–204.

Christi, Y. & Moo-Young, M. (1990) Large-scale protein separations: engineering aspects of chromatography. *Biotechnological Advances* 8, 699–708.

Datar, R. V., Cartwright, T. & Rosen, C. G. (1993) Review: Process economics of animal cell culture and bacterial fermentations: a case study analysis of tissue plasminogen activator. *Bio/Technology* 11, 349–357.

Daugulis, A. J. (1994) Integrated fermentation and recovery processes. *Current Opinion in Biotechnology* 5, 192–195.

Foster, D. (1995) Optimizing recombinant product recovery through improvements in cell-disruption technologies. *Current Opinion in Biotechnology* 6, 523–526.

Harrison, S. T. L. (1991) Bacterial cell disruption: a key unit operation in the recovery of intracellular proteins. *Biotechnological Advances* 9, 217–240.

Harrison, S. T. L., Dennis, J. S. & Chase, H. A. (1991) Combined chemical and mechanical processes for the disruption of bacteria. *Bioseparation* 2, 95–105.

Marston, F. A. O., Angal, S., Lowe, P. A., Chan, M. & Hill, C. R. (1988) Scale-up of the recovery and reactivation of recombinant proteins. *Biochemical Society Transactions* 16, 112–115.

Sassanfeld, H. M. (1990) Engineering proteins for purification. *Trends in Biotechnology* 8, 88–92.

Books

Biotol Series (1992) *Product Recovery in Bioprocess Technology*. Butterworth-Heinemann, Oxford.

Atkinson, B. & Mavituna, F. (1991) *Biochemical Engineering and Biotechnology Handbook*, 2nd edition. Macmillan, Basingstoke.

Blanch, H. W. & Clark, D. S. (1997) *Biochemical Engineering*. Marcel Dekker, New York.

Brown, C. M., Campbell, I. & Priest, F. G. (1987) *Introduction to Biotechnology*. Blackwell Scientific Publications, Oxford.

Bu'lock, J. & Kristiansen, B. (1987) *Basic Biotechnology*. Academic Press, London.

Doran, P. M. (1995) *Bioprocess Engineering Principles*. Academic Press, London.

Hambleton, P., Melling, J. & Salusbury, T. T. (1994) *Biosafety in Industrial Biotechnology*. Blackie Academic and Professional, Glasgow.

Harrison, R. G. (ed.) (1993) *Protein Purification Applications. A Practical Approach*. Marcel Dekker, New York.

Jackson, A. T. (1990) *Process Engineering in Biotechnology*. Open University, Milton Keynes.

Stanbury, P. F., Whitaker, A. & Hall, S. J. (1995) *Principles of Fermentation Technology*, 2nd edition. Butterworth–Heinemann (Pergamon), Oxford.

Wheelwright, S. M. (1991) *Protein Purification: Design and Scale-up of Downstream Processing*. Hanser Publications, New York.

8 Product development, regulation and safety

The development of any new fermentation product depends on many factors, including the **market**, the current level of **scientific knowledge** and the **regulatory environment**. Some of these factors are out of the control of the developer. The market is obviously affected by product uniqueness and where a market is perceived to exist for a new product, referred to as 'market pull', there will be an incentive to try and fill that niche, e.g. drugs to treat cancers, HIV infection, dementia, etc. In the case of medical products, a new treatment for a previously untreatable disease or a more effective drug can be extremely profitable for a company. However, the market size is a major consideration. For example, the development of drugs targeted at rare diseases may not be economically viable for commercial companies. In the USA and Japan, and soon in the EU, their development can be achieved through government-sponsored '**orphan drug**' programmes. An additional problem arises when the potential market cannot afford the product, as in developing countries. In these situations, international agencies such as the United Nations may be involved in supplying the products.

The level of scientific knowledge is a further key factor. Some advances in science result in the rapid development of industrial processes, termed 'scientific push'. For example, Fleming's discovery of penicillin in 1928 ultimately led to the establishment of the antibiotic industry in the 1940s. A particularly rapid development was seen with monoclonal antibodies. Their production was first developed in the early 1970s and led to a multimillion dollar industry within just a few years. The rapid progress of recombinant DNA technology in recent years has also generated a 'push' for many new products and, importantly, this technology can be readily applied to established fermentation processes. Nevertheless, some major markets exist, but at present the science cannot produce the necessary product, e.g. an effective HIV vaccine and drugs for the treatment of cancers, new variant Creutzfeldt–Jakob disease (nvCJD), mental disorders, heart disease, etc.

In drafting a proposal for the production of any new fermentation product, and throughout its development, a great deal of input must come from disciplines other than microbiology. Major contributions are required from, among others, biochemists, geneticists, molecular biologists, chemists, chemical and process engineers, mathematicians and computer technologists. Transition from initial discovery through to a marketable new product can be a long and costly process. A typical pharmaceutical, for example, currently takes 10–12 years to come to the market and incurs development costs in excess of $200 million (Fig. 8.1). However, the target is now to reduce this to a maximum of 8 years. A major factor influencing the long development time is the regulatory environment, involving safety of the product, employees and the environment, along with legal aspects, patents, product licences, and social and ethical aspects.

Patent protection of an invention is normally sought at an early stage, in order to give the patentee the legal right to exclude others from its commercial use for a period of usually 17–20 years, depending upon the country. Patents can be separated into three distinct types. *Product patents* (substances, composition of matter and devices) such as bioinsecticides, recombinant proteins, monoclonal antibodies, plasmids, etc. are most readily protected as patent infringement can be determined by product analysis. *Manufacturing process patents*, e.g. DNA isolation, purification of a recombinant protein, etc., presuppose an improved method of production and are rather more difficult to enforce. *Methods of use patents* involve a novel role for a product, such as mode of bioinsecticide application, drug delivery, etc.

Patenting of microorganisms, where they are part of a process, has been carried out for over a hundred years. However, the patenting of gene sequences and whole organisms, particularly plants and animals, is highly controversial as it raises enormous legal, economic, environmental and not least ethical and moral issues. Social and ethical aspects of biotechnological product

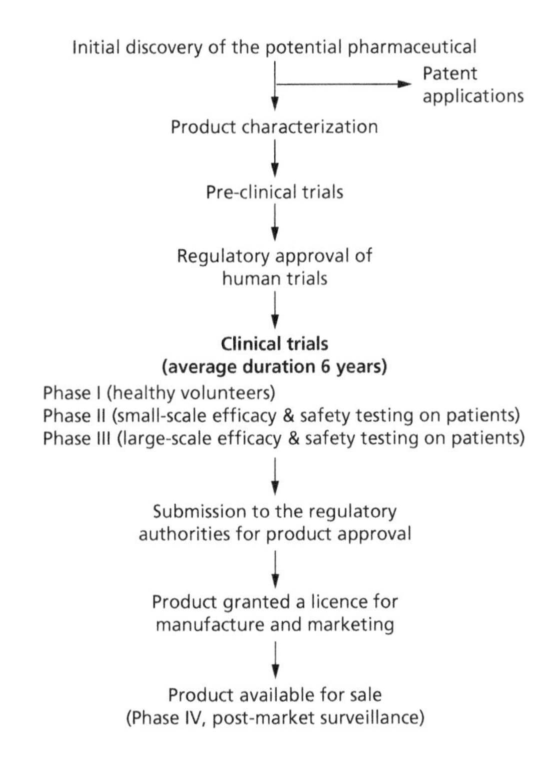

Fig. 8.1 Development of a pharmaceutical product.

development are becoming increasingly important, as witnessed in recent years, by the destruction of genetically manipulated (GM) agricultural crops in Europe and Asia. Also, there are concerns over breeding pigs with part human immune systems for use as sources of organs for transplants (xenotransplantation), patenting of human DNA sequences and the release of recombinant DNA systems into the environment.

Product quality and safety

Prior to approval for marketing, any new food or health care product that is to be used by humans and domestic or farm animals must be thoroughly tested. This involves aspects of safety and efficacy, and required specifications for purity and activity. Product licences are granted only after the product has passed all the requirements of the regulatory system. The necessary testing is lengthy and can be extremely expensive, but is considerably easier where the production organism has GRAS status (generally regarded as safe). In the case of pharmaceuticals, for example, the process of manufacture must be validated and the product 'well characterized'. This involves a rigorous examination of its identity (structure, physicochemical and immunochemical characteristics), quantification of potency and purity, and the identification and quantification of the impurities. In addition, pharmaceuticals must pass through three levels of trials prior to marketing and undergo postmarketing safety surveillance (Fig. 8.1). Approval for non-food and non-drug products is rather less costly and the risk assessments will depend on the microorganism involved, the specific product, and the mode of its production and downstream processing. However, this is obviously more complex for genetically modified microorganisms (GMMs) and pathogens (see p. 138).

Once approved, the product must be manufactured to specified quality standards. Stipulated levels of purity and activity will depend on the individual product. Importantly, the product must not be contaminated by potentially dangerous substances or pathogens. This means that a product, such as an injectable drug for human use, would have a somewhat different and more stringent set of quality criteria than silage for feeding to farm animals. Nevertheless, the silage would still have to be produced to specified quality criteria.

Any product that is to be released on to the market has to conform to quality procedures, which are usually made up from at least three sources, depending upon the specific product. These are national and international regulatory requirements, which are legal and compulsory; guideline requirements, often devised by manufacturers associations, which are not necessarily compulsory; and 'in-house' company standards.

The approach to manufacture should involve an overall ethos of **quality assurance (QA)**. This is defined as the sum total of the organized arrangements made with the object of ensuring that products will be consistently manufactured to a quality appropriate to their intended use (Fig. 8.2). Within this, appropriate **good manufacturing practices (GMP)** must be adopted, for example, those imposed by the US Food and Drug Administration (FDA) (GMP issued by the US Code of Federal Regulations on design, validation, and operation of the pharmaceutical/biotechnology manufacturing facility). Validated **standard operating procedures (SOPs)** are also required. Their validation involves establishing documented evidence that provides a high degree of assurance that a specific process will consistently produce a product that meets the predetermined specifications and quality attributes. Overall, the GMP framework determines the organization and functions of the production and purification processes. They involve

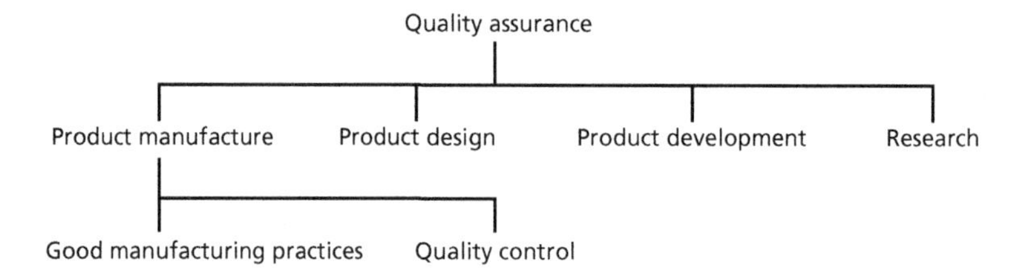

Fig. 8.2 Components of quality assurance.

both manufacturing and quality control (QC). The latter is concerned with sampling, specifications, testing, organization, documentation and product release procedures. Overall, implementation of these practices should result in the controlled production of a safe product with the elimination of significant batch-to-batch variations in quality.

Manufacturing and environmental safety

Safety of employees involved in a manufacturing process is a primary concern. They should not be exposed to potential pathogens or allergens and irritants that may cause disease. For example, dusts generated from enzyme powders have caused allergic reaction in staff working in fermentation plants manufacturing these products. Certain downstream processing unit operations, such as the use of centrifuges for cell harvesting and some mechanical methods of cell disruption, may generate bioaerosols that could potentially affect employees. These bioaerosols can be contained, but at extra cost. The GMP adopted, involving the use of appropriate containment (methods for managing the 'infectious' agent), where necessary, and safety equipment, etc., must eliminate these potential safety problems. Also, all employees must be fully trained, carry out safe working practices and be aware of potential hazards.

Microorganisms are deliberately released into the environment as pesticides and for bioremediation, but there is still considerable debate about some aspects of deliberate release, especially where GMMs are concerned. However, the population and the environment must be protected from the accidental release of potentially harmful microorganisms or microbial products from large-scale fermentation processes. Protocols must be established to ensure that this does not occur. There should also be safe disposal of waste products that arise from these processes, which may require sterilization and other treatments prior to release.

Safety within fermentation industries is particularly important where known pathogens and certain GMMs are employed. The classification of microorganisms in terms of danger to humans divides them into four catagories, a division that is regularly updated by the regulatory authorities. These categories are similar for regulations formulated by UK, US and EU authorities, and by the World Health Organization (WHO), whose classification is shown in Table 8.1. Cultivation of microorganisms necessitates the implementation of different levels of barrier/containment systems depending upon the fermentation process and the classification of the producer microorganism (Fig. 8.3). Traditional and many established 'proven' fermentation processes require a minimum level of containment and sterility, whereas cultivation of pathogens and some GMMs requires high levels of both containment and sterility (see below).

Primary containment barriers protect the personnel and immediate processing facility by preventing the escape of microorganisms from the fermenter or as aerosol generated from downstream processing equipment (Table 8.2). This protection is provided by implementing good microbiological techniques and the use of appropriate safety equipment. Some of the problematic areas in large-scale fermentations are the stirrer shaft and its seals. Usually two or three seals are now required so that if one breaks another will hold. Also, all valves, O rings, taps and pumps must be regularly checked. Secondary barriers involve the use of protective clothing, regular medical supervision and vaccination of the laboratory and manufacturing personnel. Further secondary containment barriers entail specific design criteria for manufacturing plants and laboratories. They include the use of positive and negative air pressure, high-efficiency particulate air (HEPA) filters, air locks and

Table 8.1 World Health Organization classification of microorganisms on the basis of hazard

	Containment required
Risk group 1 Low individual and community risk. A microorganism that is unlikely to cause human disease or animal disease of veterinary importance	Good industrial large-scale practices (GILSP)
Risk group 2 Moderate individual risk, limited community risk. A pathogen that can cause human disease or animal disease, but is unlikely to be a serious hazard to laboratory workers, the community, livestock or the environment. Laboratory exposures may cause serious infection, but effective treatment and preventive measures are available and the risk of spread is limited, e.g. *Salmonella* food poisoning. Vaccines and antibiotics are available	Level 1 containment
Risk group 3 High individual risk, low community risk. A pathogen that usually causes serious human disease, but does not ordinarily spread from one infected individual to another. Prophylaxis and treatment may be available	Level 2 containment
Risk group 4 High individual and community risk. A pathogen that usually causes serious human or animal disease and may be readily transmitted from one individual to another. No effective prophylaxis or treatment is available, e.g. Ebola virus	Level 3 containment

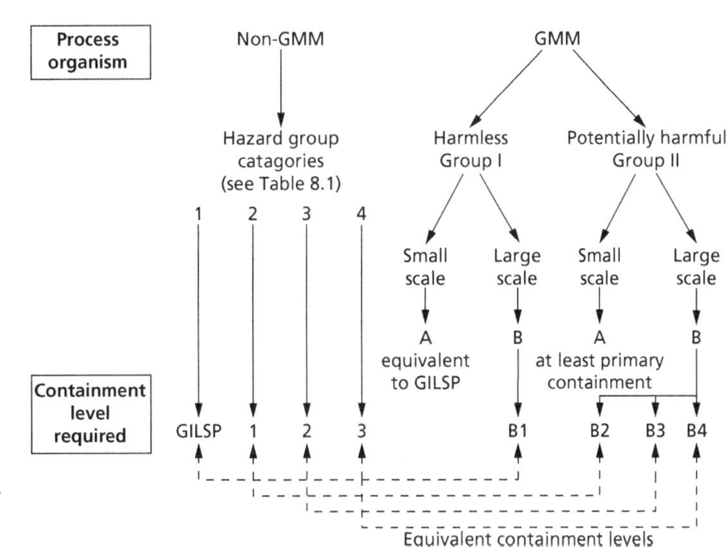

Fig. 8.3 Categories of process microorganisms and the level of containment required when used at research and industrial sites within the European Federation of Biotechnology (GILSP = good industrial large-scale practice; GMM = genetically manipulated microorganism)

changing rooms for operating personnel, along with specific protocols for the sterilization of waste before it leaves the site and its safe disposal.

Overall safety is primarily controlled by international and individual government agencies that regulate: foods, beverages and pharmaceuticals; protection of the environment; and safety in the workplace. Controls in the USA are principally imposed by the FDA, the Environmental Protection Agency (EPA) and the National Institutes of Health (NIH). In Japan this is largely overseen by the Ministry of Health and Welfare. Within the EU, the European Commission and European Parliament, as well as individual member states, frame new regulations and ensure their implementation. For example, in the UK, the Health and Safety Executive (HSE) has a major policing role. Attempts are being

Table 8.2 Examples of safety precautions required for different levels of containment, as required in the European Federation for Biotechnology

	Containment category (see Fig. 8.3)			
Non-genetically modified microorganisms	GILSP	1	2	3
Genetically modified microorganisms	B1	B2	B3	B4
Procedures				
written code of practice	+	+	+	+
biosafety manual	–	+	+	+
restricted access	–	+	+	+
medical surveillance	–	+	+	+
Primary containment				
use of closed systems for viable microorganisms	–	Minimization of release	Prevention of release	Prevention of release
treatment of exhaust gases	–	Minimization of release	Prevention of release	Prevention of release
removal of material, products and effluents	–	Minimization of release	Prevention of release	Prevention of release
penetration of the closed system, e.g. agitator shaft, monitoring devices, etc.	–	Minimization of release	Prevention of release	Prevention of release
Secondary containment				
disinfection facility	–	+	+	+
emergency shower facility	–	–	+	+
airlock + compulsory shower	–	–	–	+
decontaminated effluents	–	–	+	+
HEPA filters in air ducts	–	–	+	+
area hermetically sealable	–	–	–	+
controlled negative pressure	–	–	–	+

GILSP, good industrial large-scale practices.

made to standardize regulations to facilitate international free trade of products.

Before any new industrial process can start operating, most countries now require that an **assessment of risk analysis** be carried out. This is specifically required for processes using GMMs. Risk is a function of the estimated probability that an event will have an adverse effect, multiplied by an estimate of the magnitude (seriousness and consequences) of that effect. Risk analysis requires both quantitative risk assessment and risk management. The latter is implemented once it has been agreed that the process should go ahead as any risk is outweighed by the benefits.

Part of this risk assessment depends on whether the microorganism contains foreign DNA. If the organism is a non-GMM it can be placed in one of four categories as mentioned previously (Table 8.1). The specific categorization of a microorganism depends upon its pathogenicity, virulence, the infective dose necessary to cause disease, the route of infection, the level of incidence in the community and whether vectors are needed. Obviously, the availability of prophylaxis, such as vaccination and other treatments, is a major consideration. The techniques to be used and the scale of production are also key factors. For example, the handling and disposal of 1 L of *Escherichia coli* (a class 2 microorganism) has completely different problems, and therefore separate risks, compared with a 100 000 L fermentation of the same microorganism.

Risk group 1 microorganisms are the least likely to cause a problem and, for industrial use, simply require **good industrial large-scale practices (GILSP)**. Most industrial microorganisms are in this category. Use of microorganisms within risk groups 2, 3 and 4 must be more strictly controlled, necessitating containment levels 1, 2 and 3, respectively (Table 8.1). Microorganisms classified in these groups are used industrially, e.g. for vaccine production, but have very specialized manufacturing requirements.

Large-scale fermentation of GMMs has specific potential problems, should they escape from the fermenter. For example, they often contain antibiotic resistance selection markers that could pose problems were they to be released into the environment. In some cases, the microorganism can be engineered so that it is unable to survive outside the fermentation environment. Nevertheless, fermentations involving GMMs require case-by-case analysis, but can be essentially classified into organisms that are harmless (group I) or potentially harmful (group II). The level of their containment depends upon the size of the process: less than 10 dm^3 constitutes class A (small-scale) and greater than 10 dm^3 is class B (large-scale). Risk assessment of large-scale production of a group I GMM requires GILSP, whereas cultivation of a class II organism on this scale, and also on a small scale, would demand further risk analysis. This is required to determine the necessary level of containment, B2, B3 or B4; these are equivalent to levels 1, 2 and 3 for non-GMM microorganisms (Fig. 8.3 and Table 8.2). Those GMMs classified in these higher risk groups must have secondary or integral containment of equipment, particularly where aerosols are generated. Equipment is also validated (i.e. tested to specific specifications before use and regularly retested) and special validated SOPs are also required. Inspectors from the regulatory agencies must give permission for a process to start and for its continued operation. They inspect the manufacturing plant and laboratories on a regular basis, to determine whether the manufacturer is operating in a state of control, and in compliance with the laws and regulations. Although desirable, as yet, there is neither a global unified categorization for the fermentation of GMMs nor standardization of SOPs.

Further reading

Papers and reviews

Chiu, Y. H. (1988) Validation of the fermentation process for the production of a recombinant DNA drug. *Pharmaceutical Technology* **12**, 132–138.

Claude, J.-R. (1992) Difficulties in conceiving and applying guidelines for the safety evaluation of biotechnologically produced drugs: some examples. *Toxicology Letters* **64/65**, 349–355.

Crespi, R. S. (1997) Biotechnology patents and morality. *Trends in Biotechnology* **15**, 123–129.

Jeffcoate, S. L. (1992) New biotechnologies: challenges for the regulatory authorities. *Journal of Pharmacy and Pharmacology* **44** (Supplement 1), 191–194.

Kirsop, B. (1993) European standardization in biotechnology. *Trends in Biotechnology* **11**, 375–378.

Lunel, J. (1995) Biotechnology regulations and guidelines in Europe. *Current Opinion in Microbiology* **6**, 267–272.

Seamon, K. B. (1998) Specifications for biotechnology-derived protein drugs. *Current Opinion in Microbiology* **9**, 319–325.

Wilson, M. & Lindow, S. E. (1993) Release of recombinant microorganisms. *Annual Review of Microbiology* **47**, 913–944.

Books

Brown, F. & Fernandez, J. (1997) *Developments in Biological Standardization*. Karger, New York.

Chiu, Y.-Y. H. & Gueriguian, J. L. (eds) (1991) *Drug Biotechnology Regulation: Scientific Basis and Practices*. Marcel Dekker, New York.

Collins, C. H. & Beale, A. J. (eds) (1992) *Safety in Industrial Microbiology and Biotechnology*. Butterworth-Heinemann, Oxford.

Crespi, R. S. (1988) *Patents: A Basic Guide to Patenting in Biotechnology*. Cambridge University Press, Cambridge.

Hacking, A. J. (1986) *Economic Aspects of Biotechnology*. Cambridge University Press, Cambridge.

Hambleton, P., Melling, J. & Salusbury, T. T. (1994) *Biosafety in Industrial Biotechnology*. Blackie Academic and Professional, Glasgow.

Moses, V. & Capes, R. E. (1991) *Biotechnology: The Science and the Business*. Harwood Academic, Chur.

Ono, R.D. (1991) *The Business of Biotechnology*. Butterworth-Heinemann, Boston.

Peters, M. S. & Timmerhaus, K. D. (1991) *Plant Design and Economics for Chemical Engineers*, 4th edition. McGraw–Hill, New York.

Thomas, J. A. & Myers, L. A. (eds) (1999) *Biotechnology and Safety Assessment*, 2nd edition. Raven Press (Taylor & Francis), London.

Part **3**

Industrial processes and products

9 Microbial enzymes

Commercial microbial enzymes are increasingly replacing conventional chemical catalysts in many industrial processes. Enzymes have several advantages over chemical catalysts, including the ability to function under relatively mild conditions of temperature, pH and pressure. This results in the consumption of less energy and there is usually no requirement for expensive corrosion-resistant equipment. Enzymes are specific, often stereoselective, catalysts, which do not produce unwanted byproducts. Consequently, there is less need for extensive refining and purification of the target product. Also, compared with chemical processes, enzyme-based processes are 'environmentally friendly' as enzymes are biodegradable and there are fewer associated waste disposal problems. Certain enzymes are not restricted to aqueous environments and can operate in two-phase water–organic solvent systems and in non-aqueous organic media, particularly hydrophobic solvents. Operation under such conditions can often improve enzyme performance, especially where substrates have limited water solubility.

Enzyme classification is based on a system originally established by the Commission on Enzymes of the International Union of Biochemistry (1979). There are six main classes, grouped according to the type of reaction catalysed:

1 **oxidoreductases** (class 1) catalyse oxidation/reduction reactions, the transfer of H atoms, O atoms or electrons;
2 **transferases** (class 2) catalyse transfer of a group from one molecule to another;
3 **hydrolases** (class 3) catalyse hydrolysis, the cleavage of bonds by addition of a water molecule;
4 **lyases** (class 4) catalyse splitting bonds, other than via hydrolysis or oxidation;
5 **isomerases** (class 5) catalyse structural rearrangements of molecules; and
6 **ligases** or **synthetases** (class 6) catalyse the formation of new bonds, e.g. C–N, C–O, C–C and C–S, with breakdown of ATP.

Each enzyme has a four-figure code for class, subclass, etc. For example, invertase, a hydrolase, is classified as EC 3.2.1.26.

Suspended or immobilized microbial cells and spores may be employed as biocatalysts in some industrial bioconversions, particularly where coenzymes are necessary. However, in many cases this has limitations because:

1 the cells 'waste' energy and resources in growth and/or maintenance activities;
2 side reactions may lead to a reduction in the potential yield of the target product;
3 conditions for microbial growth, where required, may be different from those necessary for optimum product formation; and
4 difficulties may be encountered in the isolation and purification of the product from the cells or spent fermentation medium.

To overcome such problems, the use of partially purified 'bulk' microbial enzyme preparations is often preferred for numerous and varied industrial processes. In some cases, whole conventional microbial fermentation processes may eventually be replaced by multienzyme systems that could provide more efficient substrate utilization, higher yields and greater product uniformity.

Several thousand tonnes of commercial enzymes are currently produced each year, which have a value in excess of US$1500 million. A few animal and plant enzymes are used, but most commercial enzymes are now obtained from microbial sources. Many of these are extracellular enzymes, with the majority being derived from various species of *Bacillus*. Their proteases and amylases are the most widely used, and there is a particular demand for the thermostable enzymes that are available from several members of this genus. The greatest proportion of all commercial enzymes, some 34%, are used as detergent enzymes, 14% for dairy-related uses, with 12% for starch processing and 11% for textile applications. The remaining 29% are divided

Table 9.1 Applications of a range of bulk microbial enzymes

Proteases, mostly neutral proteases or zinc metalloprotease from species of *Aspergillus* and *Bacillus*, along with some alkaline serine proteases
(a) biological detergents, e.g. the alkaline protease subtilisin EC 3.4.4.16 from *Bacillus licheniformis* and *B. subtilis*
(b) baking—dough modification/gluten weakening and flavour improvement
(c) beer brewing, e.g. 'chillproofing' of beer to remove protein haze
(d) leather baiting and tendering
(e) cheese manufacture—clotting of milk protein and promotion of ripening, e.g. an aspartic protease (EC 3.4.23.6) from *Rhizomucor* species and recombinant *Kluyveromyces* species that produces calf chymosin. Aminopeptidase from *Lactococcus lactis* are also used to enhance flavour development
(f) meat tenderization and removal of meat from bones
(g) flavour control and production in food products
(h) waste treatment, e.g. recovery of silver from spent photographic film

Lipases EC 3.1.1.3, primarily from species of *Bacillus, Aspergillus, Rhizopus* and *Rhodotorula*
(a) biological detergents, also available from recombinant *Aspergillus oryzae* that produces *Humicola* lipase
(b) leather processing—fat removal
(c) production of flavour compounds and acceleration of ripening in dairy and meat products

Carbohydrases
α-Amylase EC 3.2.1.1, mostly derived from species of *Aspergillus* and *Bacillus*
(a) starch processing—liquefaction of starch in the production of sugar syrups
(b) baking—partial starch degradation in flour modification and the generation of fermentable sugars, and for improved crust colour
(c) brewing—starch hydrolysis during wort preparation
(d) biological detergents—starch removal from food stains
(e) textiles manufacture—desizing
β-Amylase EC 3.2.1.2 from *Bacillus* species such as *B. polymyxa*, and species of *Streptomyces* and *Rhizopus*
(a) maltose syrup production
(b) brewing—increasing wort fermentability
Amyloglucosidase (glucoamylase) EC 3.2.1.3 from *Aspergillus niger* and *Rhizopus niveus*
(a) glucose syrup production—complete starch saccharification
(b) baking—improved bread crust colour
(c) brewing—dextrin saccharification in the production of low-carbohydrate beer
(d) wine and fruit juice—starch removal
β-Galactosidase (lactase) EC 3.2.1.23 from species of *Bacillus, Kluyveromyces and Candida*
(a) whey syrup production—hydrolysis of lactose to glucose and galactose to give greater sweetness
(b) milk and dairy product processing—lactose reduction/removal for those who are lactose intolerant
(c) ice-cream manufacture, prevention of 'sandy' texture caused by lactose crystals
Glucose isomerase EC 5.3.1.5 from species of *Actinoplanes, Arthrobacter* and *Streptomyces*
(a) manufacture of high-fructose syrups as 'high sweeteners'—glucose conversion to fructose
Glucose oxidase EC 1.1.3.4 from *Aspergillus niger and Penicillium notatum*
(a) antioxidant, often used along with catalase to remove oxygen from food products
Inulinase EC 3.2.1.7 from *Aspergillus niger*, *Candida* species and *Kluyveromyces* species
(a) processing of Jerusalem artichoke tubers—hydrolysis of polyfructans and levans
Invertase EC 3.2.1.26 from species of *Saccharomyces* and *Kluyveromyces*
(a) sweets and confectionery production—liquefaction of sucrose, e.g. soft-centred chocolates
(b) invert sugar production—sucrose conversion to glucose and fructose
Pullulanase EC 3.2.1.41 from *Klebsiella pneumoniae* and *Bacillus acidopullulyticus*
(a) starch processing—used for the debranching of starch in sugar syrup manufacture and brewing
Pectinases (a mixture of enzymes such as polygalacturonase EC 3.2.1.15, produced by *Aspergillus niger* and *Penicillium* species, often by solid-substrate fermentation)
(a) plant juice and oil extraction
(b) fruit juice and wine clarification
(c) coffee bean fermentation

Continued

Table 9.1 *continued*

Hemicellulases and β-glucanases/cellulases
β-Glucanases, e.g. cellulases EC 3.2.1.4 from *Trichoderma reesei*, *Penicillium funiculosum*, *Aureobasidium pullulans* and *Bacillus* species, and specific β-glucanases from *Penicillium emersonii*
(a) fruit juice and olive production and processing
(b) wine and beer production and processing—hydrolysis of β-glucan gums to improve filtration rates
(c) malting—speeds modification of grains
(d) textile processing—biopolishing of cellulose fibres
(e) wood pulp processing
Hemicellulases from the yeasts *Cryptococcus* and *Trichosporon*, e.g. xylanase EC 3.2.1.32
(a) baking—pentosanases are used to alter dough consistency
(b) brewing
(c) animal feedstuffs
(d) nutraceutics
(e) wood pulp processing

Miscellaneous enzymes and their uses
Catalase EC 1.11.1.6 from *Aspergillus niger*, *Corynebacterium glutamicum* and *Micrococcus lysodeikticus*
(a) bleaching textiles
(b) cheese manufacture
Phytase EC 3.1.3.8
(a) addition to feeds for monogastric animals, in order to release phosphate from phytic acid, e.g. from recombinant *Pichia angusta*
Urease EC 3.5.1.5 from *Lactobacillus fermentum*
(a) wine production
(b) ceramics manufacture, to enhance the manufacture and quality of modern ceramics which are of importance for the computer and aeronautics industries or for orthopaedics

Enzymes in miscellaneous biotransformation processes
(a) **Penicillin and cephalosporin acylases** (amidases) EC 3.5.2.6, e.g. *Bacillus megaterium* and *Escherichia coli* produce penicillin acylase and *Pseudomonas* species produce cephalosporin acylase. Used for antibiotic conversion to remove the side groups of penicillins or cephalosporins to form 6-aminopenicillanic acid or 7-aminocephalosporanic acid, respectively, the base molecules for the synthesis of semisynthetic penicillins and cephalosporins
(b) **L-amino acid acylases** EC 3.5.1.14 from *Aspergillus* species, used for the resolution of L-amino acids from acylated racemic mixtures of D,L-amino acid mixtures, e.g. see Chapter 10, L-lysine production
(c) **Steroid biotransformations**, e.g. 11-β-hydroxylase (monooxygenase) EC 1.14.15.4 from *Curvularia lunata* for the production of prednisolone from 11-deoxycortisone (see Chapter 11)

among a vast array of applications (Fig. 9.1 and Table 9.1). Besides their role as aids in traditional processes, 'bulk' enzymes are at the centre of many novel processes and biotechnological innovation continues to expand the range of applications. Bulk enzymes are normally quite crude preparations that are purified only sufficiently to fulfil customer requirements for activity and stability. They often contain many other enzymes, which in some instances are beneficial to overall application performance.

Smaller quantities of 'fine' high-purity enzyme preparations are also required for numerous applications. These include roles as therapeutic agents, many of which are now recombinant products (see Chapter 11), components of food and medical diagnostic test kits, biosensors, as research tools and many other purposes

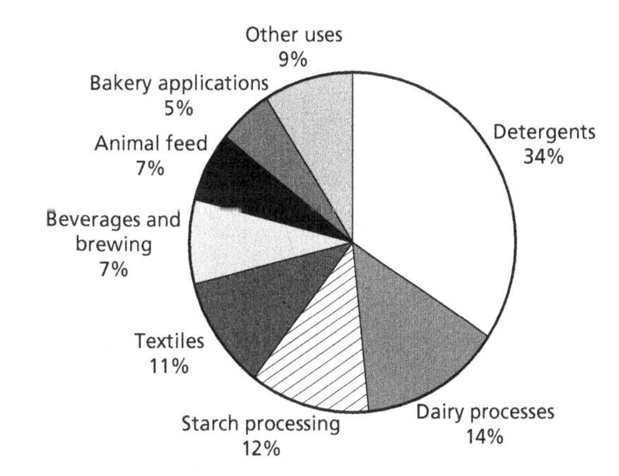

Fig. 9.1 Applications of bulk microbial enzymes.

(Table 9.2). There is an increasing demand for fine enzymes used in molecular biology, particularly restriction endonucleases and DNA polymerases, e.g. DNA polymerase from *Thermus aquaticus* (Taq DNA polymerase) or *Pyrococcus furiosus* (Pfu DNA polymerase). These enzymes are used in polymerase chain reaction (PCR) for DNA fragment amplification. The purification of native and recombinant Taq DNA polymerase is outlined in Fig. 9.2, as an example of the protocols needed to deliver a highly purified enzyme. Most other fine enzymes are also intracellular and are subjected to high levels of purification, particularly those for clinical use.

Enzymes that are more expensive to produce, including many intracellular microbial enzymes, and those employed in biosensors and analytical test systems, are often used in an immobilized form. In industrial processes (Table 9.3), immobilization facilitates confinement and recovery, and these enzyme preparations have several other beneficial features, including:

1 reuse;
2 suitability for application within continuous operations;
3 their product is enzyme free, therefore further processing to remove or inactivate the enzyme is not required;
4 improved enzyme stability; and
5 reduced effluent disposal problems.

Enzymes can be immobilized by attachment through physical adsorption, ionic bonding or covalent

Table 9.2 Examples of microbial enzymes used for analytical, diagnostic and therapeutic purposes

Enzyme	Source	Use
α-Amylase EC 3.2.1.1	*Aspergillus* & *Bacillus* species	Digestive aids
Alcohol dehydrogenases EC 1.1.1.1 or alcohol oxidases EC 1.1.3.13	*Saccharomyces cerevisiae* *Candida boidinii* & *Pichia pastoris*	Ethanol testing
Asparaginase EC 3.5.1.1	*Escherichia coli*, *Serratia marcescens* & *Erwinia carotovora*	Maintains low asparagine levels in treating cancers (lymphomas and leukaemias) whose cells cannot synthesize the amino acid
Catalase EC 1.11.1.6	*Aspergillus niger*, *Corynebacterium glutamicum* & *Micrococcus lysodeikticus*	Contact lens cleaning systems
Cholesterol esterase EC 3.1.1.13	*Pseudomonas fluorescens*	Monitoring serum cholesterol levels
Creatininase EC 3.5.2.10 (creatinine amidohydrolase)	*Pseudomonas putida* & recombinant *Escherichia coli*	Determination of serum creatinine levels
Glucose oxidase EC 1.1.3.4	*Aspergillus niger*	Analysis of blood glucose levels
β-Lactamase EC 3.5.2.6	*Bacillus cereus* & *Escherichia coli*	Treatment of penicillin allergy
Proteases (various)	Several bacterial & fungal sources	Used as digestive aids, for wound debridement and contact lens cleaning
Rhodanase EC 2.8.1.1 (thiosulphate sulphur transferase)	*Trichoderma* species	Treatment for cyanide poisoning
Streptokinase EC 3.4.22.10 (a cysteine protease)	Various haemolytic streptocci	To break down blood clots
Urease EC 3.5.1.5	*Lactobacillus fermentum*	Removal of urea from blood in renal failure
Uricase EC 1.7.3.3 (urate oxidase)	*Arthrobacter globiformis* & *Candida utilis*	Used in the treatment and diagnosis of gout
Some microbial enzymes used in molecular biology (*DNA directed enzymes)		
DNA ligases (e.g. ATP, EC 6.5.1.1)	*Escherichia coli*	Joining pieces of DNA
DNA polymerase* EC 2.7.7.7	*Thermus aquaticus*	DNA synthesis, e.g. in PCR for forensic analysis
Restriction endonucleases e.g. *Bam* HI and *Eco* RI	*Bacillus amyloliquefaciens* H & *Escherichia coli* RY13	'Cutting' DNA at specific base sequences
RNA polymerase* EC 2.7.7.6	*Salmonella typhimurium*	RNA synthesis

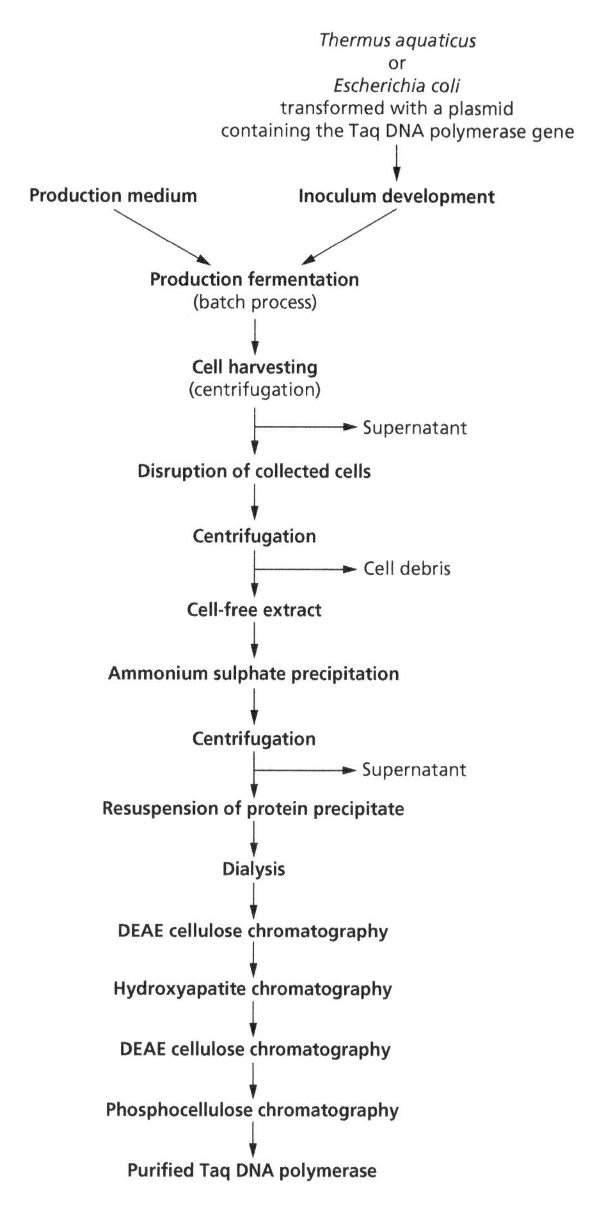

Fig. 9.2 An example of a purification strategy for DNA polymerase (EC 2.7.7.7) from *Thermus aquaticus* (Taq DNA polymerase) and genetically modified *Escherichia coli*.

bonding, to inert inorganic or organic solid support materials, such as nylon, bentonite, cellulose and dextran. Alternatively, they can be microencapsulated within semipermeable membranes, entrapped within gels or fibres, and intracellular enzymes may be immobilized within their producer cells.

Commercial microbial enzyme production

Microbial enzymes are predominantly produced by submerged fermentations, although some solid-substrate fermentations are used, particularly for the production of extracellular fungal enzymes (see Chapter 6, Solid-substrate fermentations). In fact, the first commercial microbial enzyme preparation was produced via solid-substrate fermentation. This enzyme, 'Takadiastase', a fungal amylase, was produced by culturing *Aspergillus oryzae* on moist rice or wheat bran. The process was initially developed by Dr Jokichi Takamine and patented in the USA in 1884. However, large-scale production of microbial enzymes was not generally feasible until after the middle of the 20th century. It became possible only after the vast improvements to submerged fermentation technology that followed the development of penicillin fermentations in the 1940s. Most industrial enzymes are products of batch processes and few are currently produced via continuous fermentation. The fermenters for bulk enzyme production are up to 100 m^3 capacity, but fine enzymes may be produced on smaller scales of a few hundred litres or less. Most fermenters are stirred tank reactors that are operated under aseptic conditions and use low-cost undefined complex media.

As in the development of any fermentation process, enzyme production processes traditionally begin with the search for a suitable producer organism. Use of GRAS-listed (generally regarded as safe) organisms (Table 4.2) is an important consideration for enzymes that are to be used in food or medical applications. A programme of microorganism screening and selection is necessary, to determine enzyme properties, such as optimum pH and heat resistance, and examination of the ability to secrete the target enzyme. Enzymes from thermophiles generally provide several advantages. They are thermostable, able to operate at higher temperatures than enzymes of mesophiles, thereby increasing diffusion rates and solubility, and decreasing both viscosity and the risk of microbial contamination. The fermentation system and conditions for maximum production of the enzyme per unit of biomass, using inexpensive carbon and nitrogen feedstocks, must then be determined. Apart from enzyme productivity, a further consideration is enzyme stability, which can influence the timing, and operations used in, downstream processing. The level of purification applied varies considerably depending on whether the enzyme is intracellular or extracellular, and on its end use. Downstream processing involves separation, purification, stabilization and preservation

Table 9.3 Some important industrial microbial enzymes that are used in an immobilized form

Enzyme	EC number	Source	Product/role
Aminoacylase	3.5.1.14	*Aspergillus oryzae*	L-amino acids
Amyloglucosidase (glucoamylase)	3.2.1.3	*Aspergillus niger* *Rhizopus niveus*	glucose production from starch
Glucose isomerase	5.3.1.5	*Actinomyces missouriensis* *Bacillus coagulans*	high fructose corn syrup
Hydantoinase	4.5.2.2	*Flavobacterium ammoniagenes*	D- and L-amino acids
Invertase	3.2.1.26	*Saccharomyces cerevisiae* *Aspergillus niger*	invert sugar (glucose + fructose)
Lactase (β-galactosidase)	3.2.1.23	*Aspergillus oryzae* *Kluyveromyces fragilis*	lactose-free milk and whey
Lipase	3.1.1.3	*Rhizopus arrhizus*	cocoa butter substitutes
Naringinase* (hesperidinase)	3.2.1.40 3.2.1.21	*Penicillium decumbens*	debittering of citrus fruit juice
Nitrile hydratase	4.2.1.84	*Rhodococcus rhodochrous*	acrylamide
Penicillin acylase (penicillin amidase)	3.5.2.6	*Escherichia coli* *Bacillus subtilis*	penicillin side chain cleavage
Melibiase (raffinase) (α-galactosidase)	3.2.1.32	*Aspergillus niger* *Saccharomyces cerevisiae*	removal of raffinose from sugar beet extracts
Thermolysin (a zinc protease)	3.4.24.27	*Bacillus thermoproteolyticus*	Aspartame (L-aspartyl-L-phenylalanine methyl ester) a low-calorie sweetener

* Consists of two enzymes, an α-rhamnosidase and a β-glucosidase

(Fig. 9.3; also see Chapter 7 for specific methods of protein purification).

Strain improvement may be attempted to further enhance enzyme productivity. In the past, this has often involved cycles of random mutagenesis and screening. However, other strategies are now available for both organism and protein engineering. Targets for improvement often include increased secretion efficiency for extracellular enzymes and overcoming of the organism's own regulatory mechanisms. The latter may involve attempts to relieve catabolite repression, a common regulatory mechanism for many hydrolytic enzymes. Other improvements may be achieved by enhancing mRNA half-life and increasing gene dosage through chromosomal amplification or by plasmid amplification, if the enzyme is plasmid encoded. Targets for enzyme (protein) engineering are enhancement of enzyme activity, improved stability, altered pH optima or temperature tolerance, modified specificity and general improvements to their industrial performance. However, such changes involve the manipulation of specific amino acid constituents and are dependent on prior knowledge of the amino acid sequence of the protein.

Today, if a useful enzyme is identified in a microorganism, plant or animal, which is itself difficult to culture, or if little is known of its physiology and biochemistry, other strategies for commercial production can be adopted. Rather than proceeding with extensive research and development programmes to facilitate the production of the enzyme by the natural producer organism, the structural gene for the enzyme can be transferred into a selected, readily cultivated, 'host' microorganism, along with appropriate mechanisms for control. Genetic engineering now makes possible the expression of the gene coding for almost any enzyme, no matter what its origin. This enables enzyme manufacturers to develop rapidly a process for producing large quantities of the enzyme.

Expression in GRAS-listed organisms provides obvious advantages. Prime candidates for this role are species within three genera of microorganisms, namely *Bacillus*, *Aspergillus* and *Saccharomyces*. Their biology is well understood and they have proved safe to handle, quick to grow and can produce high yields of enzymes, many of which can be excreted into the fermentation medium. A further advantage is that the medium and conditions under which they grow and perform well are already known, thus minimizing further costly experimentation to optimize fermentation conditions.

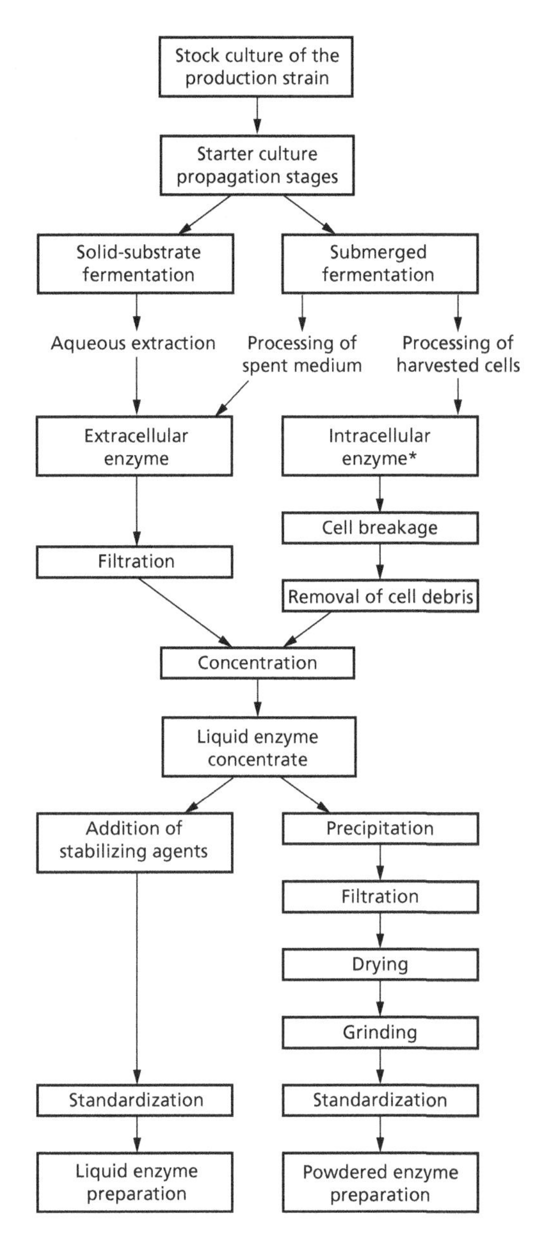

Fig. 9.3 Bulk industrial enzyme production. *Note: some intracellular enzymes, such as glucose isomerase, may not be extracted from cells but immobilized within.

Detergent enzymes

The incorporation of enzymes into detergents provides several benefits. Energy savings are made as a lower wash temperature can be used and levels of less desirable detergent chemicals can be reduced. Unlike other detergent components, enzymes do not have a negative impact on sewage treatment processes. They are totally and rapidly biodegraded to leave no harmful residues. Consequently, they are environmentally safe and are no risk to aquatic life.

The first commercial use of microbial enzymes in detergents was in 1959. A Swiss chemist Jaag, who worked for the detergent company Gebrüder Schnyder, developed a new product containing a bacterial protease, which replaced the animal trypsin that had previously been incorporated into these detergent products. From this point there was an increasing use of microbial enzymes in detergents. In the 1960s, there were further improvements through the introduction of proteases active at alkaline pH. They were relatively unaffected by other components of washing powder and functioned at the desired wash temperatures. A well-known example is subtilisin, a bacterial alkaline serine protease from *Bacillus licheniformis* and *B. subtilis*, which is used extensively in laundry detergents at levels of 0.015–0.025% (w/w). This enzyme has now been engineered to improve pH and temperature characteristics, and reduce sensitivity to peroxide.

Proteases are not the only enzymes used in detergents; since the late 1980s, amylases and lipases have been available for incorporation, e.g. Lipolase from Novo Nordisk. Lipolase was the first detergent enzyme to be produced through recombinant DNA technology. The lipase gene was isolated from a strain of the filamentous fungus *Humicola* and then transferred to *Aspergillus oryzae*, which is more readily cultivated in submerged fermentations. Superior fatty acid digesting enzymes have now been discovered, such as a cutinase from *Fusarium solani*, which naturally degrades the mixture of fatty acids that form plant cuticles. The gene for this enzyme has been transferred to *Saccharomyces cerevisiae* and commercial production is now being examined.

Starch processing enzymes and related carbohydrases

Microbial enzymes have proved to be of immense value in the processing of starch, a polysaccharide composed of amylose (linear α-1,4-linked glucose units) and amylopectin (branched polymer with both α-1,4 and α-1,6 linkages). **α-Amylase** (1,4-α-D-glucan glucanohydrolase) is one of the most important of these industrial enzymes. This endo-enzyme acts randomly on α-1,4 linkages, and ultimately generates glucose, maltose and maltotriose units. It can also catalyse the cleavage of internal α-1,4 linkages in glycogen and oligosaccharides.

Since the 1950s, fungal amylases have been used to manufacture sugar syrups containing specific mixtures of sugars that could not be produced by conventional acid hydrolysis of starch. In addition, α-amylases are employed in several industries, particularly brewing and baking, where starch liquefaction and dextrin hydrolysis are required (Table 9.1). They are also secreted by many other microorganisms, including bacteria and some yeasts. The thermostable α-amylases from *Bacillus* species, e.g. *B. licheniformis*, have proved to be especially useful.

A further success has been **amyloglucosidase (glucoamylase)** from *Aspergillus niger* and *Rhizopus* species, which first became available in the early 1960s. This enzyme can completely break down starch and dextrins into glucose. Within a few years of its introduction, almost all conventional glucose production via acid hydrolysis was replaced by enzyme-based processes, because of the greater yield, higher degree of purity and easier product crystallization. The process was further improved by using highly thermostable bacterial amylase in starch pretreatment (liquefaction).

Glucose isomerase has also been a notable success in the starch processing industry. This enzyme can be obtained from many bacteria, including species of *Bacillus* and *Streptomyces*, and is usually immobilized for use in the conversion of glucose to fructose. The product, where conversion to fructose is completed, has the same calorific value as glucose, but its sweetening effect is approximately twice as high. Substrate glucose is normally derived from hydrolysed corn starch (see above), and is then converted to a mixture of glucose and fructose by glucose isomerase. Practical conversion yields are in the range of 40–50% and the products are referred to as 'high fructose corn syrup' (HFCS) or 'isosyrup'. Fructose concentration can be increased to 55% by chromatographic enrichment. Alternatively, the fructose can now be separated from the glucose–fructose mixture and the remaining glucose fraction is isomerized to produce a final combined product containing up to 80% fructose. These syrups, derived from starch, are somewhat cheaper than sucrose from cane or beet sources, but have the same sweetening power. Consequently, they have become major food ingredients, particularly in North America, and annual production of HFCSs is now in excess of 8×10^9 kg.

Invertase (β-fructofuranosidase) was the first enzyme to be immobilized for use on an industrial scale. It was developed in the UK by Tate and Lyle during the early 1940s for syrup production. The conversion of sucrose to glucose and fructose using this immobilized invertase replaced conventional acid hydrolysis methods. This enzyme was originally prepared from autolysed *S. cerevisiae* preparations, which were adjusted to pH 4.7, clarified by filtration through calcium sulphate and then adsorbed onto bone char. Invertase is now also obtained from other yeasts or filamentous fungi, e.g. *A. niger* and *A. oryzae*. Apart from syrup production, it is employed in confectionery manufacture. For example, soft-centred chocolates are prepared by taking a solid sucrose-based filling, containing some invertase, and coating with chocolate. Within 2 weeks the centre becomes converted into a fructose/glucose syrup.

Lactase (β-galactosidase) is employed in several industrial processes that require the hydrolysis of lactose from milk, where the disaccharide is present at a concentration of 4.7% (w/v). Hydrolysis of lactose to glucose and galactose is performed on milk and milk products for infants who are lactose-intolerant, and in regions of the world where a large proportion of the adult population exhibit lactose-intolerance, e.g. regions of Africa and South-East Asia. Lactose hydrolysis is also useful in the manufacture of products such as ice cream, as the low solubility of this disaccharide leads to crystal formation that may give an unpleasant sandy texture. In addition, its conversion to glucose and galactose increases the relative sweetness by about four-fold.

Commercial lactase is obtained from *Kluyveromyces marxianus* (formerly *K. fragilis*), *A. niger* or *A. oryzae*. The yeast enzymes have pH optima of 6.0–7.0, whereas the *Aspergillus* enzymes have optima as low as pH 3.0–4.0. These enzymes may be used free or immobilized, the latter being particularly useful for whey treatment. Yeast enzymes have been immobilized by incorporation into cellulose acetate fibres and the *Aspergillus* enzymes have been immobilized on 0.5 mm diameter porous silica for use in packed bed reactors.

Enzymes in cheese production

Rennet preparations from the stomachs of calves, lambs and kids have been used in cheese production for thousands of years. These rennets contain the enzyme rennin (chymosin, an aspartic protease), which performs limited proteolysis of milk protein (casein) to form curds at the start of cheese making (see Chapter 12, Dairy fermentations). Fig sap, containing the proteolytic enzyme ficin, has sometimes been used for the same purpose, but many other proteases carry out too much proteolysis.

Animal rennets, particularly from calves, predominated in cheese manufacture until the 1960s when an increase in cheese production coincided with a shortage of available animal rennet. Specific fungal proteases, which have very similar properties to calf chymosin, were then developed as microbial rennets, such as proteases from *Rhizomucor miehei* and *R. pusillus*. This overcame the shortfall and facilitated the production of so-called 'vegetarian' cheese. Some microbial enzyme preparations are also more temperature-sensitive (thermolabile) than calf rennet, which is useful in certain cheese-making processes.

More recently, the calf chymosin gene has been introduced into several microorganisms, including *E. coli*, *K. lactis*, *Aspergillus nidulans* and *A. niger* var. *awamori*. These genetically engineered microorganisms are capable of expressing and secreting the enzyme. Recombinant chymosin preparations now have approval for use as food additives in over 20 countries. Animal and microbial chymosins (constituting about one-third of the market) have combined world sales worth in excess of $75 million. Microbial lipases are also used in dairy products, especially cheeses, for the hydrolysis of fatty acid esters to accelerate flavour development.

Enzymes in plant juice production

Several microbial enzymes are employed in fruit juice processing, but probably the most important are pectinases. These processing aids are often produced by solid-substrate fermentation of *A. niger* or *Penicillium* species. Fruits and berries contain varying amounts of pectin, which acts as a binding layer between plant cells to hold adjacent cell walls together. Pectin is a heteropolymer of galacturonic acid, methyl esters of galacturonic acid and other sugar residues. In plant juice production, some of this pectin is extracted during pressing. It causes an increase in juice viscosity, leading to difficulties in obtaining optimal juice yields, and in juice clarification and filtration. These problems can be overcome by adding pectinase preparations to the fruit pulp before pressing. Similar enzyme treatment is used to increase the yield of oils from olive pulp, palm fruit and coconut flesh.

The commercial 'pectinase' is not one enzyme, but is usually a complex cocktail of enzymes (pectin methyl esterases, polygalacturonases and pectin lyases) capable of attacking a variety of bonds in correspondingly diverse pectin molecules. Compositions of commercial pectinase preparations vary considerably in the proportions of these different enzymes. Fungal pectinases are also available that remain active in very acidic juices from citrus fruits, where the pH can be as low as 2.2–2.8. They may be used in the enzymic peeling of citrus fruit for canning.

The polysaccharide araban, a polymer of the pentose arabinose, is also an important component of fruit cell walls. Like pectin, it is often extracted during pressing of some fruits, especially pears. This may lead to haze formation, but it can now be eliminated by using commercial arabanases. Also, some extracts, such as apple juice, contain starch, which must be degraded to produce clear juices and concentrates. This is achieved by adding amylases, along with the pectinase, during depectinization of the juice.

In wine making, commercial enzymes are used for several purposes, as well as for juice extraction. For red wines, colour extraction from the grape skins during pressing can be promoted by the addition of commercial cellulases, e.g. from the fungus *Trichoderma reesei*. In addition, glycosidases can be employed to hydrolyse terpenyl glycosides, releasing the terpenes that are important constituents of the wine bouquet. Wines prepared from grapes allowed to undergo attack by the fungus *Botrytis cinerea* before harvesting ('noble rot') are often difficult to clarify and filter, due to the presence of fungal β-glucans. These polymers can be degraded by adding a specific microbial β-glucanase to the wine (see also Chapter 12, Beer brewing).

Citrus fruits, such as grapefruit and bitter oranges, contain the bitter-tasting fruit flavonoid called naringin. The bitterness of products derived from these fruits can be adjusted using naringinase (α-rhamnosidase + β-glucosidase) from *A. niger*. It first converts naringin to the less bitter compound prunin, then on to non-bitter naringenin. The level of naringin hydrolysis may be controlled by regulating the flow rate of fruit juice through a column of immobilized naringinase.

Glucose oxidase from *A. niger* or *Penicillium* species, often coupled with catalase, can be employed to remove molecular oxygen from wine, beer, fruit juices and soft drinks. This prevents potentially damaging oxidation that otherwise affects product quality.

$$2\,\text{glucose} + O_2 \xrightarrow{\textit{glucose oxidase + catalase}} 2\,\text{gluconic acid}$$

Enzymes in textile manufacture

Enzymes are being increasingly used in textile processing for the finishing of fabrics and garments, especially in desizing, biopolishing and denim washing.

Prior to weaving fabrics from cotton, or blends of

cotton and synthetic fibres, the threads are reinforced with an adhesive (termed 'size') to prevent breakage of warp threads. The size may be composed of starch, starch derivatives, vegetable gum and water-soluble cellulose derivatives, such as methyl- and carboxymethylcellulose. Before dyeing, bleaching and printing, the size must be removed from the cloth. This has previously been accomplished by treatment with acids, alkalis and oxidizing agents that may damage the fibres. Consequently, enzymatic desizing with amylases or cellulases is now widely used, as they are non-corrosive and produce no harmful effluent wastes.

Cellulases are also employed in bio-polishing cotton and other cellulose fibres to produce fabrics with a smoother and glossier appearance. Similar processes using microbial proteases have been developed for treating wool fibres, which are composed of keratin.

Treatment of denim garments to give a worn look before sale is called 'stonewashing'. Traditionally, this has been accomplished by washing the garments in a tumbling washing machine with pumice stones. Cellulases are now used to accelerate abrasion and aid the loosening of the indigo dye. This 'bio-stonewashing' is less damaging to the garments, reduces wear on machines and generates less pumice dust.

Enzymes in leather manufacture

Proteases and lipases are now extensively used in the processing of hides and skins. These enzymes are easier to use, more pleasant to handle and safer than the harsh chemicals that were previously employed. Their most important applications are in soaking, dehairing, degreasing and baiting. Soaking is the first important operation of leather processing. Apart from cleaning the hides and skins by removing debris derived from blood, flesh, grease and dung, it rehydrates them. Proteases enhance water uptake by dissolving intrafibrillary proteins that cement the fibres together and prevent water penetration. Lipases are also used to disperse fats as an alternative to degreasing with solvents. These enzymes hydrolyse surface fats and those within the structure of the hides and skins, but without causing any physical damage.

Dehairing of hides and skins has traditionally employed chemicals such as slaked lime and sodium sulphide, which have in the past made tannery effluent a severe pollution problem. Enzyme-assisted dehairing involves proteases, often the alkaline protease from *Aspergillus flavus*. Their use reduces or totally replaces the requirements for chemical processing and provides a major cost reduction in effluent treatment.

Leather baiting prepares the leather for tanning. It involves removal of any remaining unwanted protein to make the grain surface of the finished leather clean, smooth and fine. Traditional methods employed dog, pigeon or chicken manures. These were very unpleasant to use, unreliable and slow, and have been replaced by microbial proteases.

Enzymes used in the treatment of wood pulps

Paper manufacture is a major world industry. In the USA alone over 70 million tonnes of paper and paperboard are manufactured each year, which have a value in excess of US$50 billion. Paper manufacture involves the processing of mostly virgin fibres, but there is increasing utilization of some recycled materials or secondary fibres. Recycling saves both trees and energy, decreases waste effluents and reduces landfill loads.

Traditionally, wood pulp processing has involved the extensive use of chemicals, which can lead to problems with effluent treatment and environmental pollution. Thus, the development of enzyme-based technology for pulp processing and paper manufacture has major advantages. Microbial enzymes can be used in several stages of pulp and paper processing to:

1 enhance pulp digestion;
2 improve drainage rates in water removal during paper formation;
3 increase fibre flexibility;
4 selectively remove xylan without affecting other components;
5 remove resins;
6 enhance bleaching;
7 remove contaminants, such as in the de-inking of high-quality waste paper; and
8 fibrillate or increase interfibre bonding in chemical pulps and herbaceous fibres.

Those microbial enzymes used in these processes include a wide range of cellulases, hemicellulases, pectinases and lipases. However, they cannot yet replace all mechanical and chemical treatments.

Enzymes as catalysts in organic synthesis

There is a rapidly growing market for microbial enzymes used in the synthesis of high-value organic compounds for the chemical, food and pharmaceutical industries. One commercially valuable area involves the application of cyclodextrin glycosyltransferases, often obtained from extreme thermophiles, to create cyclodextrins from simple starch molecules. These cyclodextrins are useful carriers of fragile compounds such as vitamins and flavours. No chemical means exists for creating this class of substances, so the enzymatic route is unique.

An increasingly important field is the synthesis of chiral compounds. Chiral compounds exist in two forms called enantiomers, which are non-superimposable mirror images of each other and are therefore asymmetrical. In virtually all respects, the two enantiomers are physically and chemically identical. However, solutions of one enantiomer rotate the plane of polarized light in a clockwise direction (+), whereas solutions of the other enantiomer rotate it anticlockwise (−), but by exactly the same amount. This phenomenon is known as optical activity and enantiomers are sometimes referred to as optical isomers.

Chirality is vitally important in biology as most natural organic compounds are chiral. For example, most amino acids are enantiomeric, with only one enatiomer usually being found in nature. However, when the compounds are chemically synthesized an almost 50 : 50 mixture of enantiomers is made, which is called a racemic mixture. Usually, for most biological molecules, as for amino acids, only one of its variants or enantiomers is biologically effective as an enzyme substrate, pharmaceutical drug, nutrient, etc. Its other enantiomer may be useless/futile or even damaging to health. The reason that identical molecules of opposite chirality can have such different biological behaviour is that proteins, especially enzymes, and nucleic acids are themselves chiral molecules. Consequently, any drug or nutrient must have the proper chirality if it is to interact with a protein of specific chirality. In the industrial synthesis of such compounds, there have to be controls to guarantee the production of only the desired enantiomer, often at a purity higher than 99%. Not only is it economically disadvantageous to make a large quantity of chemicals when only half (one enantiomer) is biologically functional, but for some drugs the other enantiomer may be toxic. Certain chiral syntheses can be performed chemically, but enzymic routes are usually preferred.

Enzyme-based processes can be used to prepare specific enantiomers and resolve enantiomers from racemic mixtures. These processes utilize racemases, which are a family of enzymes that accept either enantiomer, although they are themselves chiral like all other proteins. Their biological function is to turn a substrate of one chirality into its opposite form. They can be used to synthesize chiral amines, alcohols and many other compounds.

Further reading

Papers and reviews

Adams, M. W. W. & Kelly, R. M. (1995) Enzymes isolated from microorganisms that grow in extreme environments. *Chemical and Engineering News* 73, 32–42.

Benkovic, S. J. & Ballesteros, A. (1997) Biocatalysts—the next generation. *Trends in Biotechnology* **15**, 385–386.

Headon, D. & Walsh, G. (1994) The industrial production of enzymes. *Biotechnology Advances* **12**, 636–646.

Jaeger, K.-E., Dijkstra, B. W. & Reetz, M. T. (1999) Bacterial biocatalysts: molecular biology, three-dimensional structures, and biotechnological applications of lipases. *Annual Review of Microbiology* **53**, 315–351.

Marrs, B., Delagrave, S. & Murphy, D. (1999) Novel approaches for discovering industrial enzymes. *Current Opinion in Microbiology* **2**, 241–245.

Ogawa, J. & Shimizu, S. (1999) Microbial enzymes: new industrial applications from traditional screening methods. *Trends in Biotechnology* **17**, 13–19.

Shulze, B. & Wubbolts, M. G. (1997) Biocatalysis for industrial production of fine chemicals. *Current Opinion in Biotechnology* **10**, 609–615.

Books

Bornscheuer, U. & Kaszlauskas, R. (1998) *Biotechnology. Biotransformations*, Vol. 8. VCH, Weinheim.

Chaplin, M. F. & Bucke, C. (1990) *Enzyme Technology*. Cambridge University Press, Cambridge.

Drauz, K. & Waldmann, H. (eds) (1995) *Enzyme Catalysis in Organic Synthesis*. VCH, Weinheim.

Faber, K. (1996) *Biotransformation in Organic Chemistry*. Springer Verlag, Berlin.

Fogarty, W. M. & Kelly, C. T. (1990) *Microbial Enzymes and Biotechnology*. Elsevier, London.

Gerhartz, W. (1990) *Enzymes in Industry*. VCH, Weinheim.

Koskinen, A. M. P. & Klibanov, A. M. (eds) (1996) *Enzymatic Reactions in Organic Media*. Blackie Academic & Professional, London.

Nagodawithana, T. & Reed, G. (1993) *Enzymes in Food Processing*. Academic Press, San Diego.

Sheldon, R. & Kuster, A. (1993) *Chirotechnology*. Marcel Dekker, New York.

Tucker, G. A. & Woods, L. F. J. (1995) *Enzymes in Food Processing*, 2nd edition. Blackie Academic and Professional, London.

10 Fuels and industrial chemicals

Within the next 50–100 years some fossil fuel supplies, particularly oil, are likely to become depleted. Consequently, there is an urgent need to develop alternate energy sources. Most requirements will be met from geothermal, nuclear, solar, water and wind sources. However, biological fuel generation is likely to become increasingly important, especially as it can provide both liquid and gaseous fuels. Importantly, these fuels are produced from renewable resources, primarily plant biomass, in the form of cultivated energy crops, natural vegetation, and agricultural, domestic and industrial wastes. The two main microbial fuel products currently derived from these resources are methane and ethanol, but these are not the only fuels that can be formed. Other liquid and gaseous examples include hydrogen, propane, methanol and butanol; electricity can also be generated by microbiological systems.

Numerous other important chemical compounds are now most economically produced by microbial fermentation and biotransformation processes. The majority are primary metabolites that include organic acids, amino acids, industrial solvents and a wide range of biopolymers. Many of these microbial products are used as chemical feedstocks and functional ingredients in a wide range of industrial and food products.

Alkanes

Methane is used for both domestic and industrial fuel. At present, supplies mostly come from gas and oil fields or the gasification of coal. Consequently, methane production via fermentation is attractive only in limited small-scale local situations. However, this mode of production may become increasingly important later in the 21st century, when supplies from the non-renewable sources begin to dwindle.

Methane production by microorganisms is a very complex process, which involves a mixture of anaerobic microorganisms found naturally in marshes, organic sediments and in the stomachs (rumen) of ruminant animals. Current microbial production of methane for combustion may be via the anaerobic digestion of agricultural, industrial and urban wastes (see Chapter 15). These wastes are predominantly plant biomass (lignocellulosic materials) that have high collection costs. A potentially attractive, but costly, mode of large-scale production is via landfilling (see Chapter 15), provided that stable long-term gas production can be developed. In contrast, biogas fermenters use low technology in the small-scale local production of methane. They often use animal excreta and are particularly valuable in locations where other fuels are not available. The biogas generated is primarily composed of 50–80% methane and 15–45% CO_2, along with some trace gases.

Mixed microbial populations associated with methane generation are very versatile with regard to the range of substrates that they can utilize. Methane production from organic materials involves three specific phases. First, a group of microorganisms hydrolyse organic polymers, including fats, proteins and polysaccharides, to their respective soluble monomers. These compounds are then metabolized to organic acids by anaerobic acidogenic organisms. In the final phase, the organic acids are converted to alkanes and carbon dioxide (Fig. 10.1). Methanogenic bacteria produce methane from acetate, which is the major product. However, methane has a lower energy yield than the longer chain alkanes, ethane and propane (Table 10.1), which are derived from propionate and butyrate, respectively (Fig. 10.1). Normally, only small amounts of these two organic acids are produced during acidogenesis, but the quantities generated depend upon the specific conditions. Therefore, there is potential for the future manipulation of such fermentations to produce a greater proportion of the more attractive fuels ethane and propane.

Butanol

Acetone, butanol, butyric acid and isopropanol, along

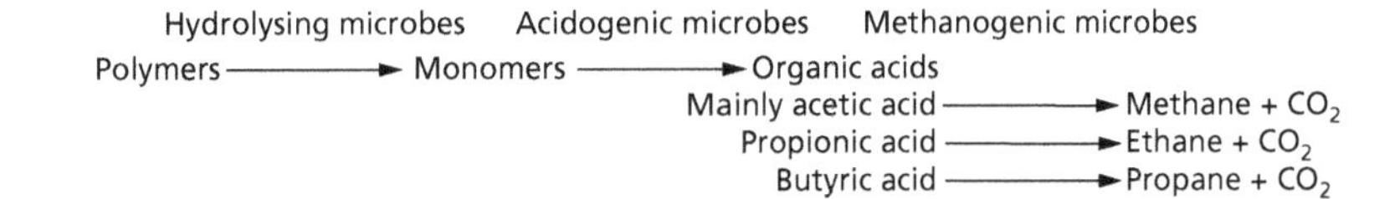

Fig. 10.1 Production of short chain alkanes by fermentation.

Table 10.1 Energy yield of short chain alkanes

Alkane	Energy yield (kJ/m^3)
Methane (CH_4)	37
Ethane (C_2H_6)	64
Propane (C_3H_8)	94

with other organic acids and alcohols, may be obtained by clostridial fermentation of a range of raw materials, including starch, molasses and hydrolysed cellulosic materials. The relative amount of each fermentation product is dependent upon the bacterial species and the specific strain used, and the environmental conditions of the fermentation. There are three main fermentation types:

1 **acetone–butanol** (*Clostridium acetobutylicum*), additional products: butyric acid, acetic acid, acetoin, ethanol, CO_2 and H_2;

2 **butanol–isopropanol** (*Clostridium butylicum*), additional products: butyric acid, acetic acid, CO_2 and H_2; and

3 **butyric acid–acetic acid** (*Clostridium butyricum*), additional products: CO_2 and H_2.

Members of the genus *Clostridium* are Gram-positive rods with peritrichous flagella and are accordingly very motile. They are characterized by their ability to form heat-resistant spores, a highly fermentative metabolism and their response to oxygen. All are anaerobic, but range from obligate anaerobes to aerotolerant species. Normally, they contain no haem derivatives, such as cytochromes and catalase. However, some species can produce cytochromes if supplied with appropriate precursors. Most are mesophiles, although several thermophilic species are known, e.g. *C. thermoaceticum*. They grow well at neutral and alkaline pH, but are inhibited under acidic conditions and vary widely in the range of substrates that they can utilize.

The acetone–butanol fermentation has a long history as a successful industrial fermentation process. It was Weizmann in the UK, early in the 20th century, who conducted much of the basic research into the production of acetone, butanol and ethanol by *C. acetobutylicum*. This was invaluable during World War I, particularly for the production of acetone, which was needed in the manufacture of explosives. Acetone was specifically used as a gelatinizing agent for nitrocellulose in the production of cordite. The Weizmann process also produced riboflavin (vitamin B_2) as a byproduct.

Following World War I, butanol became the main product of interest. It was used extensively as a feedstock chemical in the production of lacquers, rayon, plasticizers, coatings, detergents, brake fluids and butadiene for synthetic rubber manufacture. Butanol was also used as a solvent for fats, waxes, resins, shellac and varnish, and as a valuable extractant and solvent in the food industry. The annual production of fermentation-derived butanol was over 20 000 tonnes in 1945, but in the western world the process began to decline by the late 1940s. This was due to changes in supply of fermentation raw materials (molasses, sugar cane, etc.) and the increase in availability of inexpensive petrochemical feedstocks for its chemical synthesis.

Butanol is now predominantly manufactured from petroleum-based raw materials. In the USA, for example, the current production of chemically synthesized butanol is more than 500 000 tonnes/year, with annual growth of 3–4%. Nevertheless, butanol fermentations are still operated in certain countries. In the former states of the Soviet Union some processes were based on beet molasses, whereas fermentations using sugar cane molasses continued until relatively recently in South Africa, and China still maintains some fermentation-based manufacturing plants. The future for fermentation-based production looks quite bright, particularly as worldwide consumption of butanol has surged over the past few years. This provides an opportunity for the introduction of new and more efficient fermentation process technology, especially as the supply of petrochemicals dwindles.

Besides the existing role as a solvent and chemical feedstock, butanol has several properties that are favourable for motor fuel use, either alone or when blended with gasoline (Table 10.2). Butanol has good octane-enhancing properties, a relatively high heat of combustion and a much lower vapour pressure than both methanol and ethanol. These characteristics make

Table 10.2 Characteristics of alcohols for use as cosolvents and octane enhancers with motor fuels

Physical property	Methanol	Ethanol	Butanol	Gasoline	Diesel
Heat of combustion (kJ/g)	23.9	30.6	36.7	43.8	42.7
Fuel values*: RON	110	108	100	92	15
(octane number) MON	91	90	87	83	–
Miscibility with					
Gasoline	Poor	Fair	Good	–	–
Diesel	Poor	Poor	Good	–	–
Water	High	High	Low	Low	Low

*Octane number defines a fuel's antiknock quality in an internal combustion engine, i.e. its resistance to preignition. The performance value for iso-octane is given the arbitrary value of 100 and *n*-heptane a value of zero, and is used for the standardization of fuels by comparing a fuel with mixtures of these two alkanes. There are two laboratory tests, used RON (research octane number), which correlates best with low-speed, mild-knocking conditions, and MON (motor octane number), which correlates with high-speed, high-temperature conditions.

butanol an even better liquid fuel extender than ethanol, which is currently used in the formulation of gasohol. Furthermore, butanol has low miscibility with water, but high miscibility with both diesel and gasoline. Owing to its high heat of combustion, butanol solutions containing as much as 20% (v/v) water have the same combustion value as anhydrous ethanol. This can indirectly reduce nitrogen oxide (NO_x) emissions by lowering the operating temperature of internal combustion engines. As fuel additives, alcohols such as butanol also have the potential for reducing carbon monoxide emissions.

Butanol production process

In the past, the economically viable production of butanol has normally required a fermenter volume of at least $1000\,m^3$. The fermenters were not stirred, as the evolution of gases provided sufficient mixing. These fermentations were operated as batch processes, often using 5–7% (w/v) starch or molasses as the carbon substrate. More recently, with the increasing demand for butanol, advanced processes based on corn, corn processing byproducts and other cellulosic wastes have been proposed, particularly in the USA (see below).

In the conventional process, prior to a fermentation, the medium and fermenter are sterilized and purged with CO_2. The fermenter is then inoculated with a relatively low level of inoculum, 0.03% (v/v) *C. acetobutylicum*. Over the first 18–24 h the pH falls from an initial level of 5.8–6.0 to pH 5.2, due to the production of butyric and acetic acids during this rapid growth phase. Over the following 20–24 h the pH rises back to pH 5.8–6.0, as these acids are metabolized to form the neutral solvents acetone, butanol and ethanol. Their total concentration reaches around 2% (v/v) in a ratio of 6:3:1 for acetone, butanol and ethanol, respectively. Overall yields of up to 37% (w/w), based on the initial carbohydrate, can be achieved. Product recovery has traditionally involved fractional distillation. Gases generated during the fermentation, particularly CO_2, may be recovered for sale as byproducts and the distillation residues may be sold as animal feed.

A proportion of butanol production in China and a few other countries may still be via similar fermentation processes, but the last industrial acetone–butanol–ethanol fermentation operated in the western world closed in the early 1990s. This was carried out by National Chemical Products in Germiston, South Africa, using *C. acetobutylicum* P262. It involved batch fermentations with molasses as the substrate, which ran for 40–60 h and generated average product concentrations of 15–18 g/L. However, economic viability of these fermentations now demands a solvent yield of 22–28 g/L within this time, otherwise they cannot compete with chemical processes. The exact economically competitive concentration is dependent upon the prevailing price of oil.

The traditional batch methods have suffered from several problems, including contamination by lactobacilli, bacteriophage attack, product inhibition, high energy costs for distillation and the fact that a mixture of fermentation products is obtained. More efficient processes are now being developed with improved strains that predominantly produce one fermentation product and use cheaper substrates, including domestic and agricultural wastes. They may involve temperature-programmed, multistage, continuous suspension cultures or immobilized cells incorporated into fluidized-bed reactors. Such fermentation processes are likely to have integrated product recovery systems, such as pervaporation. This particular method involves selective membrane separation of solvent components into a low-pressure chamber (e.g. using poly dimethylsilane membranes) followed by product condensation. These improved procedures have the potential for producing a total solvent concentration greater than 30 g/L and allow simultaneous recovery of solvents from the broth during fermentation. This eliminates product inhibition and allows complete utilization of the carbon substrate at a relatively high rate during continuous fermentation.

Industrial ethanol

Most regions of the world have traditionally produced alcoholic beverages from locally available substrates (see Chapter 12). Similar alcoholic fermentations are now used in some countries to produce fuel grade or chemical feedstock ethanol. The annual world production of ethanol is over 30 billion litres, approximately 70% of which is produced by fermentation, the remainder being mostly manufactured by the catalytic hydration of ethylene. Almost 12% of the fermentation ethanol is beverage alcohol, 20% is for various industrial uses and the remaining 68% is fuel ethanol.

Ethanol is an attractive fuel because it may be be used alone or mixed with other liquid fuels, e.g. 'gasohol', a blend of 10–22% (v/v) ethanol with gasoline (see Table 10.2). In the 1970s, Brazil and a few other countries undertook the full-scale production of ethanol from indigenous renewable biomass resources to offset the growing costs of oil imports. The ethanol was produced by fermentation of sucrose, derived from sugar cane, using *Saccharomyces cerevisiae*. Brazil is now responsible for over 46% of annual world production, some 14.5 billion litres of ethanol. However, this is not sufficient to keep up with the growing demand for fuel. Failure to develop their production processes has resulted in the need to import ethanol from the USA and other producing countries.

Apart from sucrose, other conventional fermentation substrates for ethanol fermentations include simple sugars derived from plants and dairy wastes. These require relatively little processing (Figs 10.2 and 10.3). However, use of root and tuber starch (cassava, potato, etc.) or grain starch (maize, wheat, rice, etc.) demands energy-consuming processing operations to achieve hydrolysis. Even greater processing is necessary prior to the utilization of lignocellulosic plant materials (see p. 149).

Bioconversion of maize starch

In North America, wet or dry milling processes have been developed for maize processing to separate corn oil from the starch. This also generates byproducts that can be used for animal feed. Extracted starch is subjected to gelatinization and saccharification, and the resultant sugars can then undergo alcoholic fermentation. The technology initially used was largely based on that previously developed for alcoholic beverage production, but now the processes are much more efficient. Enzymatic saccharification is used for converting starch into fermentable sugars employing various thermostable amylases, including glucoamylases. Fermentation of the resulting sugars, mostly glucose, is carried out by selected strains of *S. cerevisiae* at 32–38°C and pH 4.5–5.0. Fermentations may be batch or continuous processes, often with some form of cell recycle, which reduces both the fermentation time and the amount of substrate 'wasted' in conversion to unwanted biomass. Operation under vacuum, facilitating continuous removal of ethanol to reduce ethanol inhibition, and even cell immobilization have been trialled (Table 10.3).

These alcoholic fermentations generate a 'beer' containing approximately 10% (v/v) ethanol from which the yeast is usually separated prior to distillation. Recovered ethanol may then be dehydrated (see Chapter 7,

Direct sugar sources (cane, beet, whey, etc.) require less energy consuming processing and are readily fermented

starches —*saccharification*→ readily fermented products (glucose, maltose)

lignocellulose —*delignification*→ cellulose and hemicellulose —*saccharification*→ hexoses & pentoses*

Fig. 10.2 Substrates for ethanol production (*much less readily fermented products).

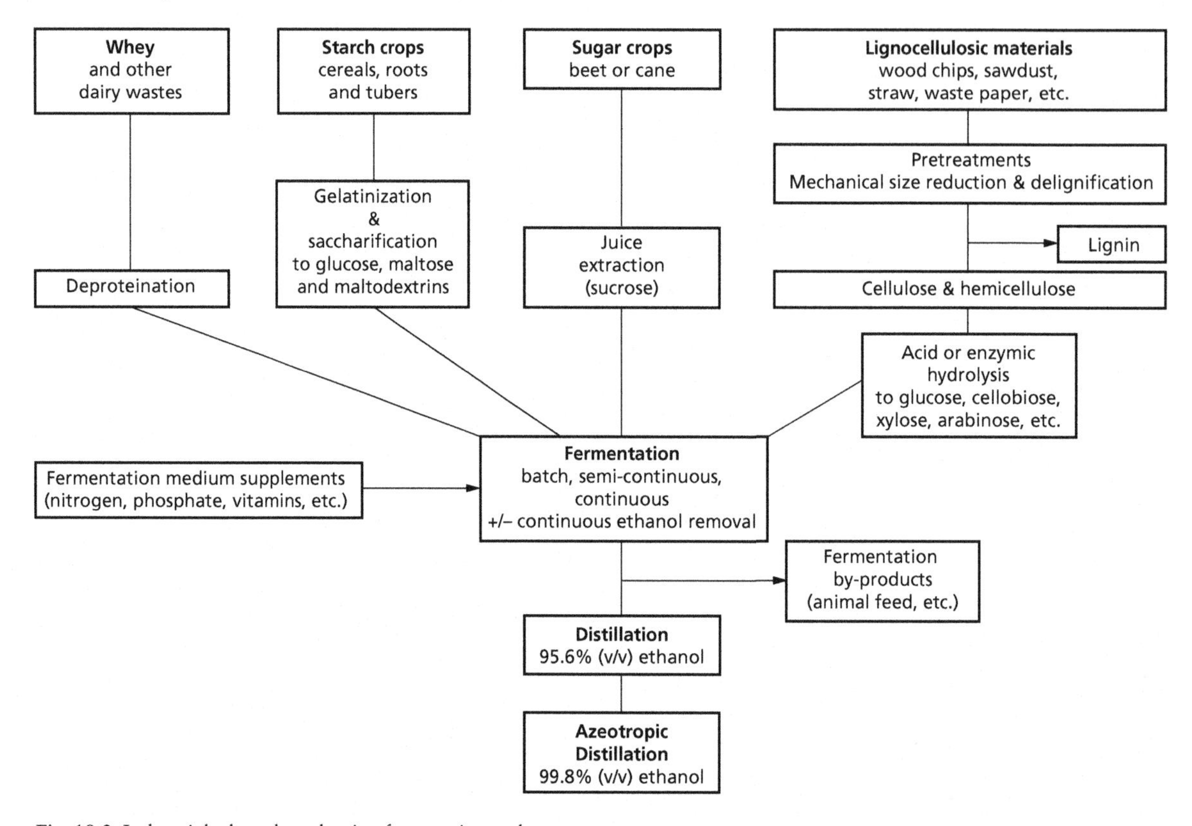

Fig. 10.3 Industrial ethanol production from various substrates.

Table 10.3 Ethanol production by *Saccharomyces cerevisiae* in various fermentation systems

Fermentation type	Ethanol production rate (g/L/h)
Batch	2.5–3.5
Batch with yeast recycle	5.0–6.5
Continuous (stirred tank reactor)	12–14
Continuous (under a vacuum of 35 mm Hg and yeast recycle)	60–80*
Continuous (immobilized yeast cells)	> 100

* In similar systems *Zymomonas mobilis* produces up to 120 g ethanol per litre per hour.

Distillation). The cost of this ethanol recovery is often up to 50% of the total process expenditure. Process byproducts include methanol, glycerol and higher alcohols, such as amyl, butyl and propyl alcohols.

Possible alternate approaches to ethanol recovery include the use of continuous extractive fermentation processes using non-volatile, non-toxic solvents, such as oleyl alcohol, which have a high affinity for ethanol. This stategy is useful in overcoming end-product inhibition. The solvents employed are continuously introduced into the fermenter and rise through the medium to form a layer that is continuously removed. Passage through a centrifuge results in the separation of ethanol-laden solvent from media and cells, which are returned to the fermenter. The ethanol may be recovered by flash vaporization and the non-volatile solvent is reused.

Although *S. cerevisiae* is still extensively used for the

alcoholic fermentation of simple sugar substrates, there are other organisms with commercial potential (Table 10.4). These include species of the bacterial genus *Zymomonas*, such as *Z. mobilis*, which are Gram-negative facultative anaerobes that normally ferment only glucose, fructose or sucrose. They afford greater ethanol yields than does *S. cerevisiae*, but are not as ethanol tolerant. In future, alternate routes are likely to involve genetically engineered organisms that have the ability to utilize a wider range of carbon sources and have better fermentation properties. For example, *Escherichia coli*, which normally produces only relatively small amounts of ethanol, has been transformed with a plasmid that encodes the alcohol dehydrogenase and pyruvate decarboxylase from *Z. mobilis*. Such transformants produce ethanol under both aerobic and anaerobic conditions.

Bioconversion of lignocellulosic materials

Unlike energy crops (cereals, sugar cane and beet, etc.), lignocellulosic plant wastes (sawdust, wood chips, straw, bagasse, waste paper, etc.) have no direct food use. They are renewable resources yet to be fully exploited. Billions of tonnes of these cellulosic materials currently go to waste each year, which could be converted into chemical energy or other useful fermentation products. Lignocellulose is composed of the following polymers.

1 **Lignin** (10–35%, w/w), a polymer of three phenolic alcohols (*p*-coumaryl, sinapyl and coniferyl alcohols) that encrusts the cellulose. This material cannot be degraded by microorganisms under anaerobic conditions, but may be used as sources of vanillin, catechol, dimethylsulphide (DMS) and dimethyl sulphoxide (DMSO) via chemical processes.
2 **Cellulose** (15–55%, w/w), a linear homopolymer of β-1,4-linked glucose units. Once hydrolysed, the resultant glucose is readily fermented by many microorganisms, but few can directly utilize the native polymer.
3 **Hemicellulose** (25–85%, w/w), a class of heteropolymers containing various hexoses (D-glucose, D-galactose and D-mannose) and pentoses (L-arabinose and D-xylose). Xylose is the second most abundant sugar in nature after D-glucose and may constitute up to 25% of the dry weight of some woody trees, but only a few microorganisms ferment pentoses to ethanol. Importantly, ethanol production from lignocellulose

Table 10.4 Ethanol-producing microorganisms

Bacteria	
Clostridium thermohydrosulfuricum	Extreme thermophile
Clostridium thermocellum	Thermophilic, hydrolyses cellulose
Thermoanaerobacter ethanolicus	Ferments xylose and starch*
Zymomonas mobilis	The wild-type ferments only glucose, fructose and sucrose, but with high productivity
Yeasts	
Candida pseudotropicalis, C. tropicalis	Ferments xylose*
Candida species	Ferment xylose and cellobiose*
Kluyveromyces lactis†	Ferments lactose in dairy wastes*
Kluyveromyces marxianus	Hydrolyses inulin (polyfructosan)
Pachysolen tannophilus	Ferments xylose*
Pichia stipitis	Ferments xylose*
Saccharomyces cerevisiae	Most strains ferment only glucose, sucrose, fructose, maltose and maltotriose
S. cerevisiae var. *distaticus*	Hydrolyses starch
Schwanniomyces alluvius	Hydrolyses starch
Filamentous fungi	
Fusarium species	Ferment xylose*
Monilia species	Hydrolyse cellulose and xylan
Mucor species	Ferment xylose and arabinose*

* In addition to common hexose monosaccharides and disaccharides.
† Does not exhibit the Crabtree effect.

is economically viable only if both pentoses and hexoses are fermented.

Few microorganisms can utilize lignocellulose directly and those that do, such as some species of *Clostridium*, produce little or no ethanol. Therefore, direct microbial fermentation of cellulosics to ethanol is a distant opportunity. An approach more likely to prove successful in the shorter term involves several steps. First, pretreatment of the lignocellulosic material is necessary prior to the saccharification of hemicellulose and cellulose components. Sugars resulting from chemical and/or enzymic hydrolysis may then be fermented to produce ethanol, which can be separated from the aqueous phase by distillation.

Pretreatment and saccharification must be conducted in a way that maximizes subsequent bioconversion yields and minimizes the formation of potentially inhibitory compounds, especially furfurals and soluble phenolics. Most lignocellulosic materials require a pretreatment to render the cellulose and hemicellulose more amenable to acid or enzyme hydrolysis. Pretreatment requirements vary with the feedstock and are often substantially less for processed materials such as paper and card. Methods employed include mechanical size reduction by milling, chemical pulping, acid hydrolysis, alkali treatment, autohydrolysis, solvent extraction, steaming and steam explosion (explosive decompression following high-pressure steam treatment at 4000 kPa for 5–10 min).

Various combinations of pretreatment processes may be used depending upon the source of lignocellulosic materials (Fig. 10.3). Some achieve partial saccharification, but further treatment with acid or enzymes is usually necessary. Acid hydrolysis is generally carried out with dilute acid (e.g. 0.5–5% (v/v) sulphuric acid) under pressure to achieve elevated temperatures of 100–240°C. This treatment is relatively inexpensive, but also generates large quantities of degradation byproducts and undesirable inhibitory compounds. Strong acid hydrolysis often uses concentrated hydrochloric acid at ambient temperature, which gives the highest sugar yield of any acid hydrolysis process. However, such operations are highly corrosive and almost complete acid recovery is essential to make the process economically viable. Acid hydrolysis of mixtures of cellulose and hemicellulose is difficult to control. Hemicellulose is more readily hydrolysed than cellulose and generates sugars early in the process. These sugars may undergo further breakdown to inhibitory compounds, e.g. furfurals. Consequently, conditioning of hydrolysates may be necessary in order to remove these compounds, prior to fermentation.

The sugars generated by hydrolysis are primarily glucose, cellobiose (a disaccharide composed of β-1,4-linked glucose units) and xylose. Fermentation of xylose is problematic. *S. cerevisiae*, which is currently responsible for 95% of all ethanol produced by fermentation, does not ferment this monosaccharide. Those organisms that do (see Table 10.4), are not ethanol tolerant and give poor ethanol yields. There are a number of possible ways by which *S. cerevisiae* could be employed in the alcoholic fermentation of xylose (Fig. 10.4).

1 Isomerization of the aldo-sugar, D-xylose, to the keto-form D-xylulose, which *S. cerevisiae* can ferment. This may be achieved by carrying out the yeast fermentation in the presence of a bacterial xylose isomerase.

2 Genetic engineering of *S. cerevisiae*, to express genes for either:

(a) a bacterial xylose isomerase, e.g. from species of *Actinoplanes*, *Bacillus*, etc.; or

(b) xylose reductase and xylitol dehydrogenase from a pentose fermenting yeast, e.g. species of *Candida*, *Pichia*, etc. However, there are likely problems with cofactor imbalances with this option.

Z. mobilis has also been genetically engineered to ferment xylose and may play a future role in the production of ethanol from plant biomass, as may the genetically engineered *E. coli* mentioned earlier, and certain thermophilic microorganisms.

Hydrogen

Hydrogen is a very attractive fuel because of its high energy content (118.7 kJ/g), which is about four-fold greater than ethanol and over two-fold higher than methane. The technology for its use has already been developed and the product of its combustion is water. A wide range of microorganisms produce hydrogen as a part of mechanisms for disposing of electrons that are generated during metabolic reactions:

$$2e^- + \underset{\text{hydrated hydrogen ion (hydronium ion)}}{2H_3O^+} \leftrightarrow H_2 + 2H_2O$$

The generation of hydrogen using microorganisms, or cell-free systems based on microbial components, is still very much in its infancy. However, there are three possible routes of production.

1 **Biophotolysis of water** involves splitting water using light energy and does not require an exogenous sub-

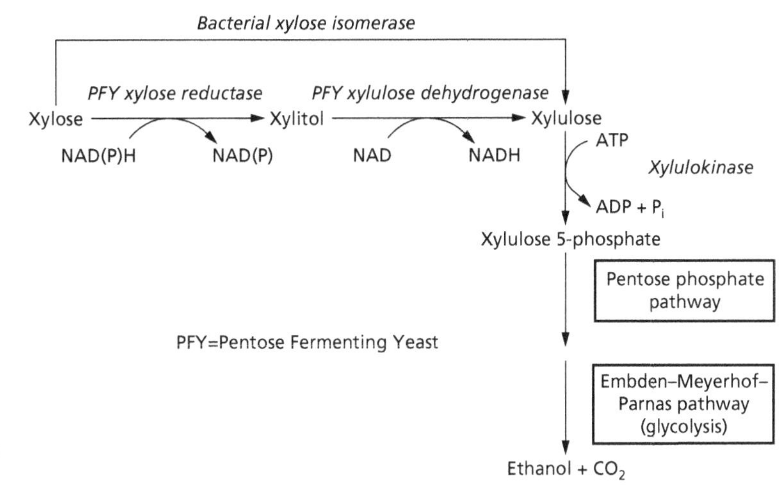

Fig. 10.4 Strategies for the fermentation of pentoses to ethanol by *Saccharomyces cerevisiae*.

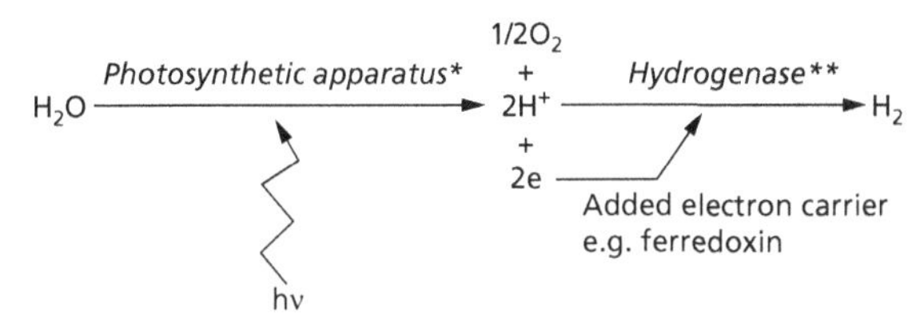

Fig. 10.5 Scheme for cell-free biophotolysis.

strate. This can be performed using photosynthetic systems, such as algal chloroplasts, which may be regarded as solar cells. *In vivo*, the energy generated is normally used to form reduced nicotinamide adenine dinucleotide phosphate (NADPH). However, in the presence of a bacterial hydrogenase and an appropriate electron carrier, molecular hydrogen can be generated (Fig. 10.5).

2 **Photoreduction**, the light-dependent decomposition of organic compounds, is performed by photosynthetic bacteria. This is an anaerobic process requiring light and an exogenous organic substrate, which is inhibited by oxygen, dinitrogen and ammonium ions. Formation of hydrogen is attributed to a nitrogenase that can reduce protons as well as dinitrogen. Members of the Chlorobiaceae, Chromatiaceae and Rhodospirillaceae carry out photoreduction. Those bacteria with most potential are probably the purple non-sulphur bacteria, such as *Rhodospirillium* species, which photometabolize organic acids.

3 **Fermentation** of organic compounds by many bacteria generates a small amount of hydrogen. For example, some enterobacteria produce hydrogen and CO_2 by cleaving formate, and in clostridia it is produced from reduced ferredoxin. Theoretically, 4 mol of hydrogen could be generated from each mole of glucose, which represents only a 33% energy yield. However, most organisms produce much less. Consequently, there is little possibility for the commercial production of hydrogen via this route in the near future.

$$C_6H_{12}O_6 + 2H_2O \rightarrow 2CH_3COO^- + 2H^+ + 2CO_2 + 4H_2$$

Electricity

The role of microorganisms in electricity generation may involve microbially produced gaseous and liquid fuels, such as ethanol or methane, being used to drive conventional mechanical generators. Alternatively, direct generation may be used, but this is still in the early stages of development. Possible routes are via intact microorganisms or microbial enzymes incorporated within fuel cells (Fig. 10.6). Enzyme-based systems are preferred, as electron transfer between whole cells and electrodes is generally less efficient. In some cases, immobilized enzymes may be used. Possible candidates are microbial dehydrogenases coupled to electrode systems and catalysing the interconversion of hydrogen and electricity. Also, there is the possibility that phototrophic microorganisms, or their photoactive systems, could directly convert sunlight to electricity. For example, using artificial membranes incorporating bacteriorhodopsin-based systems from archaeans, e.g.

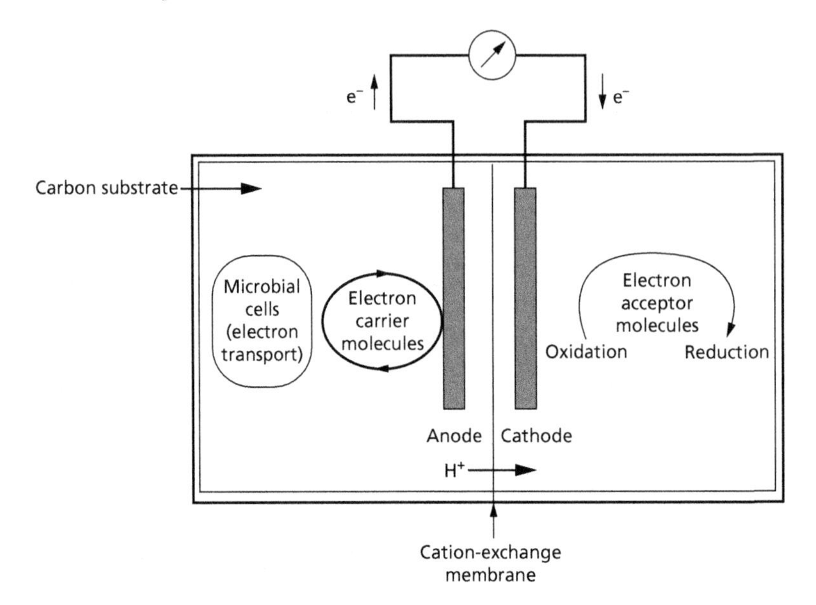

Fig. 10.6 A microbial fuel cell.

Halobacterium halobium. Such systems facilitate the light-dependent translocation of protons and the resulting transmembrane electrochemical gradient created could be used to generate electricity.

Amino acids

Several amino acids are produced in commercial quantities via direct fermentation processes using overproducing microbial strains, or by microbial biotransformation. They are mostly employed as food or animal feed supplements and flavour compounds. However, several amino acids also have uses in pharmaceuticals and cosmetics, and in the chemical industry for the manufacture of polymers.

L-Glutamic acid

Of all the amino acid production processes, that of L-glutamic acid is probably the most important in terms of quantity. Its main use is as the flavour enhancer, monosodium L-glutamate (MSG), which can heighten and intensify the flavour of foods without adding significant flavour of its own. MSG is naturally present in certain foods and was discovered to be the 'active' component of a traditional flavour-enhancing seaweed stock used in Far Eastern foods. This compound was first isolated from the seaweed, *Laminaria japonica*, in 1908. Commercial production in Japan followed almost immediately, using extracts of soya protein and wheat gluten. In 1959 the US Food and Drug Administration (FDA) classified MSG as 'generally regarded as safe' (GRAS) due to its history of safe use, and the Joint Food and Agriculture Organization (FAO)/World Health Organization (WHO) Expert Committee on Food Additives (1970) gave the acceptable daily intake as 0–120 mg/kg body weight.

Since the early 1960s, the classical production methods using plant sources have largely been replaced by fermentation processes, which are now responsible for an annual production in excess of 400 000 tonnes. The price of MSG in international trade is an average of US$1.20/kg and apart from extensive use in oriental foods, it is added to a wide range of food products, particularly soups, gravies, sauces and snack foods.

Glutamic acid-producing microorganisms include species of the closely related genera *Arthrobacter*, *Brevibacterium*, *Corynebacterium*, *Microbacterium* and *Micrococcus*. These are Gram-positive, biotin-requiring, non-motile bacteria that have intense glutamate dehydrogenase activity.

$$\text{Oxoglutaric acid} + NH_4^+ + NAD(P)H + H^+ \xrightarrow{\textit{Glutamate dehydrogenase}} \text{L-glutamic acid} + NAD(P)^+ + H_2O$$

Species of *Brevibacterium* and *Corynebacterium* are used for most industrial fermentations. The wild-type *Corynebacterium glutamicum*, for example, exhibits feedback inhibition when cellular glutamic acid concentrations rise to 5% on a dry weight basis. However, the production strains developed using mutagenesis and selection programmes are both regulatory and auxotrophic mutants. These strains have been developed with a high steady-state cytoplasmic amino acid concentration. They accumulate about 30% L-glutamic acid and yield 1 mol of glutamate per 1.4 mol of glucose, and importantly are phage resistant. More recently, recombinant DNA technology has been used to increase the activity of specific biosynthetic enzymes by transformation with multicopy plasmids bearing the structural genes for these enzymes. The overall strategy for achieving overproduction of the amino acid involves:

1 increasing the activity of anabolic enzymes;
2 manipulation of regulation to remove feedback control mechanisms;
3 blocking pathways that lead to unwanted byproducts;
4 blocking pathways that result in degradation of the target product; and
5 limiting the ability to process the immediate precursor of L-glutamic acid, namely oxoglutaric acid, to the next intermediate of the tricarboxylic acid (TCA) cycle, succinyl coenzyme A (CoA), i.e. use of mutants lacking oxoglutaric acid dehydrogenase. During the growth phase these mutants produce essential intermediates from isocitrate via the glyoxylate cycle (Fig. 10.7).

In addition, as these bacteria do not normally secrete glutamate, a range of treatments are employed to render the cells more permeable and aid release of the amino acid into the medium. These treatments include: biotin limitation, restriction of phospholipid biosynthesis by adding C_{16}–C_{18} saturated fatty acids during the growth phase, and inclusion of surfactants (e.g. Tween 40) and penicillin in the production media.

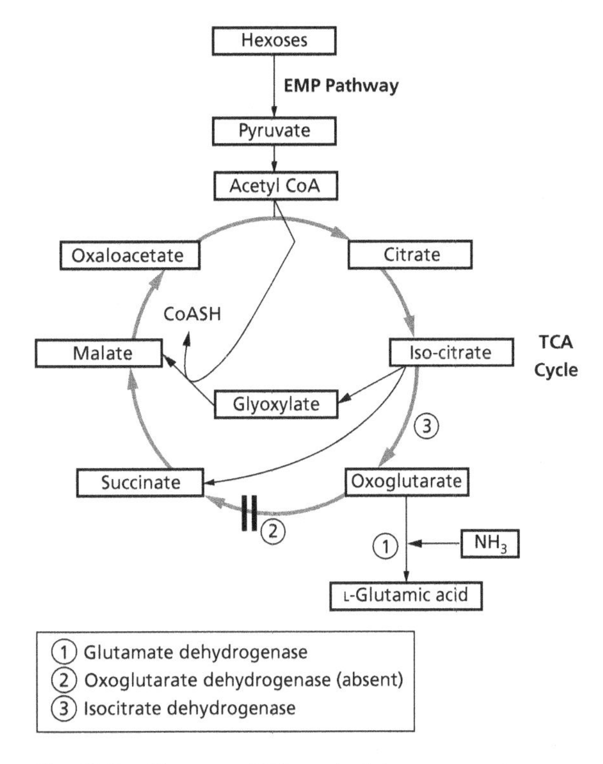

Fig. 10.7 L-Glutamic acid biosynthesis in mutant strains.

INDUSTRIAL PRODUCTION OF L-GLUTAMIC ACID

Industrial-scale fermenters are normally stainless steel stirred tank reactors of up to 450 m^3. These are batch processes, operated aerobically at 30–37°C, the specific temperature depending on the microorganisms used. Apart from carbon and nitrogen sources, the fermentation medium normally contains inorganic salts, providing magnesium, manganese, phosphate and potassium, and limiting levels of biotin. Corynebacteria are nutritionally fastidious and may also require other vitamins, amino acids, purines and pyrimidines. The preferred carbon sources are carbohydrates, preferably glucose or sucrose. Cane or beet molasses can be used, but the medium requires further modification as their biotin levels tend to be too high. This can be overcome by the addition of saturated fatty acids, penicillin or surfactants which promote excretion. The nitrogen source (ammonium salts, urea or ammonia) is fed slowly to prevent inhibition of L-glutamate production. Medium pH is maintained at 7–8 by the addition of alkali, otherwise the pH progressively falls as the L-glutamate is excreted into the medium. Accumulation of L-glutamic acid does not become apparent until midway through the fermentation, which normally lasts for 35–40 h and achieves L-glutamic acid levels in the broth of 80 g/L.

Product recovery involves separation of the cells from the culture medium. The L-glutamic acid is then crystallized from the spent medium by lowering the pH to its isoelectric point of pH 3.2 using hydrochloric acid. Crystals of L-glutamic acid are then filtered off and washed. MSG is prepared by adding a solution of sodium hydroxide to the crystalline L-glutamic acid followed by recrystallization.

L-Lysine

L-Lysine is not synthesized by humans and other mammals. This 'essential' amino acid must be acquired as part of their diet. However, many cereals and vegetables are relatively low in lysine. Consequently, food products and animal feeds derived from these sources are often supplemented with this amino acid. The annual world production of L-lysine necessary to fulfil these requirements is now in excess of 380 000 tonnes. Over 90 000 tonnes of this lysine are currently produced by direct microbial fermentation and biotransformation methods. The remaining portion is produced by chemical synthesis. However, this route has the major disadvantage that a mixture of the D- and L-isomers is synthesized, but it is only L-lysine that the body utilizes. Thus, optical resolution is required following chemical synthesis, whereas microbial production has the advantage that only the L-isomer is formed.

INDUSTRIAL PRODUCTION OF L-LYSINE

Metabolic control of L-lysine production in wild-type *C. glutamicum* is shown in Fig. 10.8. The first key step of this metabolic pathway, aspartate to aspartyl phosphate, catalysed by aspartokinase, is controlled via feedback inhibition by two end-products of this branched pathway, lysine and threonine. Homoserine dehydrogenase activity is also subject to feedback inhibition by threonine and repression by methionine. However, dihydropicolinate synthetase is not inhibited by lysine accumulation, which is unusual for the first enzymes following the branch point of a pathway.

The over-producing strains of *C. glutamicum* selected for lysine production have defects in these feedback control mechanisms. They lack homoserine dehydrogenase activity and are thus homoserine auxotrophs. These auxotrophs convert all aspartate semialdehyde to lysine, and because of the lack of threonine synthesis, there is no longer feedback control (see Fig. 10.8). However, carefully measured amounts of threonine, methionine and isoleucine must be added to the culture medium to enable this auxotrophic bacterium to grow.

Most commercial L-lysine fermentations are operated as batch processes in aerated stirred tank reactors. Cane molasses is the preferred carbon source, although other carbohydrates, acetic acid or ethanol can be used, often supplemented with soya bean hydrolysates. The temperature is held at 28°C and the pH is maintained at, or near, neutrality by feeding ammonia or urea, which also act as a nitrogen source. Control of the biotin level is very important, as concentrations below 30 μg/L result in the accumulation of L-glutamate instead of L-lysine (see L-glutamic acid production above). However, cane molasses usually contains sufficient biotin to fulfil this requirement.

The lag phase is shortened by using a high concentration of inoculum, normally about 10% (v/v) of the fermentation volume. Production of lysine starts in the early exponential phase and continues through to the stationary phase. These fermentations last about 60 h and yield 40–45 g/L L-lysine from a molasses concentration of 200 g/L, containing 100 g/L sucrose.

Lysine recovery is relatively simple. Once the cells have been removed, the fermentation medium is acidified to pH 2.0 with hydrochloric acid and the L-lysine is

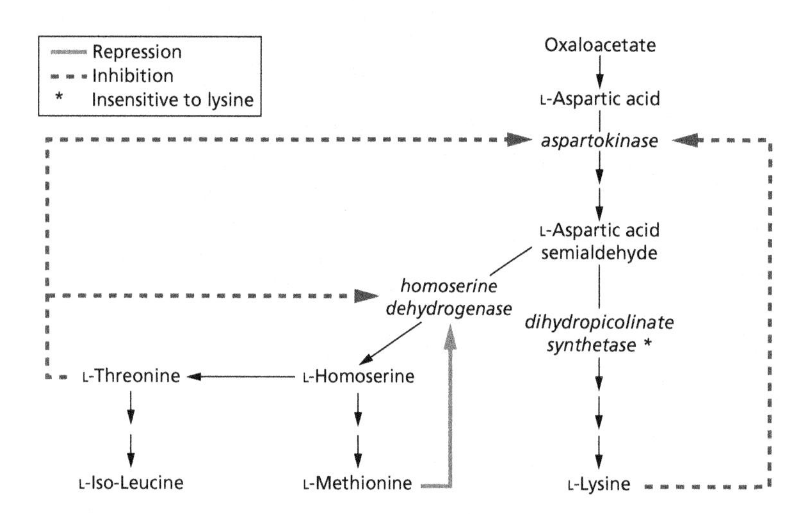

Fig. 10.8 Control of L-lysine production in *Corynebacterium glutamicum*.

adsorbed onto a cation-exchange column in the ammonium form. A dilute solution of ammonia is then used to elute L-lysine from the column. This eluate is reacidified and the product is finally crystallized as L-lysine hydrochloride.

ALTERNATIVE BIOTRANSFORMATION METHODS FOR THE PRODUCTION OF AMINO ACIDS

Production of L-amino acids may also be performed by the enantioselective hydrolysis of racemic precursors. L-Glutamic acid, for example, can be produced from chemically synthesized D-L-hydantoin 5-propionic acid. This process uses *Bacillus brevis*, which produces the necessary hydantoinase and gives a 90% yield (Fig. 10.9a).

L-Lysine can also be produced via a batch biotransformation process from D-L-α-aminocaprolactam, an inexpensive starting material chemically derived from cyclohexane. D-L-α-Aminocaprolactam is added to a reaction vessel at a concentration of 100 g/L, along with the yeast *Cryptococcus laurentii* and the bacterium *Achromobacter obae*. The result is an almost complete conversion of the substrate to L-lysine. This method exploits the stereospecificity of a hydrolase found in *C. laurenti* to convert the L-isomers of the substrate to L-lysine. The remaining D-isomer of α-aminocaprolactam is brought back into the production pathway by a racemase in *A. obae* (Fig. 10.9b). Similar methods are also available for the synthesis of D-amino acids, some of which are important side-chain precursors for semisynthetic penicillins and cephalosporins, e.g. D-*p*-hydroxyphenylglycine.

Organic acids

Acetic acid

See Chapter 12, Vinegar production.

Citric acid

Citric acid is widely used in the food industry as an acidulant and flavouring agent in beverages, confectionery and other foods, and in leavening systems for baked goods. As a food constituent, its use is unrestricted because it has GRAS status. This organic acid also has many non-food applications. They include roles in maintaining metals in solution for electroplating, as a cleaning and 'pickling' agent for metals, and as

(a) L-Glutamic acid

Starting material: a racemic mixture of D- and L-hydantoin 5-propionic acid

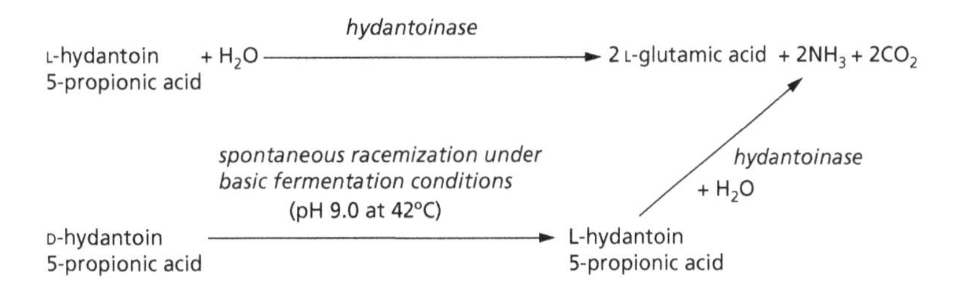

(b) L-Lysine

Starting material: a racemic mixture of D- and L-α-aminocaprolactam

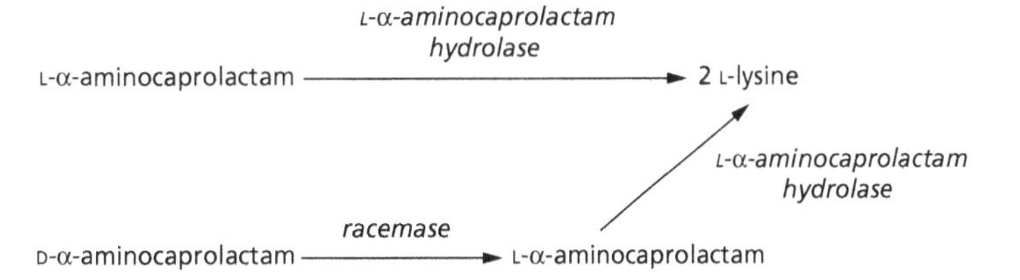

Fig. 10.9 Production of L-amino acids by biotransformation.

a replacement for polyphosphates in the detergent industry, along with several pharmaceutical uses.

Until the 1920s citric acid was mainly prepared from lemon juice, but in 1923 Pfizer began operating a fermentation-based process in the USA. The production organism was the filamentous fungus *Aspergillus niger*, an obligate aerobe, which was grown in surface culture on a medium of sucrose and mineral salts. Virtually all the worldwide output is now produced by fermentation, which is primarily located in Western Europe, the USA and China. Citric acid has become one of the world's major fermentation products, with an annual production of over 550 000 tonnes and a value approaching US$800 million. The demand for citric acid is still increasing, particularly for beverage applications.

Surface methods are still operated, but since the late 1940s, submerged fermentations have become the principal mode of production. Many microorganisms, including filamentous fungi, yeasts and bacteria, can be used to produce this primary metabolite. However, *A. niger* still remains as the predominant industrial producer. Specific strains have been developed for various types of fermentation processes, which are capable of generating high yields of citric acid, often in excess of 70% of the theoretical yield from the carbon source.

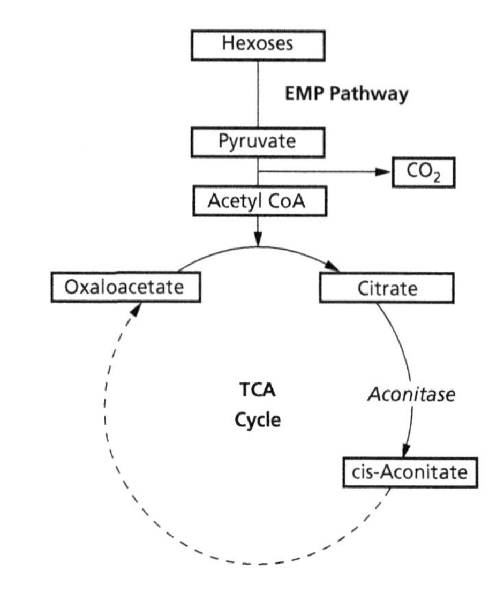

Fig. 10.10 Citric acid biosynthesis.

CITRIC ACID BIOSYNTHESIS

The metabolic pathways involved in citric acid biosynthesis are the Embden–Meyerhof–Parnas (EMP) pathway and the TCA cycle. *A. niger* also operates the pentose phosphate pathway, which can compete with glycolysis for carbon units.

The first stages of citric acid formation involve the breakdown of hexoses to pyruvate in glycolysis, followed by its decarboxylation to produce acetyl CoA (Fig. 10.10). Very importantly, the CO_2 released during this reaction is not lost, but is recycled by pyruvate carboxylase (produced constitutively in *Aspergillus*) in the anaplerotic formation of oxaloacetate (other anaplerotic routes to oxaloxacetate are also operated; see Chapter 3). Normally, oxaloacetate would largely be supplied through the completion of the TCA cycle, allowing recommencement of the cycle by condensing with acetyl CoA to form citrate, catalysed by citrate synthase. However, in order to accumulate citrate, its onward metabolism (continuation of the cycle) must be blocked. This is achieved by inhibiting aconitase, the enzyme catalysing the next step in the TCA cycle. Inhibition is accomplished by removal of iron, an activator of aconitase. Consequently, during citrate accumulation, the TCA cycle is largely inoperative beyond citrate formation, hence the importance of the anaplerotic routes of oxaloacetate formation.

Conventional strain improvement methods and the genetic engineering of elements of primary metabolism in *A. niger* are being employed in an effort to improve citric acid production. The aim is to increase metabolic flux leading directly to citric acid formation by decreasing fluxes through branches of this pathway, thus producing fewer byproducts, particularly gluconic acid and oxalic acid. Utilization of mutants lacking glucose oxidase, and consequently unable to produce gluconic acid from glucose, are examples of such an approach. Alternatively, a direct increase of the flux through the main pathway may be achieved by overproduction of the constituent enzymes.

FERMENTATION PROCESSES USED IN CITRIC ACID PRODUCTION

Surface and solid-substrate fermentations

These methods use simple technology and have low energy cost, but are more labour intensive. Liquid surface methods involve placing the sterilized medium, usually containing molasses plus various salts, into shallow (5–20 cm deep) aluminium or stainless steel trays stacked in an aseptic room. The medium is formulated with relatively low levels of iron, otherwise the citric acid yield is reduced (see above). The trays are inoculated by spraying with *A. niger* spores, either a spore suspension or dry spores. The fungus then develops on the surface of the medium. Sterile air is blown over these cultures, which is important for maintaining aerobic conditions, temperature control and in lowering the CO_2 level. Medium pH gradually falls to below 2, at which point citric acid production begins. At 30°C, the fermentation takes about 8–12 days to complete and achieves a productivity of around 1.0 kg/m^3 per day.

Solid-state fermentation processes for citric acid production are small-scale operations. Each plant generates only a few hundred tonnes per annum, and uses a solid medium of steam-sterilized wheat bran or sweet potato waste that has a 70–80% moisture content. This mash is inoculated with spores of *A. niger* and then spread on trays or a clean floor to a depth of 3–5 cm. Air circulation helps to maintain the temperature at about 28°C. This process runs for 5–8 days, after which the mash is collected and the citrate is extracted using hot water.

Various solid food processing residues are being evaluated to determine whether they too could serve as low-cost substrates for citric acid production. In addition, technological developments are being sought, such as the use of packed bed reactors. Preliminary trials have produced high levels of citric acid with low levels of fungal biomass, as these reactors retard fungal growth and promote greater substrate conversion to citric acid.

Submerged processes

More than 80% of the worldwide supply of citric acid is produced using submerged batch fermentation in stirred tanks of 40–200 m^3 capacity or larger airlift fermenters of 200–900 m^3 capacity. The fermenters are corrosion-resistant, made of stainless steel, or steel lined with special glass or plastic.

These fermentations mostly use beet or cane molasses as the carbon source. Unlike surface methods, vegetative inocula, rather than spores, are normally used. Consequently, the culture organism is taken through several propagation stages in order to generate a sufficient quantity of inoculum. Initially, spores of the production strain of *A. niger* are produced on solid medium and then used to inoculate a small-scale submerged fermentation where fungal pellet formation takes place. Quantities of stable fungal pellets are then developed for the inoculation of the production fermenter.

The structure of these pellets has a major influence on productivity. Small pellets of less than 1 mm diameter, with fluffy centres and smooth surfaces are preferred. These structural properties and their physiology are strongly dependent on medium composition and operating conditions. Pellets that produce high levels of citric acid are characterized by short forked, bulbous hyphae. The presence of even low levels of some heavy metals, particularly manganese, can be detrimental to pellet formation, resulting in hyphae that are long and unbranched. Thus, it is necessary to pretreat all raw materials to reduce manganese concentrations to below 0.02 mmol/L. Low manganese levels also limit the operation of the pentose phosphate pathway, which would otherwise divert flux away from glycolysis and reduce citrate production. Alternatively, copper ions may be added to counteract the manganese, by preventing its uptake. Citric acid yields are also improved by formulating the medium with minimum levels of iron. This reduces onward metabolism of citrate because, as mentioned previously, aconitase has a requirement for iron. Also, the addition of copper further diminishes aconitase activity, as it acts as an antagonist to iron, as well as manganese.

In order to maintain good citric acid yields, media sugar concentrations must be at least 140 g/L, which promotes the activity of both glycolytic enzymes and pyruvate carboxylase. It is also important to restrict growth through nitrogen limitation. This is normally accomplished by providing ammonium salts at levels of 0.1–0.4 g/L. The ammonium ions also stimulate citric acid production by counteracting the inhibitory effect of citrate on phosphofructokinase, a key enzyme of glycolysis. These fermentations are highly aerated and maintained at 30°C. For the initial growth phase, the pH starts at 5–7, but must then be kept below 2, otherwise oxalic and gluconic acids accumulate at the expense of citric acid, i.e. low pH inhibits glucose oxidase. Overall yields of 0.7–0.9 g citrate per gram glucose can be attained in these submerged fermentations with productivities of up to 18.0 kg/m^3 per day.

Smaller volumes of citric acid are also produced using yeasts such as *Candida guilliermondii* and *Yarrowia* (formerly *Candida*) *lipolytica*. These yeasts are free from problems with metal ions, and provide shorter and more productive fermentations than those currently available with *A. niger*.

CITRIC ACID RECOVERY

Recovery of citric acid commences with the removal of fungal mycelium from the culture medium. Further polishing filtration may be necessary to remove residual mycelia and precipitated oxalate. The resulting clarified solution is heated and lime (CaO) is added to form a precipitate of calcium citrate. This is separated by filtration and treated with sulphuric acid to generate citric acid and a precipitate of calcium sulphate (gypsum). Following filtration, the dilute citric acid solution is decolorized with activated carbon and evaporated to produce crystals of citric acid. These crystals are recovered by centrifugation, then dried and packaged. Alternate recovery methods being evaluated in order to avoid the use of lime and sulphuric acid include solvent extraction, ion-pair extraction and electrodialysis.

Gluconic acid

Calcium gluconate and ferrous gluconate are widely used as therapeutic agents to treat patients with calcium and iron deficiency. The free acid is also used as a mild acidulant in the tanning industry. Over 50 000 tonnes of gluconic acid are produced each year using *A. niger* grown in submerged fermentations on glucose and corn steep liquor, under both phosphate and nitrogen limitation. These highly aerobic fermentations are performed at pH 6–7 and 30°C. They last for 20 h and achieve yields of over 90%.

Itaconic acid

This dicarboxylic acid is used in the manufacture of adhesives, paper products and textiles. It is also incorporated into plastics as a copolymer with acrylic acid, methyl acrylate and styrene (Fig. 10.11). Itaconic acid is produced commercially by submerged culture of *Aspergillus terreus* or *A. itaconicus*, often using molasses and corn steep liquor, with product yields of up to 65%. The 3-day fermentation must be highly aerated and operated at a relatively high temperature of 35–42°C. Itaconic acid is formed in a branch of the TCA cycle via decarboxylation of cis-aconitate (Fig. 10.12), which is normally followed by its oxidation to itatartaric acid. Onward metabolism of itaconic acid must be prevented in commercial fermentations, otherwise yield is reduced. This is achieved by formulating the medium with high levels of calcium ions, thereby inhibiting itaconic acid oxidase, which catalyses the oxidation of itaconic acid to itatartaric acid.

Lactic acid

Lactic acid is primarily used in the food industry, where 30 000 tonnes are incorporated into food each year to act as a preservative, an acidulant, or in the preparation of dough conditioners. Its salts are also used in other industries, for example, antimony lactate is used as a mordant in dyeing and sodium lactate has applications as a plasticizer and corrosion inhibitor. Lactic acid is produced in 20 000–100 000 L anaerobic fermentations using *Lactobacillus delbruckii* or other homolactic

CH_2
||
C — COOH
|
CH_2 — COOH

Fig. 10.11 Itaconic acid.

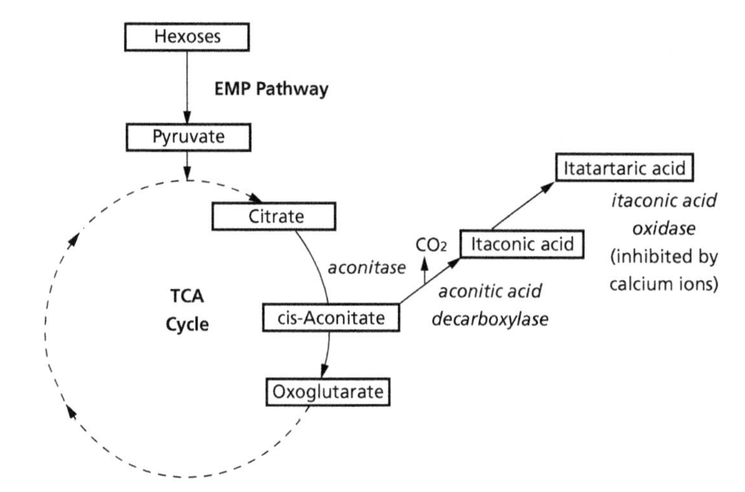

Fig. 10.12 Itaconic acid biosynthesis.

bacteria such as *L. bulgaricus*. The media normally contain a complex nitrogen source and vitamin supplements, along with up to 12% (w/v) sucrose or glucose as a carbon and energy source. Alternatively, lactose may be used, in the form of whey permeate. These carbohydrates are metabolized to pyruvate via the EMP pathway, which is then converted to L (+) lactate by L-lactate dehydrogenase (see Chapter 3). Lactic acid fermentations are operated at 45–60°C with a pH of 5–6. They last for 4–6 days and can achieve yields of over 90% based on sugar supplied.

Polyhydroxyalkanoates

Polyhydroxyalkanoates (PHAs) have considerable potential as biodegradable alternatives to petroleum-derived plastics. PHAs are linear homochiral thermoplastic polyesters produced as intracellular energy reserves by numerous microorganisms. These biopolymers accumulate as distinct 0.2–0.7 μm diameter granular inclusion bodies in response to nutrient limitation, especially in pseudomonads. The most widely encountered PHAs are poly β-hydroxybutyrate (PHB) and poly lactic acid (polyhydroxypropionate), formed from the monomers hydroxybutyric acid and lactic acid, respectively.

PHAs are produced by a distinct metabolic pathway that is divided into two stages. First is the biosynthesis of the hydroxyacyl CoA monomers, followed by their head-to-tail polymerization to form the polymer chains, which can exceed 10 000 units in length. The most fully characterized pathway is that for PHB biosynthesis in *Ralstonia eutropha* (formerly *Alcaligenes eutrophus*). This involves three enzymes: thiolase catalyses a Claisen condensation of two molecules of acetyl CoA to form acetoacetyl CoA, which is reduced to the chiral intermediate R-3-hydroxybutyryl CoA by a reductase. Polymerization is then performed by a PHA synthase (polymerase).

PHBs are the most useful of the microbially derived plastics. These biocompatible polymers are renewable resources that can be completely and quite rapidly biodegraded to carbon dioxide and water, thus providing certain advantages over conventional petroleum-based plastics. When copolymerized with polyhydroxyvalerate, as PHBV, the product has an even faster degradation time than the homo-polymer. PHBV is produced by Monsanto under the trade name Biopol. In many respects, PHB resembles polypropylene; both have a similar molecular mass, melting point, crystallinity and tensile strength, but PHB lacks the impact strength of polypropylene. Major obstacles to widespread use are their higher cost and the fact that they may become brittle with time. Currently, PHBs are used in biomedical and packaging applications, particularly for sutures that are slowly degraded by the body's enzymes, food storage materials and shampoo bottles. Derivatives of polylactic acid have also been used in medicine, as templates for tissue growth and in plastics for replacement joints.

R. eutropha is employed for the commercial production of PHBs as the polymer can constitute up to 90% of cell dry weight. In order to obtain maximum product yield with respect to the carbon source, industrial fermentations have a growth phase preceding the product formation stage. The latter operates under low oxygen concentration and with limited nitrogen, phosphate, magnesium or sulphate.

Currently, PHB commands a relatively high price of US$15–30/kg. Consequently, cheaper means of production are being sought. PHAs can now be synthesized by recombinant microorganisms, e.g. *E. coli*, which contain the genes encoding the enzymes necessary for PHA biosynthesis. Such microbial recombinants may become an economically attractive source of PHAs. Alternatively, transformation of a higher plant with these genes could provide an even cheaper means of PHA production in the longer term. Likely transgenic hosts include *Arabidopsis thaliana* (thale cress), *Brassica napus* (canola/oil-seed rape) or *Zea mays* (maize).

Polyhydric alcohols

Yeasts produce several polyhydric alcohols, including glycerol, arabitol, erythritol, mannitol and xylitol. Xylitol is becoming increasingly used as a low-calorie sweetener and is particularly useful for sweetening food products for diabetics. It can be produced by the pentose fermenting yeasts, *Candida* species, *Pachysolen tannophilus* and *Pichia stipitis* (see p. 147, Industrial ethanol production).

Glycerol has numerous medical, food and industrial applications as a plasticizer, solvent and sweetener. Probably the most commercially important is its role as a raw material for explosives manufacture. Microbial glycerol production was first noted by Louis Pasteur, who discovered that in wine and beer fermentations, yeasts form about 2.5 g for each 100 g of sugars fermented. Thus, glycerol is normally only a minor fermentation product of any yeast alcoholic fermentation.

At the beginning of the 20th century, Neuburg discovered that glycerol accumulation could be greatly enhanced by fixing the acetaldehyde formed during fermentation by adding bisulphite. This suppresses the reduction of acetaldehyde to ethanol by alcohol dehydrogenase, which is the last step in the alcoholic fermentation pathway of yeasts and normally functions to reoxidize NADH. As a result, the yeast is 'forced' to regenerate NAD via an alternate route, otherwise the EMP pathway stops. An alternate route for NADH oxidation is through the reduction of dihydroxyacetone phosphate (DAP), an earlier product of the EMP pathway. DAP is reduced to glycerol 3-phosphate and then on to glycerol.

$$\text{DAP} \xrightarrow[\ ]{\text{NADH} + \text{H}^+ \rightarrow \text{NAD}^+} \begin{array}{c}\text{glycerol}\\ \text{3-phosphate}\end{array} \longrightarrow \text{glycerol} + \text{P}_i$$

As a result, a 'steered' industrial fermentation process was developed, where 4% (w/v) sodium bisulphite was incorporated into a fermentation medium composed of 10% (w/v) sucrose, 0.5% (w/v) ammonium nitrate and 0.075% (w/v) potassium phosphate. The medium was inoculated with 1% (v/v) *S. cerevisiae* and maintained at 30°C for 48–60 h to give yields of 20–25% (v/v) glycerol. These fermentation methods were used extensively up to the mid-1940s, but currently chemical synthesis is usually preferred.

Microbial exopolysaccharides

A wide range of microorganisms produce exopolysaccharides in the form of discrete capsules or as soluble slimes located outside the cell. These are either homopolymers or heteropolymers and have several functions. They may protect the microorganism against desiccation, help in the evasion of the immune system for animal pathogens, act as a barrier to viruses and chemical agents, aid attachment to surfaces, and provide carbon and energy reserves. Their forecast commercial potential is now being realized. Microbial exopolysaccharides are beginning to replace traditional higher plant and algal polysaccharides (starch, alginate, carrageenan, gum arabic, locust bean gum, guar gum, etc.) as thickeners and stabilizers in numerous food and non-food applications. This is due to their increasing availability, varied and novel properties, ease of production and cost effectiveness. The range of cell wall and exopolysaccharides includes the following.

1 **Alginates**, which are linear heteropolymers of L-guluronic acid and D-mannuronic acid, some units of which contain *O*-acetyl groups. These polymers are formed by *Pseudomonas* species and *Azotobacter vinlandii*. They may be used as sizing agents in the paper and textile industries, or as food stabilizers.

2 **Cellulose**, a β-1,4 glucan, formed as a pellicle by strains of the acetic acid bacterium *Acetobacter xylinum*. This material can be produced in surface or submerged culture and has several potential uses. These include application as a food ingredient, as temporary artificial skin following surgery or skin burns, and for acoustic membranes.

3 **Chitin**, a polymer of *N*-acetylglucosamine residues, and its deacylated derivative, **chitosan**, are cell wall components of fungi. Commercial preparation of these polymers is currently from shellfish wastes, but in the future they may be more readily purified from fungal cell walls. These polymers can be made into fibres to make wound dressings and also have uses as chelating agents, clarifying agents and food preservatives.

4 **Curdlan**, a β-1,3 glucan from species of *Alcaligenes* and *Agrobacterium*. This polysaccharide is capable of forming a hard irreversible gel when heated in an aqueous suspension over a wide pH range of 2.0–9.5. Curdlan is manufactured in Japan where it is used in various food preparations. However, it is not currently accepted as a food ingredient in either the USA or the EU.

5 **Dextrans** are short branched glucans containing α-1,6 glycosidic linkages and α-1,3 branch points. They are produced by several microorganisms, including *Leuconostoc mesenteroides*, and are used as blood plasma supplements and adsorbants.

6 **Gellan** gum (E418) is a heteropolymer containing glucose, rhamnose and glucuronic acid in a ratio of 2 : 1 : 1, produced by *Sphingomonas paucimobilis* (formerly *Pseudomonas elodea*) in aerobic fermentations. This polysaccharide is able to form gels that exhibit different properties depending upon whether it is in substituted or unsubstituted form. The native polysaccharide forms elastic gels, whereas that deacylated by alkali treatment produces brittle gels and has separate applications. Gellan is mostly used as a replacement for the algal polymers agar and carrageenan, particularly in food applications.

7 **Glycans** and **phosphomannans** are both components of yeast cell walls. Glycans from *S. cerevisiae* have several uses in food, pharmaceuticals and cosmetics. Specific food applications include roles as thickeners, fat replacers and as animal feed supplements. Yeast glycan may also be used in skin repair cosmetics, cholesterol reduction treatments, wound healing, vaccine adjuvants,

and as an immunostimulant in animal and human health. Phosphomannans are water-soluble gums that may be obtained from *Hansenula* and *Pichia*. They exhibit several interesting properties and are resistant to microbial attack.

8 **Pullulan**, a linear α-1,4 glucan with α-1,6 linkages every third or fourth glucose unit, is produced by the yeast-like fungus *Aurobasidium pullulans*. This material has film-forming and adhesive properties that are utilized in the production of film-wrap for foods.

9 **Scleroglucan** is a β-1,3 glucan, with occasional β-1,6 branch points, produced by fungi such as *Sclerotium glucanicum*. It exhibits pseudoplasticity and is used in paints, inks and drilling muds.

Several other gelling polysaccharides, exhibiting novel characteristics, have been isolated from various microorganisms. They include polysaccharides that gel in association with monovalent or divalent cations, such as *Enterobacter* XM6 gel, beijeran from *Azotobacter beijerinckia*, a polymer from a mutant of *Rhizobium meliloti* and heteropolymer S-53 from *Klebsiella pneumoniae*.

Modification of microbial polysaccharides, to alter their functionality, may further increase their range of applications. This may be achieved by chemical and enzyme treatment of the polysaccharide, or through genetic engineering of the producer organism.

Xanthan gum

By far the most commercially successful example of a microbial exopolysaccharide is xanthan gum, which is produced by *Xanthomonas* species, e.g. *X. campestris*, *X. carotae*, *X. malvacearum* and *X. phaseoli*. These bacteria are small, motile, aerobic Gram-negative rods that produce yellow pigments. Many are phytopathogens, including *X. campestris*, the species used for commercial production of xanthan, which causes diseases of cauliflower, cabbages and rutabagas.

Much of the original work on xanthan was carried out at the US Department of Agriculture Northern Regional Research Laboratory in the late 1950s and commercial production was started in 1961 by Kelco. Approval for food use was given by the FDA in 1969 and the polysaccharide now has GRAS status. In the EU, xanthan is classified as thickener E415.

Xanthan is a high molecular weight helical heteropolymer of $1.0–2.0 \times 10^6$ Da, composed of D-glucose, D-mannose, D-glucuronic acid (in a molar ratio of 2:2:1, respectively). D-Glucose units are β-1,4-linked and form the backbone of the molecule, which is similar to cellulose. The polymer branches regularly as alternate glucose units of the backbone are linked to a trisaccharide side chain, consisting of α-D-mannose, β-D-glucuronic acid and another α-D-mannose in the terminal position (Fig. 10.13). However, there may be variations in the substituents of these side chains, which can affect various properties of the polymer. Pyruvate may be present on the terminal mannose unit and the internal mannose may be *O*-acetylated. Commercial xanthans have a degree of substitution of 30–40% for pyruvate and 60–70% for acetate. The differences

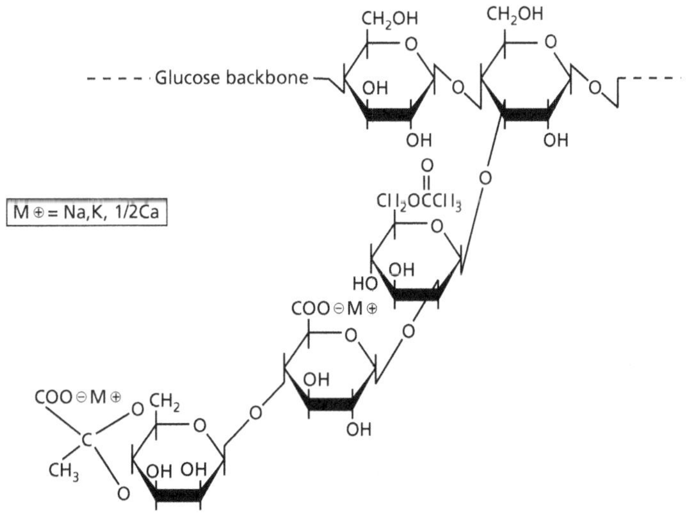

Fig. 10.13 The pentasaccharide repeating unit of xanthan gum.

depend on the strain of X. *campestris* used for production and growth media composition. Gums with abbreviated side chains, formed by mutants of X. *campestris*, exhibit very different physical properties from those produced by the wild-type bacterium.

Diverse industrial applications are based on the ability of xanthan gum to dissolve in hot or cold water and yield high viscosity, even at concentrations as low as 0.05% (w/v). Solutions of xanthan have higher viscosity than other gums at the same concentration. At a polymer concentration of 1% (w/v) in 1% (w/v) potassium chloride solution, the viscosity values for xanthan gum, guar gum, carboxymethyl cellulose and alginate are 11 300, 4000, 410 and 210 mPa s, respectively. Additional key characteristics include: translucence of xanthan solutions; compatibility with acids, bases and salts; stability at ambient temperature; and pseudoplastic rheological behaviour, i.e. xanthan solutions regain viscosity after shearing. Xanthan also interacts synergistically with other polymers. For example, it can form thermoreversible gels in combination with galactomannans or glucomannans, whereas neither component will gel alone.

Approximately 60% of the xanthan produced is used in non-food applications. These include use as a stabilizer for paint emulsions, a carrier for fertilizers and herbicides, a thickener for textile dyes, a drilling lubricant and for tertiary recovery in the oil industry, and in clay coatings for high-quality paper. Xanthan also aids the flow of pastes, e.g. facilitates toothpaste flow from containers, but recovers viscosity after the removal of the shear force.

Food applications involve roles as a thickener, an adhesive, a binder in films and coatings, an emulsifying agent and a stabilizer (Table 10.5). Xanthan can also aid rapid flavour release and provides good 'mouthfeel' characteristics. Currently, xanthan has almost a quarter of the American market for food thickeners. However, widespread use of xanthan is somewhat restricted due to its relatively high cost of US$20–25/kg, when compared with starch or some synthetic polymers. Nevertheless, its price is similar to that of other gums with comparable functionalities.

XANTHAN PRODUCTION

Approximately 20 000 tonnes of xanthan are produced each year. The production is influenced by several factors, such as the type of reactor used, mode of operation, medium composition and operational conditions. Oxygen supply, normally 1 volume of air per reactor volume per minute (vvm), is particularly important, but its maintenance is not straightforward. As xanthan is synthesized during the fermentation, the viscosity of the

Table 10.5 The role of xanthan gum in food

Function	Application
Adhesive	Icing and glazes
Binding agent	Pet foods
Coating	Confectionery
Emulsifying agent	Salad dressings
Encapsulation	Powdered flavours
Film formation	Protective coating, sausage casings
Foam stabilizer	Beer
Stabilizer	Ice cream, salad dressings
Swelling agent	Processed meat products
Syneresis inhibitor	Cheeses, frozen foods
Thickening agent	Jams, sauces, syrups and pie fillings
Synergistic xanthan–galactomannans mixtures	
Gel formation and gel stabilization	Ice cream
	Cheese and cream cheese
	Dessert gels
	Milk shakes and milk drinks
	Puddings and pie fillings
Viscosity control	Ice cream
	Instant soups
	Chocolate drinks
	Milk shakes

medium increases, which impedes mixing and leads to reduced oxygen transfer rates. Consequently, fermenter design, agitation speed and the air flow rate are key factors. Fed-batch fermentation systems are mostly used, but xanthan can also be produced in continuous processes, often under nitrogen limitation with dilution rates of 0.025–0.05/h. This offers the advantage of high yields and lower operating costs. Standardization of physiological conditions also results in a more uniform product. However, continuous operations can suffer from aeration and microbial contamination problems.

X. campestris can utilize several carbon sources, including starch, starch hydrolysates, corn syrup, molasses, glucose and sucrose. Other acceptable and cheaper substrates are whey, cereal grain hydrolysates and dry milled corn starch. The characteristics of the xanthan produced, particularly their molecular weight and rheological properties, are influenced by the composition of the substrates employed.

Initially, the bacterium is grown in a rich propagation medium to build up the inoculum within a pilot-scale fermenter. This culture in then used to inoculate mechanically agitated industrial-scale fermenters of 50–200 m^3 capacity. The production media normally contain:

1 a carbon source, commonly D-glucose, sucrose, starch or hydrolysed starch at 30–40 g/L;
2 a nitrogen source: casein or soya bean hydrolysate, ammonium salts, peptone, corn steep liquor, yeast extract or urea. The best product yields are attained with a carbon : nitrogen ratio of about 10 : 1;
3 $MgCl_2$ and other trace salts; and
4 K_2HPO_4 as a buffer.

In fed-batch mode, the fermentation is usually maintained at 28–30°C and pH 7.0; if the pH is allowed to fall gum production decreases rapidly. The bacterium starts producing xanthan during the exponential phase at rates in relation to the growth rate and production continues into the stationary phase. These fermentations are normally completed within 3 days.

A final concentration of 25 g/L is the minimum normally required for the process to be economically viable, but most industrial fermentations achieve up to 50 g/L. At the end of the fermentation, the broth is heated to 100–110°C for 10 min to kill the bacteria and improve the rheological properties of the xanthan. This is followed by a series of purification steps that are defined by the final use of the polymer (Fig. 10.14). For some applications it is necessary to remove cells by filtration or centrifugation. The xanthan is then precipitated with methanol, or isopropanol (especially when preparing food-grade product), and then separated by centrifugation. More than 50% of the production costs are incurred by these downstream processing steps and it is essential that the solvent is recovered. The product is dewatered, drum- or spray-dried, milled, sieved and finally packaged as a dispersible granulated powder.

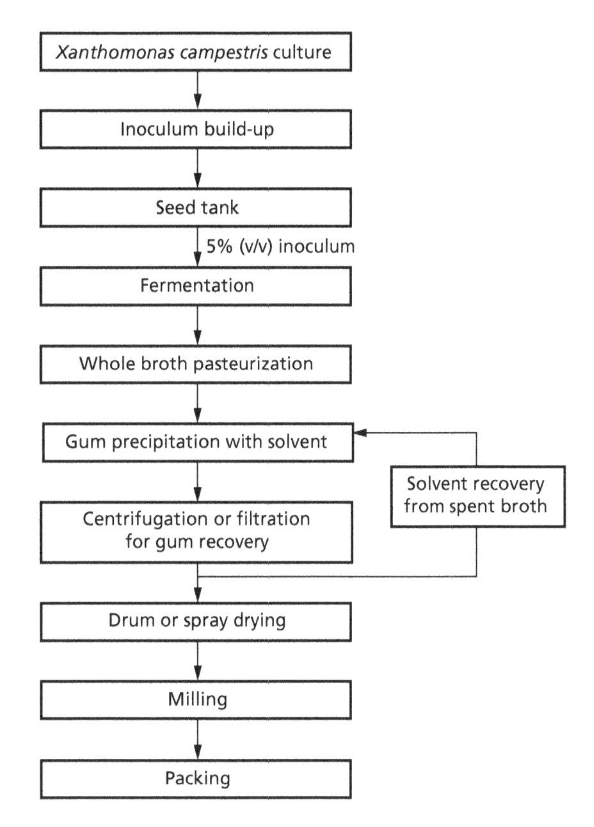

Fig. 10.14 Xanthan gum production.

Bioemulsans

Bioemulsans are amphipathic proteins and polysaccharides possessing both hydrophilic and hydrophobic properties within the same molecule. Such compounds are able to stabilize oil-in-water emulsions. They are exopolymers produced by a wide range of microorganisms, towards the end of their growth phase. However, their roles *in vivo* have not been fully elucidated. Examples include RAG-1 emulsan produced by the oil-degrading bacterium *Acinetobacter calcoaceticus*. This bioemulsan, unlike most others, contains hydrophobic long chain fatty acids covalently linked to an anionic heteropolysaccharide.

Bioemulsans have potential applications in many

industries, including the manufacture of food, paint, textiles, cosmetics and pharmaceuticals. A polysaccharide bioemulsan from *Candida utilis*, a food-grade yeast, has obvious promise for roles in food manufacture. These polymers have several advantages over the low molecular weight synthetic emulsifiers currently used in industry, as they form very stable emulsions and are biodegradable, but are relatively expensive. Nevertheless, it is predicted that they will become increasingly used as process and strain improvements lead to greater yields that should considerably lower their production costs.

Further reading

Papers and reviews

Amore, R., Kötter, P., Küster, Ciriacy, M. & Hollenberg, C. P. (1991) Cloning and expression in *Saccharomyces cerevisiae* of the NAD(P)H-dependent xylose reductase encoding gene (XYL1) from the xylose-assimilating yeast, *Pichia stipitis*. *Gene* **109**, 89–97.

Anderson, A. J. & Dawes, E. A. (1992) Occurrence, metabolism, metabolic roles, and industrial uses of bacterial polyalkanoates. *Microbiological Reviews* **54**, 450–472.

Awang, G. M., Jones, G. A. & Ingledew, W. M. (1988) The acetone–butanol fermentation. *Critical Reviews in Microbiology* **15** (Supplement 1), S33–S67.

Dahn, K. M., Davis, B. P., Pittman, P. E., Kenealy, W. R. & Jeffries, T. W. (1996) Increased xylose reductase activity in the xylose fermenting yeast *Pichia stipitis* by overexpression of XYL1. *Applied Biochemistry and Biotechnology* **57/58**, 267–276.

Daniell, H. & Guda, C. (1997) Biopolymer production in microorganisms and plants. *Chemistry and Industry*, 21 July 1997 (no. 14), 555.

Jeffries, T. W. & Kurtzman, C. P. (1994). Strain selection, taxonomy and genetics of xylose-fermenting yeasts. *Enzyme and Microbial Technology* **16**, 922–932.

Kötter, P., Amore, R., Hollenberg, C. P. & Ciriacy, M. (1990) Isolation and characterization of the *Pichia stipitis* xylitol dehydrogenase gene, XYL3, and construction of a xylose-utilizing *Saccharomyces cerevisiae* transformant. *Current Genetics* **18**, 493–500.

Kötter, P. & Ciriacy, M. (1993) Xylose fermentation by *Saccharomyces cerevisiae. Applied Microbiology and Biotechnology* **38**, 776–783.

Ladisch, M. R. (1991) Fermentation-derived butanol and scenarios for its uses in energy-related applications. *Enzyme and Microbial Technology* **13**, 280–283.

Lovitt, R. W., Kim, B. H., Shen, G.-J. & Zeikus, J. G. (1988) Solvent production by microorganisms. *CRC Critical Reviews in Biotechnology* **7**, 107–186.

Lynd, L. R. (1996) Overview and evaluation of fuel ethanol from cellulosic biomass: technology, economics, the environment and policy. *Annual Review of Energy and the Environment* **21**, 403–465.

Qureshi, N. & Maddox, I. S. (1995) Continuous production of acetone–butanol–ethanol using immobilized cells of *Clostridium acetobutylicum* and integrated with product removal by liquid–liquid extraction. *Journal of Fermentation Bioengineering* **2**, 185–189.

Roberts, I. S. (1996) The biochemistry and genetics of capsular polysaccharide production in bacteria. *Annual Review of Microbiology* **50**, 285–315.

Rosenberg, E. & Ron, E. Z. (1997) Bioemulsan: microbial polymeric emulsifiers. *CRC Critical Reviews in Biotechnology* **8**, 313–316.

Sahm, H., Eggeling, L., Eikmanns, R. & Kraemer, R. (1995) Metabolic design in amino acid producing bacterium *Corynebacterium glutamicum. FEMS Microbiology Reviews* **16**, 243–252.

Sutherland, I. W. S (1998) Novel and established applications of microbial polysaccharides. *Trends in Biotechnology* **16**, 41–46.

Warren, R. A. J. (1996) Microbial hydrolysis of polysaccharides. *Annual Review of Microbiology* **50**, 183–212.

Wheals, A. E., Bass, L. C., Alves, D. M. G. & Amorim, Y. (1999). Fuel ethanol after 25 years. *Trends in Biotechnology* **17**, 482–487.

Wood, D. R. (1995) The genetic engineering of microbial solvent production. *Trends in Biotechnology* **13**, 259–264.

Books

Bennett, J. W. & Klich, M. A. (eds) (1992) *Aspergillus—Biology and Industrial Applications*. Butterworth-Heinemann, Boston.

Crueger, W. & Crueger, A. (1990) *Biotechnology: A Textbook of Industrial Microbiology*, 2nd edition. Sinauer Associates Sunderland, Mass.

Demain, A. L., Davies, J. E. & Atlas, R. M. (1999) *Manual of Industrial Microbiology and Biotechnology*. American Society for Microbiology, Washington DC.

Ericksson, K.-E. (ed.) (1997) *Advances in Biochemical Engineering 17: Biotechnology in the Pulp and Paper Industry.* Springer-Verlag, New York.

Finkelstein, M. & Davison, B. H. (eds) (2000) *Twenty-First Symposium on Biotechnology for Fuels and Chemicals*. Humana Press, Totowa, NJ.

Glazer, A. N. & Nikaido, H. (1995) *Microbial Biotechnology*. W. H. Freeman, New York.

Nagodawithana, T. & Reed, G. (eds) (1993) *Enzymes in Food Processing*, 3rd edition. Academic Press, San Diego.

Swings, J. G. & Civerolo, E. L. (1993) Xanthomonas. Chapman & Hall, London.

Tombs, M. & Harding, S. E. (1998) *An Introduction to Polysaccharide Biotechnology.* Taylor & Francis, London.

Tucker, G. A. & Woods, L. F. J. (1995) *Enzymes in Food Processing*, 2nd edition. Blackie Academic and Professional (an imprint of Chapman & Hall, now Aspen Publishers), London.

11 Health-care products

Antibiotics are probably the most important group of compounds synthesized by industrial microorganisms. They are not produced in the greatest quantity, nor are they the most economically valuable. Nevertheless, over the last 60 years their influence in improving human health has been immense. The other major health-care products derived from microbial fermentations and/or biotransformation are alkaloids, steroids, toxins and vaccines; along with vitamins (see Chapter 13), certain enzymes (see Chapter 9), and viable microbial cell preparations used as probiotics (see Chapter 12). In addition, genetic engineering techniques have made it possible for microorganisms to produce a wide variety of mammalian proteins and peptides that have various therapeutic properties. Those of considerable medical importance and with established markets include insulin, interferons, human growth hormone and monoclonal antibodies (see Chapter 17). Apart from these therapeutic agents, which cure or reduce the incidence of disease, many diagnostic products are also derived from microorganisms. These are extensively used to test for the presence of various health and disease states.

Antibiotics

Most antibiotics are secondary metabolites produced by filamentous fungi and bacteria, particularly the actinomycetes. Well over 4000 antibiotics have been isolated from various organisms, but only about 50 are used regularly in antimicrobial chemotherapy (see Chapter 2). The best known and probably the most medically important antibiotics are the **β-lactams**, penicillins and cephalosporins; along with **aminoglycosides**, such as streptomycin, and the broad-spectrum **tetracyclines** (Table 11.1). The remainder fail to fulfil certain important criteria, particularly their lack of selectivity, exhibiting toxicity to humans or animals, or their high production costs. Some antibiotics have applications other than in antimicrobial chemotherapy. For example, actinomycin and mitomycin, produced by *Streptomyces peucetius* and *S. caepitosus*, respectively, have roles as antitumour agents; and other antibiotics are used for controlling microbial diseases of crop plants, or as tools in biochemistry and molecular biology research. Several antibiotics are also added to animal feed as growth promoters. However, worries about the development of resistance has meant that some antibiotics, used or intended for human use, may be withdrawn from use in animal feed. For example, the EU Commission voted to ban the application of bacitracin, spiromycin, tylomycin and virginiamycin as growth promoters after January 1999.

β-Lactams

Over 100 β-lactams, mostly penicillins and cephalosporins, have been approved for human use, and they account for over half of the antibiotics produced worldwide. This group is especially useful because of their wide margin of safety. They specifically target the synthesis of peptidoglycan, a vital bacterial cell wall component, which is not present in eukaryotic organisms, thus providing a high level of selectivity. They primarily inhibit the cross-linking transpeptidation reaction, resulting in the formation of incomplete peptidoglycan, severely weakening the bacterial cell wall structure (see Chapter 2, Antibiotics; and Chapter 3, Peptidoglycan synthesis).

PENICILLIN

Penicillin was discovered by Fleming in 1928 following his famous observation of an inhibitory zone surrounding a fungal contaminant, *Penicillium notatum*, on a plate of *Staphylococcus aureus*. In the late 1930s Florey, Chain and Heatley characterized the inhibitory compound responsible, penicillin, and developed a protocol that allowed it to be produced in a pure form. The discovery of penicillin and its later characterization and purification ultimately led to major advancements in

Table 11.1 Examples of important antibiotics

Antibiotic	Producer organism	Activity against	Site or mode of action
β-Lactams			
Natural penicillins	*Penicillium chrysogenum*	Gram-positive bacteria	Wall synthesis
Penicillin G (benzylpenicillin)			
Penicillin V (phenoxymethylpenicillin)			
Semi-synthetic penicillins			
Improved spectrum penicillins		Some Gram-negative rods	
Aminopenicillins			
Ampicillin			
Amoxicillin			
Hetacillin			
Penicillinase-resistant penicillins		Antistaphylococcal	
Methicillin			
Cloxacillin			
Dicloxacillin			
Nafcillin			
Oxacillin			
Extended spectrum penicillins		Antipseudomonal	
Carboxypenicillins			
Carbenicillin			
Ticarcillin			
Ureidopenicillins			
Azlocillin			
Mezlocillin			
Piperacillin			
Cephalosporins	*Cephalosporium acremonium*	Broad spectrum	Wall synthesis
Monobactams	*Chromobacterium violaceum*	Gram-negative bacteria	Wall synthesis
Polypeptide antibiotics			
Bacitracin	*Bacillus subtilis*	Gram-positive bacteria	Wall synthesis
Polymyxin B	*Paenibacillus polymyxa*	Gram-negative bacteria	Cell membrane
Macrolides			
Erythromycin	*Saccharapolyopora erythraea* (formerly *Streptomyces erythraeus*)	Gram-positive bacteria	Protein synthesis
Tylosin	*Streptomyces fradiae*	Gram-positive bacteria	Protein synthesis
Aminoglycides			
Gentamicin	*Micromonospora purpurea*	Broad spectrum	Protein synthesis
Neomycin	*Streptomyces fradiae*	Broad spectrum	Protein synthesis
Streptomycin	*Streptomyces griseus*	Gram-negative bacteria	Protein synthesis
Tetracyclines	*Streptomyces* species	Broad spectrum	Protein synthesis
Vancomycin	*Streptomyces orientalis*	Gram-positive bacteria	Peptidoglycan
Rifamycins	*Amycolatopsis mediterranei* (formerly *Streptomyces mediterranei*)	Tuberculosis	RNA synthesis
Antifungal antibiotics			
Polyenes			
Amphotericin B	*Streptomyces nodosus*	Fungi	Cell membrane
Nystatin	*Streptomyces noursei*	Fungi	Cell membrane
Griseofulvin	*Penicillium griseofulvum*	Dermatophytic fungi	Microtubules

both medicine and fermentation technology. The speed of these developments was greatly influenced by the urgent need to supply penicillin during World War II.

Penicillin exhibits the properties of a typical secondary metabolite, being formed at or near the end of exponential growth. Its formation depends on medium composition and dramatic overproduction is possible. However, *P. notatum*, the organism originally found to produce the antibiotic, generated little more than 1 mg/L from the surface cultures initially used for penicillin production. A 20–25-fold increase in yield was achieved when corn steep liquor was incorporated into the fermentation medium (see Chapter 5, Fermentation raw materials). This byproduct of maize processing contains various nitrogen sources, along with growth factors and side-chain precursors, and remains as a major ingredient of most penicillin production media. Even greater penicillin yields were obtained from a closely related species, *Penicillium chrysogenum*, which was originally isolated from a mouldy cantaloup melon. Further increases in yield were achieved when production went over to submerged fermentation. The wartime requirements for penicillin stimulated the rapid development of a large-scale submerged culture system using stirred tank reactors (STRs). Each vessel was continuously stirred via vertical shaft-driven turbine impellers and incorporated air sparging. These technological developments had a major impact on the advancement of the whole field of fermentation technology.

Since the 1940s, penicillin yield and fermentation productivity has been vastly improved by extensive mutation and selection of producer strains. The traditional approach to improving penicillin yields involved random mutation and selection of higher producing strains. Resulting mutants were grown in liquid medium and culture filtrates were assayed for penicillin. This was slow and painstaking as large numbers of strains had to be tested. Nevertheless, such methods were the key to the dramatically increased yields achieved since the discovery of penicillin. Penicillin fermentations now produce yields in excess of 50 g/L, a 50 000-fold increase from the levels first produced by Fleming's original isolate.

The contribution of classical methods of strain improvement has so far outweighed all other approaches, including more recently available genetic manipulation techniques. The latter have contributed more to our understanding of the complex mechanisms of penicillin biosynthesis, particularly the genetic arrangement of improved strains, the identification of bottlenecks in penicillin synthesis and the regulation of secondary metabolism in overproducing strains.

The basic structure of the penicillins is **6-aminopenicillanic acid (6-APA)**, composed of a **thiazolidine ring** fused with a **β-lactam ring** whose 6-amino position carries a variety of **acyl** substituents (Fig. 2.8). This β-lactam–thiazolidine structure, synthesized from **L-α-aminoadipate**, **L-cystine** and **L-valine**, is common to penicillins, cephalosporins and cephamycins. In the absence of added side-chain precursors to the fermentation medium of *P. notatum* or *P. chrysogenum*, a mixture of natural penicillins is obtained from culture filtrates, notably penicillin G (benzyl penicillin) and the more acid-resistant penicillin V (phenoxymethyl penicillin). These penicillins are most active against Gram-positive bacteria. However, an expanded role for the penicillins came from the discovery that different biosynthetic penicillins can be formed by the addition of side-chain precursors to the fermentation medium and that natural penicillins can be modified chemically to produce compounds with improved characteristics. Most penicillins are now semisynthetic (see below), produced by the chemical modification of natural penicillins, obtained by fermentation using strains of *P. chrysogenum*. Modification is achieved by removing their natural acyl group, leaving 6-APA, to which other acyl groups can be added to confer new properties. These semisynthetic penicillins, such as methicillin, carbenicillin and ampicillin (see Table 11.1), exhibit various improvements, including resistance to stomach acids to allow oral administration, a degree of resistance to penicillinase and an extended range of activity against some Gram-negative bacteria.

Commercial production of penicillin (Fig. 11.1)

Penicillin production is usually via a fed-batch process carried out aseptically in stirred tank fermenters of 40 000–200 000 L capacity, although airlift systems are sometimes used. The fermentation involves an initial vegetative growth phase followed by the antibiotic production phase. Throughout the process, the oxygen level is very important and must be maintained at 25–60 mmol/L/h. However, this is not straightforward, because the oxygen transfer rate is affected by the viscosity, which increases as the fermentation progresses. These processes are maintained at 25–27°C and pH 6.5–7.7, the specific conditions depending upon the *P. chrysogenum* strain used.

Various carbon sources have been adopted for peni-

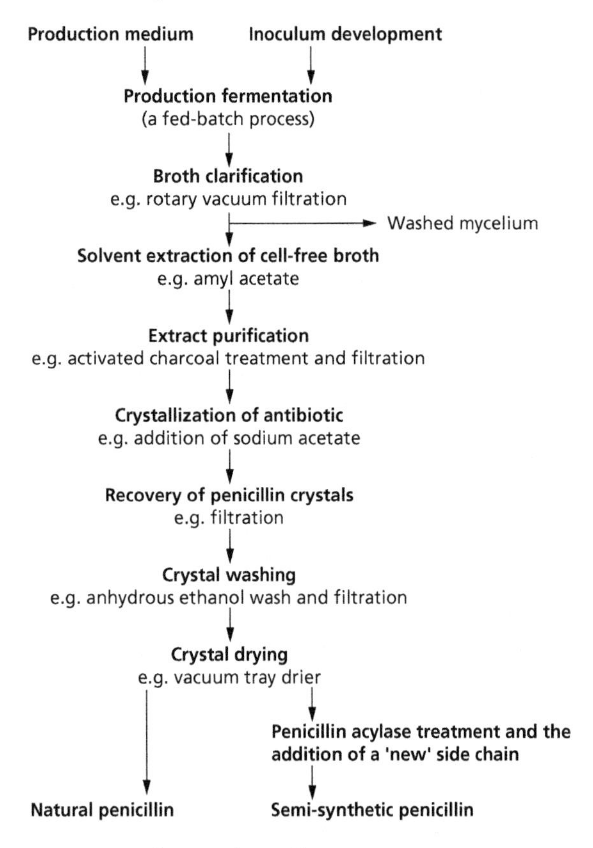

Fig. 11.1 Production of penicillin.

cillin production, including glucose, lactose, sucrose, ethanol and vegetable oils. About 65% of the carbon source is metabolized for cellular maintenance, 25% for growth and 10% for penicillin production. In the past, a mixture of glucose and lactose was used, the former producing good growth, but poor penicillin yields, whereas the latter had the opposite effect. The mode of 'feeding' of a particular carbon source is vitally important, as it can influence the production of this secondary metabolite (see Chapter 3, Secondary metabolism). Corn steep liquor is still used as a source of nitrogen, additional nutrients and side-chain precursors. Its acidic nature creates a requirement for calcium carbonate (1%, w/v) and a phosphate buffer to neutralize the medium, thereby optimizing its pH for penicillin production. Ammonia, mineral salts and specific side-chain precursors, e.g. phenyl acetic acid or phenoxyacetic acid, may also be added. However, as some precursors are toxic, they must be fed continuously at non-inhibitory concentrations.

Inoculum development is usually initiated by adding lyophilized spores to a small fermenter at a concentration of 5×10^3 spores/ml. Fungal mycelium may then be grown up through one or two further stages until there is sufficient to inoculate the production fermenter. Initially, there is a vegetative growth phase devoted to the development of biomass, which doubles every 6 h. This high growth rate is maintained for the first 2 days. To ensure an optimum yield of penicillin in the following production phase, the mycelium must develop as loose pellets, rather than compact forms. During the following production phase, the carbon source is fed at a low rate and penicillin production increases. This continues for a further 6–8 days, provided that appropriate substrate feeds are maintained.

Penicillin is excreted into the medium and is recovered at the end of fermentation. Whole broth extraction may be performed, but can lead to downstream processing problems, as additional materials leach from the mycelium. Usually, penicillin recovery follows removal of mycelium using rotary vacuum filters, the efficiency of which may be affected by the culture media composition, particularly its proteinaceous components. Recovered mycelium is then washed to remove residual penicillin, prior to its use as animal feed or fertilizer.

Antibiotic recovery is often by solvent extraction of the cell-free medium, which gives yields of up to 90%. This involves reducing the pH of the filtered medium to 2.0–2.5 by addition of sulphuric or phosphoric acid, followed by a rapid two-stage continuous countercurrent extraction at 0–3°C using amyl acetate, butyl acetate or methyl isobutyl ketone. The low temperature is necessary to reduce damage to penicillin due to the low pH. Alternatively, ion-pair extraction may be used at pH 5–7, in which range penicillin is stable. Any pigments and trace impurities are removed by treating with activated charcoal. The penicillin is then retrieved from the solvent by addition of sodium or potassium acetate. This reduces the solubility of the penicillin and it precipitates as a sodium or potassium salt. Resultant penicillin crystals are separated by rotary vacuum filtration. Solvent is recovered from the separated liquor and any other materials used, such as the charcoal, which is very important in terms of the overall economics of the process. Penicillin crystals are mixed with a volatile solvent, usually anhydrous ethanol, butanol or isopropanol, to remove further impurities. The crystals are collected by filtration and air dried. At this stage the penicillin is 99.5% pure. This product may be further processed to form a pharmaceutical grade product or is used in the production of semisynthetic penicillins.

Production of semisynthetic penicillins and cephalosporins

As mentioned previously, the objective in semisynthetic penicillin production is to generate compounds with improved properties, e.g. acid stability, resistance to enzymic degradation, broader spectrum of activity, etc. (Table 11.1). It involves removal of the side chain of the base penicillin to form 6-APA. This is achieved by passage through a column of immobilized **penicillin acylase**, usually obtained from *Escherichia coli,* at neutral pH. Penicillin G, for example, is converted to 6-APA and phenylacetic acid. The 6-APA is then chemically acylated with an appropriate side chain to produce a semisynthetic penicillin.

Yields of cephalosporins from direct fermentations are much lower than those for penicillins. Consequently, as 6-APA can also serve as a precursor of cephalosporins, it is often used as the starting material for their semisynthetic production. A base natural penicillin is converted to 6-APA, as described above, followed by its conversion to the preferred precursor, 7-amino deacetoxycephalosporic acid (7-ADCA), by ring expansion. A suitable side chain can then be readily attached.

The emergence of antibiotic resistance

The success of penicillin in the mid-1940s led to the search for other antibiotic-producing microorganisms. One of the most notable early successes was the discovery of streptomycin from a soil actinomycete, *Streptomyces griseus*. Subsequently, actinomycetes, especially *Streptomyces* species, have yielded the majority of the antibiotics used in clinical medicine today (see Table 11.1). However, the increasing development of bacterial strains that exhibit resistance to antibiotics demands the continued search for new antibiotics and alternative agents for treating microbial diseases.

Antibiotic resistance is not a recent phenomenon, it was recognized soon after the natural penicillins were introduced. The use of antibiotics creates selection pressure favouring the growth of antibiotic-resistant mutants, which is promoted by the misuse and overuse of these drugs. Over the last 10 years the situation has become alarming, due to the emergence of pathogenic bacterial strains that show multiple resistance to a broad range of antibiotics. One of the most important examples is the development of multiple-resistant strains of *S. aureus*. Certain strains, particularly methicillin-resistant *S. aureus* (MRSA) cause serious nosocomial (hospital-acquired) infections. They are resistant to virtually all antibiotics used in antimicrobial chemotherapy, including methicillin, cephalosporins and other β-lactams, the macrolide erythromycin, and the aminoglycoside antibiotics streptomycin and neomycin. The only compound that can be used effectively against these staphylococci is an older and potentially more toxic antibiotic, vancomycin. Resistance even to this antibiotic has been detected in some strains.

Many of the antibiotic-resistance genes of staphylococci are carried on plasmids that can be exchanged with species of *Bacillus* and *Streptococcus*, providing the means for acquiring additional genes and gene combinations. Some resistance genes are carried on transposons—segments of DNA that can exist either in the chromosome or within plasmids.

Ergot alkaloids

Alkaloids are a diverse group of small nitrogen-containing organic compounds produced by certain plants and microorganisms. Many are toxic, but some have various therapeutic properties. Species of the filamentous fungus *Claviceps*, which are pathogens of grasses, produce a range of alkaloids (Table 11.2). Some of the best known are the ergot alkaloids. These compounds are produced within the sclerotia (fruiting bodies) of *Claviceps purpurea* that develop naturally when this organism infects developing cereal grains. Infected grains become black and are referred to as ergots. These structures contain indole alkaloids, derived from a tetracyclic ergoline ring system (Fig. 11.2), which are classified into two groups. Members of the first group are based on clavin and contain no peptide component. Clavin-based alkaloids are also produced by other groups of fungi, including species of *Aspergillus* and *Penicillium*. Some possess antibiotic and antitumour activity, but few are produced commercially. The second group, based on lysergic acid (LSA) and containing a tripeptide or an amino alcohol, are found only in *Claviceps* species. Examples include ergometrine, ergocristine, ergosine and ergotamine. Ergotamine, for example, is a structural analogue of serotonin (a neurotransmitter) and is formed from LSA by the action of a peptide synthetase that adds alanine, proline and phenylalanine.

Table 11.2 Fungal metabolites with therapeutic activity

Class of compound	Metabolite	Producer organism
Alkaloids	Agroclavine*	*Claviceps fusiformis*
	Ergometrine	" "
	Lysergic acid	*Claviceps paspali*
	Ergocristine	*Claviceps purpurea*
	Ergosine	" "
	Ergotamine	" "
Antibiotics	Cephalosporins	*Acremonium* (*Cephalosporium*) species
	Fusidane	*Fusidium coccineum*
	Griseofulvin	*Penicillium griseofulvum*
	Penicillins	*Penicillium* species
Antitumour agent	Lentinan	*Lentinus edodes*
Antiviral agent	Funiculosin	*Penicillium funiculosum*
Cholesterol inhibitor	Zaragozic acid	*Sporomiella intermedia*
Enzymes	Lactase	*Kluyveromyces* species
Hypotensive	Fusaric acid	*Fusarium* species
Immunoregulators	Cyclosporin	*Cylindrocarpon lucidum*
		Tolypocladium inflatum
		Trichoderma polysporum
	Gliotoxin	*Aspergillus fumigatus*

* Group 1, all other alkaloids listed are group 2.

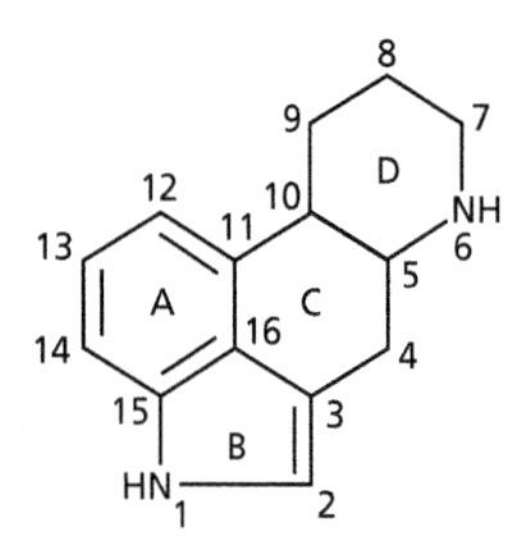

Fig. 11.2 The structure of ergoline.

These LSA-based alkaloids have medical roles as analgesics in migraine therapy, as hallucinogens and for treating circulation problems. Others have particular uses in obstetrics, for inducing the smooth muscle of the uterus to contract during labour and after childbirth.

Previously, the alkaloids were extracted from ergots that developed within infected cereal crops, usually rye, or by chemical synthesis. Most are now produced by fermentation of *Claviceps fusiformis*, *C. paspali* or *C. purpurea* in surface, submerged or immobilized cell culture. Inoculum for the production fermenter may be developed from mycelium or conidiospores. The production medium contains an organic acid of the tricarboxylic acid (TCA) cycle and a carbohydrate, such as citrate and sucrose, the specific combination depending upon the target alkaloid. In later stages, the organic acid stimulates the necessary metabolic change from the TCA cycle to the glyoxylate cycle. Alkaloid production, like that of many secondary metabolites, exhibits phosphate regulation. The synthesis is delayed until the medium phosphate has been utilized during the trophophase and the culture enters the idiophase. However, phosphate inhibition can be overcome by addition of tryptophan or a tryptophan analogue, which act as inducers and precursors.

Steroid biotransformations

The use of microorganisms in the biotransformation of compounds has been particularly successful in the manufacture of therapeutic steroids that are used for the treatment of allergies, inflammation, skin diseases and as oral contraceptives (Table 11.3). Initially, they were prepared by extraction from animal tissues or via complex chemical synthesis, both of which were extremely costly. Many steroids are now manufactured using a combination of chemical and microbial transformation steps. These processes employ relatively cheap sterols as

Table 11.3 Examples of some steroids whose production involves microbial biotransformation, and a selection of specific biotransformations

(a) **Steroids**	
Androgens	Testosterone
Corticosteroids	Cortisone, hydrocortisone, prednisone and dexamethasone
Oestrogens	Oestradiol and estrone
Gestagens	Progesterone

(b) Biotransformation	Substrate	Product	Organism
1-Dehydrogenation	Hydrocortisone	Prednisolone	*Arthrobacter simplex*
11 α-Hydroxylation	Progesterone	11 α-Hydroxyprogesterone	*Rhizopus nigricans*
11 β-Hydroxylation	Reichstein compound S	Hydrocortisone	*Curvularia lunata*
Side-chain cleavage	β-sitosterol	Androstadienedione	*Mycobacterium* species

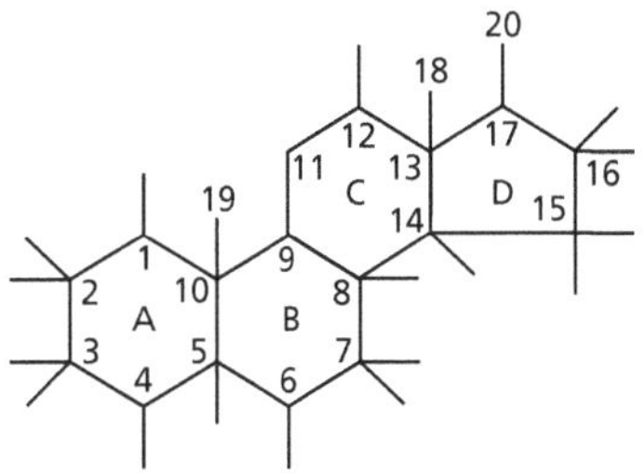

Fig. 11.3 The tetracyclic structure of a cyclopentanoperhydrophenanthrene.

the starting materials, often diosgenin extracted from the Mexican yam (*Dioscorea composita*), or stigmasterol, a byproduct of soybean oil manufacture. The microorganisms involved are mostly filamentous fungi (*Rhizopus*, *Curvularia*, *Fusarium* and *Aspergillus* species) and mycobacteria, in the form of suspensions or immobilized growing cells, resting cells, spores and cell-free extracts. They perform key reactions to modify the basic steroid structure, a cyclopentanoperhydrophenanthrene (Fig. 11.3), including hydroxylations at positions 11 and 17; various side-chain cleavages, hydrogenations and dehydrogenations; and ring expansions, from a five-membered to a six-membered ring.

When live vegetative cells are used for steroid biotransformations, the medium is kept as simple as possible in order to make later purification less problematical. Even simpler media can be formulated for use with spore preparations and there is less need for undesirable antifoam, which may otherwise affect product extraction. For vegetative cells, the culture is grown through exponential phase to obtain maximum biomass before the substrate is added, or the biomass may be harvested to set up immobilized column systems. The steroid precursors are insoluble in water and must be dissolved in solvents, e.g. methanol, ethanol or acetone. These solvents and some substrates are toxic. Consequently, substrate concentrations rarely exceed 2–5 g/L, but their conversion approaches 100%. Processing of the product depends on whether it is accumulated within the cells or excreted into the medium. Water immiscible solvents, usually methylene chloride, ethylene choride or chloroform, are used for extraction of the product from clarified medium or cell extracts. The product may be further purified if it is the end-product, or used directly in a further bioconversion step if it is an intermediate compound.

Bacterial vaccines

Early bacterial vaccines consisted of whole cultures of bacteria that had been inactivated by heat or formaldehyde, but now they can be divided into two categories, **living vaccines** and **inactivated vaccines** (Table 11.4). Living vaccines are composed of live attenuated (weakened) strains of the parent virulent strain. Inactivated forms are composed of whole bacterial cells, or a cell component or metabolic product (cell wall antigen, capsular antigens, toxin, etc.), which now may be products of recombinant DNA technology.

Microbial protein toxins can serve as vaccines following their inactivation with formaldehyde or heat to form toxoids. Vaccination with an antigenic toxoid vaccine leads to the generation of antitoxin that neutralizes the pharmacological effect of active toxin. These vaccines have been successful against Gram-positive bacteria

Table 11.4 Examples of vaccines for medical and veterinary use

Live attenuated bacterial vaccines	
Bacillus anthracis (spore vaccine)	Anthrax
Brucella abortus S19	Brucellosis
Salmonella typhi	Typhoid
Shigella sonnei	Shigellosis
Inactivated bacterial vaccines	
Capsular antigens	
Escherichia coli K88, K99 and 987P	Diarrhoea and food poisoning
Neisseria meningitidis A & C	Meningitis
Pasteurella multocida	Pasteurellosis
Cell wall antigens	
Bordetella pertussis	Whooping cough (pertussis)
Salmonella paratyphi	Paratyphoid
Salmonella typhi	Typhoid fever
Vibrio cholera	Cholera
Toxoid antigens	
Clostridium perfringens	Food poisoning
Clostridium tetani	Tetanus
Corynebacterium diphtheria	Diphtheria
Recombinant vaccines	
Bacterial	
Chlamydia species	Chlamydia (pelvic inflammatory disease)
Clostridium tetani	Tetanus
Corynebacterium diphtheria	Diphtheria
Protozoal	
Plasmodium vivax	Malaria
Viral	Hepatitis B
	Herpes simplex
	Influenza
	Measles (rubeola)
	Poliomyelitis
	Rabies
DNA vaccines	Malaria

responsible for diphtheria, tetanus and several other diseases caused by *Clostridium* species. Similar toxoid vaccines employed to counter some Gram-negative bacterial diseases have proved to be less effective. However, their surface antigens, some of which are involved in their adhesion to epithelial tissues, have been used to develop effective vaccines.

Vaccine production requires highly controlled operating conditions and strict adherence to good manufacturing practices (see Chapter 8). It normally involves the growth of bacterial cultures in sophisticated high-grade fermenters of usually no greater than 1000 L capacity. The fermentations are designed for optimized yield of antigen (cells or cell components) and containment, involving the very strict adherence to protocols that prevent bacterial release into the environment. Internal pressures never exceed atmospheric pressure, to reduce the risks of leakage, and exhaust gases from fermenters must pass through sterilizing filters, incinerators or both.

Fermentations for the production of vaccines based on whole cells aim to maximize biomass production. For inactivated whole cell vaccines, downstream processing usually follows cell inactivation by heat treatment or the addition of formaldehyde. The microbial cells, inactivated or live, are then separated from the medium by centrifugation. All harvesting equipment incorporates absolute microbial containment and is situated within a room maintained under negative pressure, so that any escape is contained. Live attenuated vaccines are usually prepared as freeze-dried products.

For the production of vaccines based on toxins or surface antigens (cell wall or capsular components), the growth conditions are aimed at producing maximum levels of these specific cellular antigens. Excreted toxins and loosely bound surface antigens that are shed into the medium are purified from the clarified culture broth and the harvested cells are safely discarded. As mentioned above, toxins are usually inactivated by treatment with heat or formaldehyde to become bacterial toxoids that have no toxicity, but retain their antigenicity. However, in some cases, notably *Clostridium botulinum* toxin, the active neurotoxin is also prepared for other therapeutic uses.

Vaccine antigens for human immunization are highly purified. Purification procedures may include conventional ammonium sulphate precipitation techniques and various chromatographic steps. Affinity chromatography is usually incorporated, utilizing specific antibodies as ligands, preferably monoclonal antibodies (see Chapter 7). For maximum effectiveness as a vaccine, the purified antigens are adsorbed onto an adjuvant that increases the immunizing power. Adjuvants include mineral salts, such as aluminium hydroxide and potassium aluminium sulphate, or water-in-oil emulsions. For some diseases with a number of serotypes, blends of antigens are used in the final vaccine.

Vaccine preparations can be injected parenterally or administered orally. Inactivated forms are usually injected and living vaccines are mostly taken orally, particularly those for enteric diseases. Injected vaccines stimulate antibody production in the bloodstream, whereas oral vaccines stimulate local production of antibody at the mucosal surface of the intestine.

Viral vaccines were previously available only via culture in live animals or from animal tissue and cell cultures. However, genetic engineering has allowed the production of recombinant viral vaccines through the cloning of viral antigens into an appropriate host microorganism. For example, the virulence factor of hepatitis B and viral protein of foot-and-mouth disease virus can be expressed in *E. coli* for the production of valuable vaccines; recombinant hepatitis B vaccine alone has worldwide sales worth over US$1000 million. Also, the safe production of recombinant vaccines for dangerous bacterial pathogens is now possible, using benign host organisms well suited to large-scale fermentation. This has the added advantage that the host can be manipulated to amplify antigen production (Table 11.4). Some lactic acid bacteria are suitable hosts and are being evaluated for use in oral immunization. These bacteria have generally recognized as safe (GRAS) status and low immunogenicity, and can be used to express antigens, such as fragments of tetanus and diphtheria toxins.

Recombinant therapeutic peptides and proteins

Previously, mammalian therapeutic proteins and peptides could be prepared only from animal or human tissues and body fluids, and were available in very limited quantities. These preparations were extremely costly to produce, some had unwanted side-effects, and in certain cases there were unfortunate problems with virus and prion contamination. Recombinant DNA technology has allowed the production of many recombinant therapeutic proteins from various sources, including over 400 human proteins and peptides with potential medical applications. At present, only about 10% have received approval for use by the US Food and Drug Administration (FDA) (Table 11.5). Overall, the worldwide market for recombinant pharmaceuticals is worth around US$20 billion and is rising at a rate of over 10% per annum. However, the fermentation volume for industrial-scale production of human therapeutic proteins is usually no greater than 2000–5000 L. Unlike recombinant vaccines, where it is essential to retain or

Table 11.5 Examples of recombinant proteins for medical use

Antibodies
tissue necrosis factor-α antibody (rheumatoid arthritis treatment)
Cancer and viral diseases
interferons
interleukins
tissue necrosis factors
transforming growth factors
Cardiovascular diseases
erythropoietin (boosts red blood cell proliferation)
hirudin (thrombin inhibitor from leech)
urokinase (thrombosis treatment)
tissue plasminogen activator (clot dissolution)
Hormones
human growth hormone
insulin
Neurological diseases
endorphins
nerve growth factors
neuropeptides
Vaccines (also see Table 11.4)
foot-and-mouth disease
hepatitis B
Wound healing and blood clotting factors
epidermal growth factor
fibroblast growth factor
clotting factor VIII (haemophilia treatment)

even enhance antigenicity, these recombinant therapeutic products must be free from antigenicity.

DNase

Cystic fibrosis is a fatal genetic disorder involving a malfunction in epithelial tissue. This condition is characterized by the presence of a thick mucus which is produced in a number of organs, particularly the lungs, where it impairs breathing and increases the risk of microbial infection. As a consequence of infection, part of the immune response involves phagocytes attacking the microorganisms, which often include *Pseudomonas aeruginosa*, *Burkholderia* (formerly *Pseudomonas*) *cepacia*, *Staphylococcus aureus* and *Haemophilus influenzae*. This results in the release of free DNA from both bacteria and phagocytes into the lungs. The DNA is very viscous and further thicken the mucus.

Genetically engineered DNase preparations are now available that can help clear these secretions by breaking up the long DNA strands into smaller sections to reduce the viscosity of the mucus. Pulmozyme, a genetically engineered DNase developed by Genentech, received approval from the US FDA in 1996. This 37000 Da human glycoprotein, consisting of 260 amino acid residues, is produced in cell lines from Chinese hamster ovary (CHO) and can be administered in an aerosol. The annual sales of this product are now in excess of US$110 million.

Erythropoietin

Erythropoietin is a 34000 Da glycoprotein hormone produced mainly in the kidneys. It is a haemopoietic growth factor, i.e. a regulatory factor involved in the control of mammalian erythrocyte production in the bone marrow. Without sufficient amounts of erythropoietin, the number of red blood cells in the body is dramatically reduced, resulting in anaemia. Recombinant erythropoietin production has been developed using cell lines from CHO and baby hamster kidney (BHK), or human B-lymphoblastic cells (see Chapter 17, Animal cell culture). Amgen, for example, has had its recombinant product, Epogen, on the market since 1989. These products are used to treat anaemia often associated with kidney disease, cancer and HIV infection, and now have an annual market value in excess of US$1 billion.

Human growth hormone (somatotrophin)

Human growth hormone (hGH) is a protein hormone that is synthesized in the pituitary gland at the base of the brain. This hormone is involved in controlling both growth and stature. Preparations of hGH are used to treat children with 'hypopituitary dwarfism', a congenital disease in which the pituitary fails to secrete sufficient hGH for normal growth. This hormone cannot be administered orally, but must be injected. The standard dose is 0.5–0.9 IU/kg body weight per week. In addition, hGH has therapeutic value in the treatment of a range of other diseases, particularly those associated with ageing, and in wound healing.

This hormone is a single protein chain that is synthesized in the body as a precursor, prehormone, composed of 217 amino acids. The prehormone contains a signal sequence of 26 amino acids that is enzymatically cleaved to give the biologically active protein. Active hGH has a molecular weight of 21500 Da and consists of 191 amino acids with two disulphide bonds linking cysteines at positions 53 and 165, and 182 and 189.

The first clinical application of hGH to treat dwarfism was in 1958, using hormone extracted from the pituitary glands of human cadavers. Its manufacture required almost 40 pituitary glands per patient per year, and consequently, the product was in short supply. In 1985 this product was withdrawn from the market, as use of these hGH preparations was directly linked with the development of a human encephalopathy called Creutzfeldt–Jakob disease (CJD). The causative agent of CJD is still not known for certain and at least two theories have been proposed: an unidentified slow virus, or a prion protein that copurifies with hGH. The prion theory involves a protein that changes the conformational shape of other proteins, whereupon they become resistant to the natural proteases in the human body, resulting in the formation of characteristic myeloid patches in brain tissue. A similar problem happened in the early 1980s when haemophiliacs were infected with HIV, which had copurified with blood clotting factor VIII from human blood. Both problems illustrate the potential dangers of animal and human products, and the vital importance of full characterization of such products, especially for the presence of contaminants.

A search for alternative sources of hGH started with animal derived products, but bovine and porcine growth hormones were found to be ineffective. Greater quantities of a safer supply are now provided through recombinant hGH, which began development in the late 1970s. Two companies, Kabivitrum in Sweden and Genentech in the USA, formed a joint venture to develop a recombinant DNA-derived hGH. However, at that time production of recombinant products was not permitted in Sweden, so the work was carried out in the USA.

The original recombinant DNA systems used the biologically active hormone DNA sequence inserted into a plasmid vector that was used to transform the bacterial host *E. coli*. The resulting protein contains 192 amino acids, its additional amino acid being a *N*-formylmethionine located at the *N*-(amino)-terminal end of the protein (i.e. *N*-formylmethionyl-hGH). This amino acid is used as the initiator of protein synthesis in most bacteria. In some bacterially synthesized proteins this extra amino acid is cleaved post-translationally, but in this case it is not removed. There was some initial concern that the extra amino acid might be immunogenic and could cause an immune reponse within patients. As a result, attempts were made to develop enzymatic steps to remove the extra amino acid, but this proved difficult. However, this product, *N*-formylmethionyl-hGH, had no apparent clinical differences to pituitary-derived hGH and Genentech began marketing their product (Protropin) in 1985. It was the second recombinant pharmaceutical to be approved for clinical use after recombinant insulin. An outline scheme for the production of recombinant hGH is shown in Fig. 11.4.

An alternate approach to recombinant hGH production is to clone the prehormone, containing 217 amino acids, and then enzymically cleave the 26 amino acid signal peptide sequence to yield the biologically active product. Other researchers have also used different hosts, including *Bacillus subtilis* and *P. aeruginosa*, which produce soluble product in the periplasmic space. Animal cell culture using mouse and human cell lines is also a possible route of hGH production.

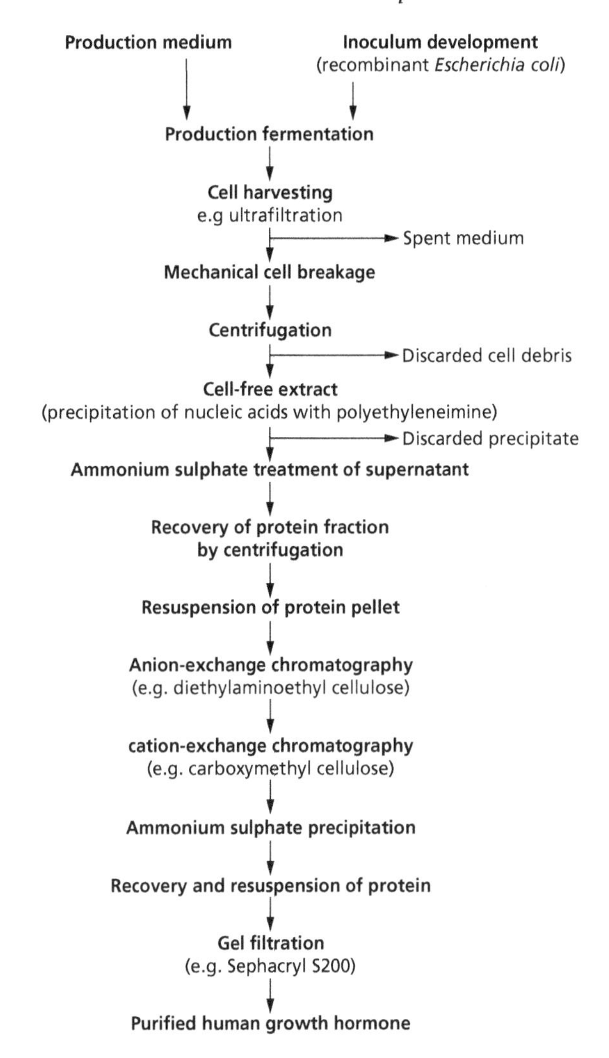

Fig. 11.4 Production of recombinant human growth hormone.

Insulin

Insulin is a 5808 Da polypeptide hormone produced in the pancreas by the islets of Langerhans. This hormone is involved in the regulation of blood glucose levels, and in the metabolism of carbohydrate, fat and starch. It consists of two peptide chains, an 'A' chain composed of 21 amino acids and a 'B' chain containing 30 amino acids.

Insulin extracted from animal pancreases, particularly bovine and porcine insulin, has previously been used to fulfil the needs of diabetic patients who suffer from insulin dependent diabetes mellitus. However, this has now been partly replaced by recombinant human insulin. The recombinant product was first developed by Genentech and marketed by Eli Lilly as Humulin. It became the first recombinant product to be approved for the treatment of human disease, following the granting of a marketing licence by the FDA in 1981. Insulin chains A and B were cloned separately and each host is cultivated independently. Each chain is purified from its respective fermentation and then combined in a chemical step to form the complete insulin molecule. The recombinant product is identical to that produced by the human body. Recombinant insulin is now available from either *E. coli* or *Saccharomyces cerevisiae*.

Interferon

Interferon (IFN) is a member of the **cytokines**, a large family of small signalling proteins involved in regulation of cell-mediated immunity, which also includes interleukins (see below), tumour necrosis factor (TNF), colony-stimulating factor (CSF), erythropoietin (see above) and thrombopoietin. All vertebrates produce a variety of interferons, and mammals, including humans, produce three types; α, β and γ. **IFN-α** forms are primarily produced by leucocytes and they consist of a single polypeptide chain of 165–166 amino acid residues. Some are glycosylated with varying amounts of carbohydrate moieties and their molecular mass is in the range of 16 000–26 000 Da. The carbohydrate portion does not appear to confer any functionality on IFN-α and may be removed without affecting their activity. This property allows recombinant IFN-α to be produced in prokaryotic systems, such as *E. coli*, which are not capable of the post-translational modifications necessary to form glycosylated polypeptides (Fig. 11.5).

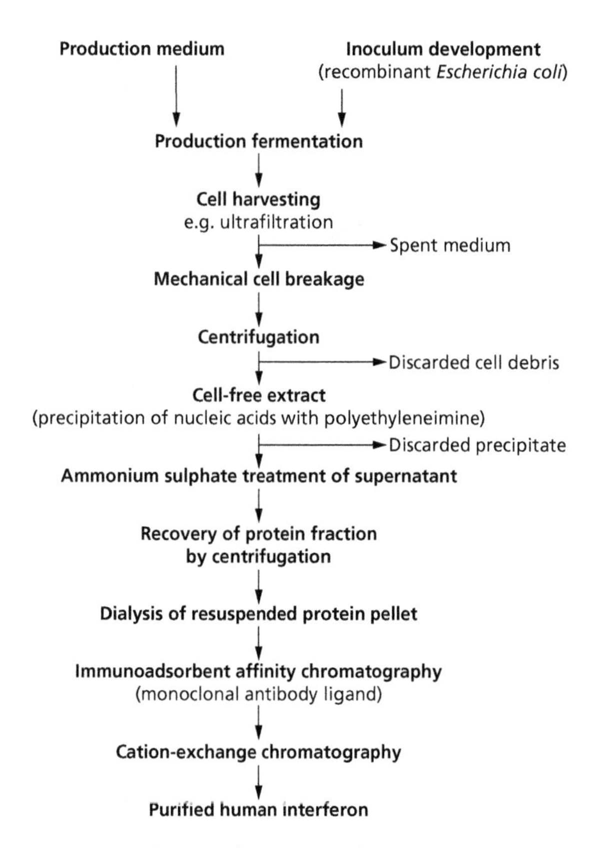

Fig. 11.5 Production of recombinant human interferon.

Recombinant IFN-α received approval from the FDA in 1991 for use in the treatment of hepatitis C. Subsequently, it has been approved for the treatment of a number of conditions, including hairy cell leukaemia, chronic myeloid leukaemia, renal cancer, melanoma, multiple myelomas and genital warts, and in a nasal spray to provide protection against colds caused by rhinoviruses. Recombinant IFN-α is now manufactured by a number of companies, for example, Hoffmann-La Roche produce Roferon and Schering Plough make Intron, which have combined sales worth in excess of US$750 million.

IFN-β is naturally synthesized by mammalian fibroblast cells and may be produced recombinantly from *E. coli*. These products, such as Betaseron and Rebif, are used in the treatment of relapsing multiple sclerosis. **IFN-γ**, sometimes referred to as immune interferon, is produced naturally by activated T-lymphocytes. It is the most important interferon involved in the immune system, where it activates helper T cells (TH1), natural killer cells (NK), cytotoxic T-lymphocytes (CTL) and macrophages, and also exhibits antiviral and oncostatic properties. Preparations of recombinant IFN-γ from *E. coli*, e.g. Genentech's Actimmune, are used in the treatment of chronic myeloid leukaemia and renal cancer.

Interleukins

The interleukins are also a subclass of the cytokines. They are usually single-chain glycosylated proteins with molecular masses of 8000–30 000 Da. There are at least 15 different members of the interleukin family (IL1–IL15). In 1992 the Chiron Corporation received approval from the FDA for a recombinant IL2 preparation, marketed as Proleukin, which is used in the treatment of metastatic renal cell carcinoma. Several other interleukins exhibit therapeutic potential and are at various stages of drug development.

Tissue plasminogen activator

Tissue plasminogen activator (tPA) is a serine protease composed of 572 amino acid residues with a molecular mass of 70 000 Da, naturally produced in the vascular endothelium. It is part of the cascade of serine proteases involved in fibrinolysis (blood clot dissolution), acting as an activator of plasminogen, a glycoprotein precursor of plasmin, which breaks down the fibrin blood clots. Genentech's recombinant tPA, Activase, was first approved in 1987, and is used in the treatment of diseases involving blockage of blood vessels by blood clots, including heart attacks caused by blocked cardiac blood vessels, pulmonary embolisms (blood clot in the lungs), deep vein thrombosis and strokes. Recombinant tPA is manufactured by animal cell culture, but recently it has been produced in the milk of transgenic animals, which is currently undergoing clinical trials.

Collagen

Collagen is the most abundant protein in the human body and is used by surgeons for suturing and repair. It is currently obtained for this purpose from cattle or human cadavers. Consequently, there are concerns about its safety with regard to potential contamination with viruses and prions. The yeasts *Pichia augusta* and *Pichia pastoris* have now been genetically engineered to produce human type I and III collagen fragments, respectively, and may be safe future sources of this protein for medical and surgical use.

Bacteriophages as therapeutic agents

Due to the increasing problems caused by antibiotic-resistant bacterial pathogens, alternatives to antibacterial chemotherapy using antibiotics are now being sought. One option may be bacteriophage (bacterial virus) therapy, which until relatively recently was little known outside the former states of the Soviet Union. Much of the research in this field has been performed over the last 80 years at the Bacteriophage Institute in Tbilisi, the capital of Georgia. The benefits of bacteriophage therapy are that bacteria do not develop resistance to the viruses, as they too evolve to continue invading their host bacteria. Also, they are specific and do not kill beneficial bacteria of the human or animal natural microflora. This therapy appears to be particularly useful for the treatment of cholera, dysentery, meningitis, septicaemia, tuberculosis, and infections caused by *S. aureus* and species of *Campylobacter* and *Pseudomonas*. Should this form of therapy become more accepted in western medicine, there will be a need for the large-scale culture of a range of bacteriophages within suitable host bacteria.

Further reading

Papers and reviews

Alisky, J., Iczkowski, K., Rapoport, A. & Troitsky, N. (1998) Bacteriophages show promise as antimicrobial agents. *Journal of Infection* **36**, 5–15.

Bruggin, K. A., Roos, E.C. & Vroom, E. (1998) Penicillin acylase in the production of β-lactam antibiotics. *Organic Process Research and Development* **2**, 128–133.

Datar, R. V., Cartwright, T. & Rosen, C.-G. (1993) Process economics of animal cell and bacterial fermentations: a case study analysis of tissue plasminogen activator. *Bio/Technology* **11**, 349–357.

Hutchinson, C. R. & Fujii, I. (1995) Polyketide synthase gene manipulation: a structure–function approach in engineering novel antibiotics. *Annual Review of Microbiology* **49**, 201–238.

Johnson, I. (1983) Human insulin from recombinant DNA technology. *Science* **219**, 632–637.

Johnson, E. A. (1999) Clostridial toxins as therapeutic agents: benefits of nature's most toxic proteins. *Annual Review of Microbiology* **53**, 551–575.

Khachatourians, G. G. (1998) Agricultural use of antibiotics and the evolution and transfer of antibiotic-resistant bacteria. *Canadian Medical Association Journal* **159**, 1129–1136.

Koths, K. (1995) Recombinant proteins for medical use: the attractions and challenges. *Current Opinion in Biotechnology* **6**, 681–687.

Marks, T. & Sharp, R. (2000) Bacteriophages and biotechnology: a review. *Journal of Chemical Technology and Biotechnology* **75**, 6–17.

Martineau, P. & Rambaud, J.-C. (1993) Potential of using lactic acid bacteria for therapy and immunodulation in man. *FEMS Microbiology Review* **12**, 207–220.

Nielsen, J. (1998) The role of metabolic engineering in the production of secondary metabolites. *Current Opinion in Biotechnology* **10**, 330–336.

Penalva, M. A., Rowlands, R. T. & Turner, G. (1998) Optimization of penicillin biosynthesis in fungi. *Trends in Biotechnology* **16**, 483–489.

Rallabhandi, P. & Yu, P.-L. (1996) Production of therapeutic proteins in yeasts: a review. *Australian Biotechnology* **6**, 230–237.

Robinson, A. & Melling, J. (1993) Envelope structure and the development of new vaccines. *Journal of Applied Bacteriology: Supplement* **74**, 43S–51S.

Roessner, C. A. & Scott, A. I. (1996) Genetically engineered synthesis of natural products: from alkaloids to corrins. *Annual Review of Microbiology* **50**, 467–490.

Seaman, K. B. (1998) Specifications for biotechnology-derived protein drugs. *Current Opinion in Microbiology* **9**, 319–325.

Wiseman, A. (1996) Therapeutic proteins and enzymes from genetically engineered yeasts. *Endeavour* **20**, 130–132.

Books

Crueger, W. & Crueger, A. (1990) *Biotechnology: A Textbook of Industrial Microbiology*, 2nd edition. Sinauer Associates, Sunderland, MA.

Demain, A. L., Davies, J. E. & Atlas, R. M. (1999) *Manual of Industrial Microbiology and Biotechnology*. American Society for Microbiology, Washington, DC.

Ellis, R. W. (ed.) (1992) *Vaccines: New Approaches to Immunological Problems*. Butterworth-Heinemann, London.

Finkelstein, D. B. & Ball, C. (eds) (1992) *Biotechnology of Filamentous Fungi*. Butterworth-Heinemann, Boston.

Glazer, A. N. & Nikaido, H. (1995) *Microbial Biotechnology*. W. H. Freeman, New York.

Strohl, W. R. (ed.) (1997) *Biotechnology of Antibiotics*. Marcel Dekker, New York.

Wainwright, M. (1992) *An Introduction to Fungal Biotechnology*. Wiley, Chichester.

Walsh, G. (1998) *Biopharmaceuticals: Biochemistry and Biotechnology*. Wiley, Chichester.

12 Food and beverage fermentations

Microorganisms have long played a major role in the production of food and beverages. Traditional fermented foodstuffs include:

1 **alcoholic beverages**, especially beers, wines and distilled spirits, which are derived from sugars and starches;
2 **dairy products**, particularly cheeses, yoghurt, sour cream and kefir;
3 **fish and meat products**, such as fish sauce and fermented sausages; and
4 **plant products**, notably cereal-based breads and fermented rice products; along with fermented fruits, vegetables and legumes, including preserved olives and gherkins, sauerkraut, soy sauce, tofu, fermented cassava, cocoa and coffee beans.

Many of these products originally evolved as a means of food preservation. The stabilizing microbial activity may result in lower water activity, modified pH, generation of inhibitory compounds (alcohol, bacteriocins, etc.) and removal of nutrients readily utilized by potential spoilage organisms. Importantly, besides providing long-term stability, these fermentation processes also generate desirable flavour, aroma and texture.

The role of microorganisms in this field is now even more diverse. Microbial biomass is used as novel food, such as single cell protein products, speciality mushrooms and probiotic preparations; numerous microbial fermentation products are also incorporated as food additives and supplements (see Chapters 13 and 14). In addition, microbial enzymes are utilized extensively as food processing aids, and some may be added to animal feed to improve its nutritional value (see Chapter 9).

Alcoholic beverages

Alcoholic beverages have been produced throughout recorded human history. They are manufactured worldwide from locally available fermentable materials, which are sugars derived either from fruit juices, plant sap and honey, or from hydrolysed grain and root starch (Table 12.1). Some alcoholic beverages are drunk fresh, but more commonly they are aged to modify their flavour, whereas others are distilled to increase alcoholic strength. Although bacteria such as *Zymomonas* species may be involved in the production of certain products, yeasts are primarily used, either in single or mixed cultures. Their fermentation products are ethanol, a range of desirable organoleptic (flavour and aroma) compounds and CO_2 (provides carbonation for some products). The yeasts involved in these alcoholic fermentations are mostly strains of *Saccharomyces cerevisiae*, which cannot directly ferment starch. They require prior hydrolysis of the polysaccharide to simple sugars and small dextrins (not greater than three glucose units). Traditionally, this is achieved by using fungal or plant amylases. These enzymes may be inherent elements of the carbohydrate source or added during processing.

Beer brewing

The term beer is given to non-distilled alcoholic beverages made from partially germinated cereal grains, referred to as malt. They include ales, lagers and stouts, which normally contain 3–8% (v/v) ethanol. Their other main ingredients are hops (giving beer a characteristic flavour and aroma), water and yeast. The brewing process is essentially divided into four main stages (Fig. 12.1).

1 **Malting** is the partial germination of cereal grain for 6–9 days to form malt. This is the primary beer ingredient and contains mostly starch, some protein and hydrolytic enzymes. Malted barley is predominantly used, but beers are also made with malted wheat, occasionally malted oats and even malted sorghum.
2 **Mashing and wort preparation** involves the production of the aqueous fermentation medium, otherwise known as **wort**. It contains fermentable sugars, amino acids and other nutrients, and is prepared by solubilizing malt components through the action of endogenous hydrolytic enzymes. In most countries, a proportion of

Table 12.1 Some examples of alcoholic beverages from around the world

Substrate	Non-distilled beverage	Product of alcoholic fermentation distilled to form	Location
Fruit juices and plant saps			
Apples	Cider	Apple brandy, calvados, etc.	N. Europe, N. America
Cacti/succulents	Pulque	Tequila	Mexico, Central America
Grapes	Wine	Grape brandy, cognac, armagnac, etc.	S. Europe, N. & S. America, Australia, New Zealand
Palmyra	Toddy	Arak	India
Pears	Perry	Pear brandy, williams, etc.	N. Europe
Honey	Mead		UK
Sugar cane or molasses	–	Rum	West Indies
Starches			
Barley, plus other cereals*	Beer –	Whisk(e)y†	Beer: ales—UK mostly; lager—worldwide Whisk(e)y: Scotland, Ireland, USA, Canada, Japan
Rice‡	Sake Pachwai Sonti	Shochu	Japan India
Sorghum*‡	Kaffir		Central and South Africa

* Saccharified primarily by malt amylases.
† Essentially prepared from an unhopped beer.
‡ Saccharified by fungal amylases.
Note: Neutral spirits, gin, vodka, etc., can be prepared from the fermentation of virtually any carbohydrate, including lactose. In northern Europe and the USA, starch is mostly employed in the form of wheat and maize, whereas in eastern Europe and Russia, potatoes are often used. These starches are usually hydrolysed using malt enzymes.

adjuncts are now also added, which are unmalted cereal and non-cereal starch sources, and sugar syrups. The resulting liquid wort is then 'sterilized' by boiling; at the same time hops are added to impart their bitter flavour and characteristic aroma. Overall, wort preparation takes approximately 5–8 h.

3 **Yeast fermentation** is a non-aseptic batch process that uses a starter culture of a selected brewing strain of *S. cerevisiae*. The inoculated wort undergoes an alcoholic fermentation to produce ethanol, CO_2 and minor metabolites that contribute to flavour and aroma. Fermentations usually last for 2–7 days depending upon the type of beer being produced.

4 **Post-fermentation** treatments are conducted to mature or condition the new beer to make it ready for consumption, which may take from one to several weeks.

The raw materials, excluding hops, the preparation of the wort and the yeast fermentation, are essentially the same for the production of both whisky and malt vinegar (see pp. 200 and 201). However, the major objectives are somewhat different. For whisky and vinegar production, the main aim is to maximize alcohol production prior to the respective steps of distillation and acetification. In beer brewing, ethanol production *per se* is rather less crucial, as development of sound flavour profile and other quality factors is equally important.

Traditionally, all necessary enzyme activities for the process were provided by the malt, which generates the wort, along with yeast enzymes, that convert wort components to ethanol, CO_2 and flavour compounds. However, commercial enzymes are now employed when raw materials are enzyme deficient, or to produce novel products, or act as aids in processing and in product stabilization.

MALT AND MALTING

Malting involves the controlled partial germination of barley grain. This modifies the hard vitreous grain into a friable (easily crushed) form containing more readily degradable starch and generates hydrolytic enzymes, especially amylases, β-glucanases and proteases.

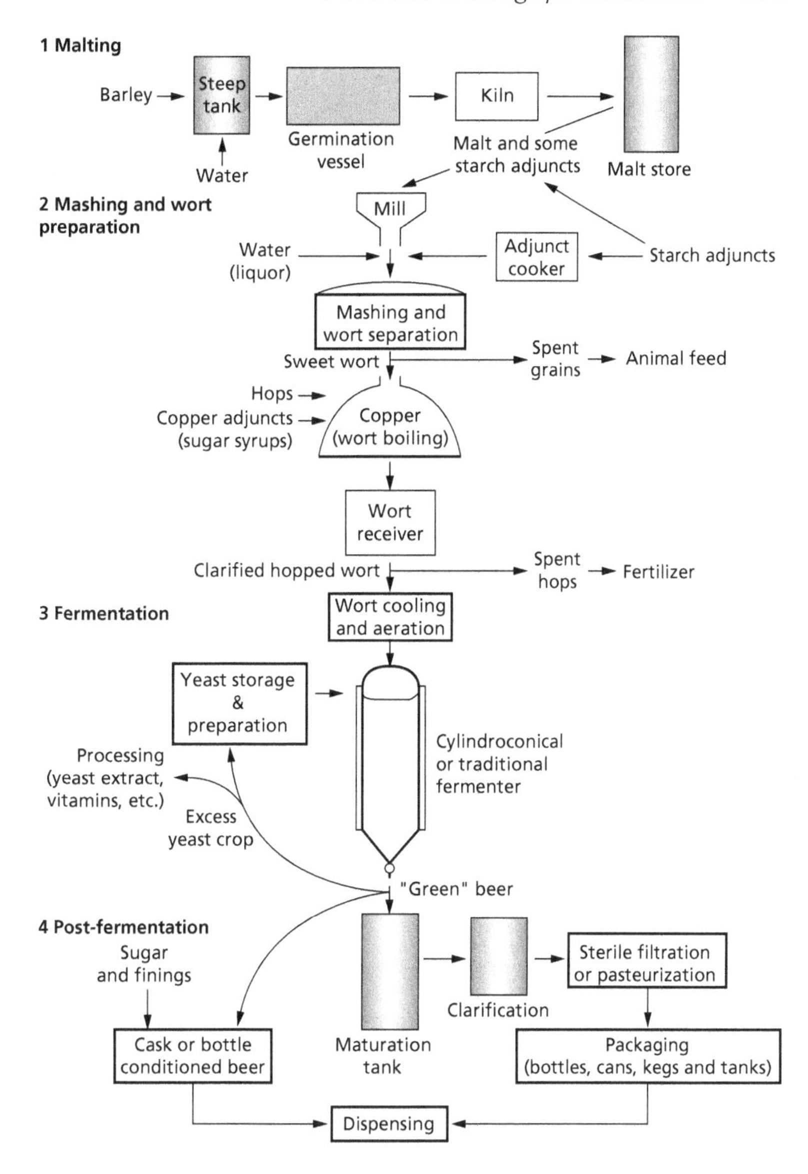

Fig. 12.1 The beer brewing process.

Barley grains contain approximately 65% starch, located within the endosperm region (Fig. 12.2). Endosperm cells are filled with starch granules embedded in a protein matrix, and their walls are composed of a mixed linkage β-glucan (β-1,3; 1,4), hemicellulosic pentosans and protein. The starch granules cannot be accessed until the cell walls have been breached and the protein matrix has been at least partially degraded. Hence, the requirement for adequate levels of β-glucanases and proteases.

Malting begins by soaking or steeping the barley in water for 2 days at 10–16°C, in order to increase the moisture content to around 45% (w/w). Periodically, the water is temporarily drained off and aeration is provided, thus preventing anaerobic conditions that can cause embryo damage. The water used for steeping is often reused to save on costs of both water and effluent treatment.

After steeping, the barley is partially germinated for 3–5 days at 16–19°C. Traditionally, this simply involved spreading the grain on malting floors to a depth of 10–20 cm. However, various mechanized systems are

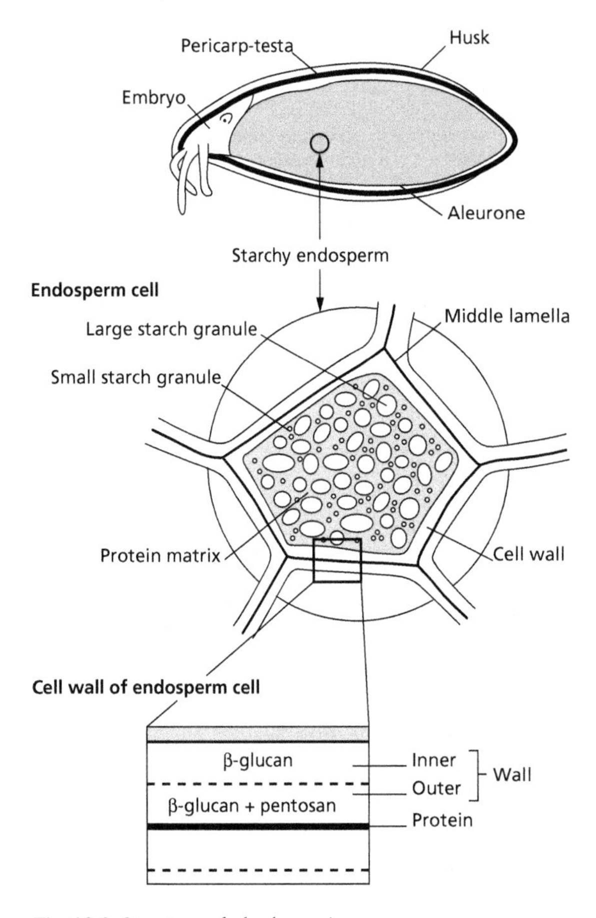

Fig. 12.2 Structure of a barley grain.

now operated, which have grain beds of about 1 m in depth. These are aerated with moist cool air and turned mechanically every 8–12 h to aid respiration by the grain and prevent the build-up of heat, otherwise the embryo may become damaged.

Grain modification can be promoted in several ways. Abrasion (the controlled damage of the husk before steeping) improves access of water and any additives. The application of the plant hormone gibberellic acid, at levels of 0.1–0.5 mg/kg barley, may be made to augment the natural embryo-produced hormone that stimulates *de novo* synthesis of certain hydrolytic enzymes, including α-amylase. Cellulase supplements, such as enzymes from *Trichoderma reesei,* at levels of 24–48 μg/kg barley, also speed germination by aiding endosperm cell wall breakage. Suppression of unnecessary root and shoot growth can be achieved by adding bromate at levels of 100–200 mg/kg barley.

When sufficiently modified, the malt is 'kilned' via a two-stage process. First it is dried at 50–60°C and then 'cured' at 80–110°C. Kilning takes about 2 days and has several functions. It arrests embryo growth and enzyme activity, while minimizing enzyme denaturation, and develops flavour and colour (melanoidin compounds). Pale lager malts that require little colour development are subjected to mild conditions. Consequently, they retain more enzyme activity than do coloured malts. Highly coloured malts required for flavouring and colouring dark beers have low enzyme activity. At a final moisture content of 2–3% (w/w), the malt is biologically stable for several months.

WORT PREPARATION

The objectives of wort preparation are to form, and extract into solution, fermentable sugars, amino acids, vitamins, etc., from malt and other solid ingredients. Malt normally provides most of the potential fermentable materials and sufficient enzymes to generate a well-balanced fermentation medium. Typically, British beer is prepared from 75% malt, and 25% unmalted cereal and non-cereal starch sources, referred to as 'adjunct'. In some cases, a portion of the adjunct may be in the form of sugar syrups that may be added later during wort boiling. In the USA, up to 60% adjuncts may be used, while a maximum of 40% is recommended in the EU. However, in Germany the beer purity law known as the Reinheitsgebot of 1516 still forbids the use of any adjuncts.

Replacement of some malt with adjuncts is mainly for economic reasons. Nevertheless, certain adjuncts, such as rice products, may improve specific beer properties, but contribute little flavour and enzyme activity. Other mash adjuncts include raw barley, wheat flour, maize grits, and other flaked or micronized cereals. Those with a high starch gelatinization temperature may require cooking before the main mash. Traditionally, 5–10% (w/w) malt was added during cooking to provide amylases that hasten the process. However, thermostable microbial α-amylases are now preferred.

Malt and those adjuncts requiring milling are usually roller-milled prior to mashing. This is often performed in such a way as to largely retain the husk intact, while reducing the remainder to a coarse powder. The husk later acts as a filtration aid in wort separation. Milled ingredients are transported to the mash vessel and mixed with hot water, the brewing liquor, whose composition can influence final beer quality. Consequently, it may be necessary to add or remove certain ions and adjust the pH. Following mashing the liquid extract is separated from residual solids to form the wort, which must then be stabilized by boiling. Resulting wort should be a well-balanced liquid medium that will supply the yeasts with all nutritional requirements for the subsequent fermentation.

Mashing systems

Three main mashing systems are operated.

1 **Infusion mashing** is the classic British method for ales and stouts, using well-modified malt and relatively simple equipment. Solid mash ingredients are mixed with hot brewing liquor to achieve a mash temperature of 62–65°C. Mixing is performed during passage to an insulated unstirred vessel, the **mash tun** (Fig. 12.3), which is 7–9 m in diameter and 2 m deep. In some breweries these traditional vessels have been superseded by mash mixers.

The mash has a liquor to solids (grist) ratio of around 3:1 and this thick mash helps to stabilize some malt enzymes. Infusion mash temperatures are suitable for starch degradation, but malt β-glucanases and proteases are rapidly inactivated; hence the requirement for well-modified malt whose cell walls and proteins have already been substantially degraded during malting. Mashing lasts for 1–2 h and is followed by mash sparging. This involves spraying water at 70–75°C onto the surface, which percolates through the grain bed and washes out the soluble extract. The grain bed sits on the false bottom of the mash tun, which has slots approximately 1 mm wide. These are kept clear by the large husk fragments in the mash that act as a filter aid. Resultant liquid extract is called 'sweet wort'. Sparging is continued until the specific gravity of the wort falls to a specified level indicating that little or no more soluble sugars remain in the mash.

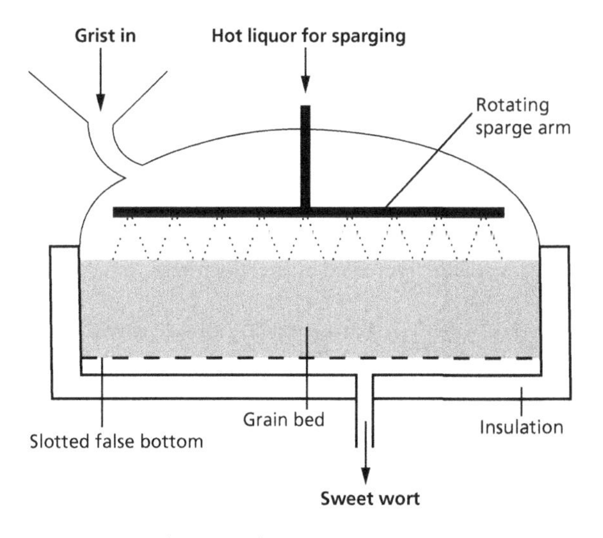

Fig. 12.3 An infusion mash tun.

2 **Decoction mashing** has been traditionally used in Europe and is suitable for less well-modified malts. The initial mash temperature starts at around 35–40°C, with a liquor to grist ratio of up to 5:1. This initial lower temperature facilitates more extensive hydrolysis of β-glucan and protein, and is often called a 'protein rest'. The temperature is then raised to 50–55°C, by removing a portion of the mash, which is boiled and immediately returned to the mash vessel. This may be repeated once or twice more, with intermediate holding periods. The aim is to raise the temperature stepwise to about 65°C, to achieve starch degradation, and finally to 75°C. Wort separation is usually via a **lauter system**, aided by rakes or knives that cut the grain bed; alternatively, **mash filter systems** may be used.

3 **Temperature programmed mashing** systems are operated by many modern breweries. The mash temperature is raised using a heating jacket from around 40°C to above 70°C through any number of holding stages. This allows much greater control over the mashing process and the programme can be adapted for variously modified malts. Mash separation is usually through a modern lautering device or mash filter.

The result of mashing, irrespective of the system used, is an aqueous extract, the **sweet wort** (for typical composition, see Table 12.2) and insoluble **spent grains**. The latter byproducts are highly perishable and are often quickly transported away for direct use as cattle feed. Attempts have also been made to generate further fermentable brewing sugars from the cellulosic components of spent grains using acid or enzyme hydrolysis (see Chapter 10, Bioconversion of lignocellulose).

Biochemistry of mashing

Starch hydrolysis. The objective in mashing is to convert as much of the malt and adjunct starch as possible to fermentable sugars. Starch is composed of 25% amylose, a linear polymer of α-1,4-linked glucose units, and 75% amylopectin, which is a branched polymer containing both α-1,4 and α-1,6 linkages. Malt enzymes involved

Table 12.2 Typical ale wort

	g/L
Carbohydrate	
Fermentable carbohydrate	71
glucose + fructose	10
sucrose	4
maltose	42
maltotriose	15
Higher dextrins	23 (not fermented by most brewing yeasts)
Total carbohydrate	94
Amino acids	1.5
Vitamins and growth factors	mg/L
biotin	< 0.01
inositol	60
nicotinic acid	10
pantothenic acid	1
pyridoxine	0.6
riboflavin	0.45
thiamine	0.3
Minerals	mg/L
sodium	20
potassium	450
calcium	40
iron	< 1
magnesium	100
chloride	350
phosphate	900
Organic acids	(total approximately 200 mg/L, wort pH 5.2–5.4)

Note: In the extraction (mashing) of grist (milled malt and adjuncts) approximately 75% is solubilized and 25% remains as spent grains.

in starch degradation are collectively referred to as diastase, consisting of a mixture of β-amylase, α-amylase, limit dextrinase (a debranching enzyme) and α-glucosidase (Table 12.3).

β-Amylase is an exo-enzyme that hydrolyses alternate α-1,4 linkages from the non-reducing ends of polymers to release disaccharide maltose units, but is unable to bypass α-1,6 branch points. This enzyme operates mainly after the initial dextrinization of starch by α-amylase, which is an endo-enzyme that acts randomly on α-1,4 linkages, and ultimately generates glucose, maltose and maltotriose units. Limit dextrinase attacks α-1,6 branch points in branched dextrins. A relatively minor role is played by α-glucosidase, which releases single glucose units.

Malt enzymes that are most likely to be limiting are β-amylase and limit dextrinase, which are more thermolabile than α-amylase. Well-modified malt normally contains 3–4 times more α-amylase than is required for mashing. Supplements of microbial α-amylases are needed only when using very poorly modified malt or low levels of malt. Mash supplements of commercial β-amylase and an enzyme that can hydrolyse α-1,6 branch points, such as a microbial pullulanase, may improve the fermentability of wort and help to control the spectrum of sugars produced (Table 12.4).

Non-fermentable residual dextrins normally constitute approximately 20–25% of the original starch. Their degradation, to further improve conversion efficiency, or allow the production of low-carbohydrate beers, is often facilitated through the addition of commercial debranching enzymes, e.g. isoamylases, pullulanases or amyloglucosidases (glucoamylases). Amyloglucosidases from *Aspergillus niger*, *Rhizopus* species or *Schwanniomyces castellii* are most frequently used for this purpose.

β-Glucan hydrolysis. Cell wall β-glucan is solubilized during malting and mashing by β-glucan solubilase and then degraded by malt endo-β-glucanases (Table 12.3). Its hydrolysis is crucial, as undegraded β-glucan may cause slow wort separation, beer filtration problems and hazes. In addition, failure to degrade β-glucan reduces potential extract yield from the mash as amylases cannot access the starch and β-glucan is itself a source of glucose.

Wort separation and beer filtration is aided by supplementing the mash with *Trichoderma* cellulase-complex or a thermostable β-glucanase from *Bacillus subtilis* (Table 12.4). Added pentosanases may be useful in the degradation of cell wall derived pentosans, particularly when using wheat adjuncts. However, resultant pentose sugars are not fermented by brewing yeast.

Protein hydrolysis. Barley malt has an extensive range of proteases (Table 12.3). Their role in mashing is to degrade wall and matrix proteins of endosperm cells. This allows amylases access to starch granules and generates a well-balanced spectrum of amino acids for subsequent yeast fermentation. The proteases are relatively labile, and as discussed above, some mashing systems have low

Table 12.3 Barley malt enzymes (after Bamforth, C. W. (1985) *Brewers' Guardian* **114**, 21–26)

Enzyme	Action	Product	pH optimum	Production	Mash stability
A. Starch degrading enzymes					
α-Amylase (EC 3.2.1.1)	Endo—splits α-1,4 links randomly, mainly dextrinizing	oligosaccharide mixture	5.5	*de novo* synthesis during malting	destroyed only after 2 h at 67°C, stabilized by Ca^{2+}
β-Amylase (EC 3.2.1.2)	exo—splits of maltose from non-reducing ends of chains (cannot bypass α-1,6 links)	maltose + β-limit dextrins	5.2	inactive form present in barley, activated during malting	largely destroyed after 40–60 min at 65°C
Limit dextrinase (EC 3.2.1.10)	hydrolyses α-1,6 links in β-limit dextrin	straight chain α-1,4 dextrins	5.0	*de novo* synthesis during malting	relatively unstable
α-Glucosidase (EC 3.2.1.3)	exo—hydrolyses terminal α-1,4 links	β-D-glucose	4.8	present in barley, levels increase during malting	relatively unstable
B. β-Glucan degrading enzymes					
β-Glucan solubilase (acidic carboxypeptidase)	esterolytic—splits bonds between β-glucan and cell wall peptides	soluble β-glucan	6.3	present in raw barley	survives mashing at 65°C, fairly thermostable
β-Glucanase (EC 3.2.1.73)	endo—on β-1,4 bonds adjacent to β-1,3 bonds	tri- and tetra-saccharides	4.7	synthesized during malting	little survives mashing at 65°C stabilized by GSH
C. Proteolytic enzymes					
Endopeptidases a. SH (90%) b. Metal (10%) (EC 3.4.4.—).	Endo	peptides	a. 3.9, 5.5 b. 5.5, 6.9, 8.5	some present in barley, mainly synthesized during malting	destroyed in 15 min at 70°C
Carboxypeptidases (EC 3.4.2.—)	exo—from C-terminal	amino acids	4.8, 5.2, 5.6	present in barley, but increases during malting	relatively stable
Neutral amino peptidases (EC 3.4.1.—).	exo—from N-terminal	amino acids	7.0, 7.2		
Leucine amino peptidases (EC 3.4.1.—).	exo—from N-terminal	amino acids	8.0–10.0		rapidly inactivated above 50°C
Dipeptidase (EC 3.4.3.—)	exo—from N-terminal	amino acids	8.0–10.0		rapidly inactivated above 50°C

initial starting temperatures (40–50°C) to promote protein degradation. Addition of commercial proteases can aid wort separation and ensure sufficient levels of amino acids for yeast metabolism (Table 12.4). This is vital, as amino acid deficiencies can lead to problems with beer flavour.

Wort boiling

Sweet wort obtained from the mash is transferred to a 'copper' ('kettle') for boiling along with dried hops or hop extracts. Hops are the flower cones of the female hop vine (*Humulus lupulus*), which contain α and β

Table 12.4 Commercial enzymes in brewing (after Bamforth, C. W. (1985) *Brewers' Guardian* **114**, 21–26)

Enzyme	Source	Stage of addition	Role
α-Amylase (EC 3.2.1.1)	*Bacillus subtilis* *Bacillus licheniformis* *Aspergillus oryzae* *Bacillus stearothermophilus* (thermostable, does not require Ca^{2+})	Mashing, adjunct cooking	Liquefaction of starch, especially maize or rice grits, has superior properties to malt enzyme
β-Amylase (EC 3.2.1.2)	Barley (*Hordeum vulgare*) Wheat (*Triticum aestivum*) Soybean (*Glycine max*) *Bacillus cereus*	Mashing	Maltogenic, improves wort fermentability and sugar spectrum
Pullulanase (EC 3.2.1.41)	*Bacillus acidopullulyticus* *Thermus aquaticus* (thermostable, pH optimum 6.4)	Mashing, fermentation	Debrancher, hydrolysis of α-1,6 links in dextrins to produce 'diet-beers', often used with added α- or β-amylase
Amyloglucosidase (EC 3.2.1.3)	*Aspergillus niger* *Rhizopus niveus* *Rhizopus delemar*	Mashing, fermentation, beer	Yields glucose, increases wort fermentability, used to produce 'diet-beers', and in enzymic priming (hydrolyses α-1,4 and α-1,6 linkages)
Endo-β-glucanase (EC 3.2.1.73)	*Bacillus subtilis* *Penicillium emersonii* *Aspergillus niger*	Mashing, especially if using barley adjuncts	Improves extract recovery in mashing, wort separation and beer filtration rates, and prevents gels and hazes
Cellulase (EC 3.2.1.4)	*Trichoderma viride* *Trichoderma reesei* *Penicillium funiculosum*	Malting, mashing	Cleaves β-1,4 links, promotes barley modification in malting, similar role in mashing as endo-β-glucanase
Proteases (Acid) (Neutral) (Papain) (Bromelain) (Ficin)	*Aspergillus niger* *Bacillus subtilis* Paw paw (*Carica papaya* L.) Pineapple (*Ananas comosus*) Fig (*Ficus* species)	Mashing, beer	Prevents protein–polyphenol haze, aids balance of wort amino acid spectrum, also activates β-amylase
Glucose oxidase (EC 1.1.3.4)/**Catalase** (EC 1.11.1.6)	*Aspergillus niger*	Beer	Removal of O_2 from beer, aiding flavour stability and haze life
Acetolactate decarboxylase (EC 4.1.1.5)	*Bacillus licheniformis* *Lactobacillus* species	Fermentation, beer	Speeds maturation by eliminating diacetyl

acids, primarily humulones and lupulones. It is predominantly the α acids that, after isomerization to iso-α acids during boiling, give beer its bitter flavour (Fig. 12.4). Hop constituents also help to inhibit certain beer spoilage bacteria and maintain foam stability. Sometimes, additional sugar (copper/kettle adjuncts) may be added to the wort prior to boiling.

Boiling is conducted for several reasons, including:

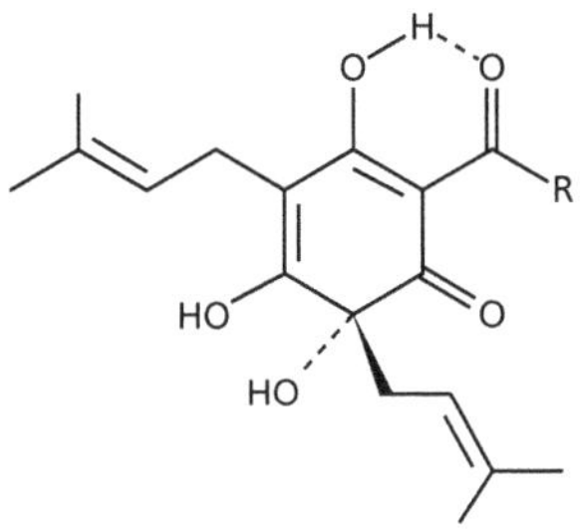

α-Acids	R
Humulone	$CH_2CH(CH_3)_2$
Cohumulone	$CH(CH_3)_2$
Adhumulone	$CH(CH_3)CH_2CH_3$

Fig. 12.4 Structural formulae of α acids (humulones) of hops.

1 isomerization of hop α acids to the more bitter tasting iso-α acids;
2 'sterilization' of the wort;
3 concentration of the wort;
4 termination of enzyme activity (the denaturation of malt enzymes and any enzyme supplements);
5 precipitation of unwanted proteins;
6 removal of volatile compounds that can impair flavour; and
7 development of some flavour compounds and colour, through the formation of melanoidins and oxidation of phenolic compounds.

Following boiling, the wort is clarified using a whirlpool separator or centrifuge to remove positively charged denatured protein and hop debris. Separation of these suspended solids, which are referred to as 'trub', may be aided by the addition of negatively charged polysaccharides such as carrageenan. Before fermentation, the clarified hopped wort is cooled and then aerated.

FERMENTATION

Yeast characteristics

Brewer's yeast must be effective in taking up nutrients from the wort and imparting the required flavour to the beer, and should be readily removed from the completed fermentation. Over the years, brewing yeast strains have evolved that are rather different from 'wild' strains, exhibiting higher fermentation rates and greater tolerance to alcohol. These yeasts have also been selected by brewers on the basis of their performance in a specific brewery environment. Brewers have their own yeast strains that produce particular flavour profiles and they endeavour to maintain the genetic purity of these strains, as certain mutations can produce changes in beer flavour. It is fortunate that beer brewing strains of *S. cerevisiae* are polyploid or aneuploid, as they are much more stable and 'less prone to mutation effects' than haploid and diploid strains. However, this makes application of conventional strain development methods problematical; fortunately, alternate genetic engineering techniques are now available.

In beer brewing, both top-fermenting strains and bottom-fermenting strains of *S. cerevisiae* are employed; the latter were formerly classified as *S. carlsbergensis* or *S. uvarum*. Top-fermenting yeasts exhibit flotational-flocculation behaviour and have been primarily used for making ales and stouts. Bottom-fermenting yeast perform sedimentary-flocculation and their traditional role is in the production of lagers in cooler fermentations (see Chapter 1, Yeasts). Flotation and sedimentation behaviour of flocculated yeast cells largely depends upon the dimensions of the fermentation vessel and physical conditions rather than the strain of yeast. This phenomenon is important for removing yeast from beer at the end of the primary fermentation. However, early separation must be avoided, otherwise the fermentation does not go to completion.

Top or 'ale' yeasts are more clearly identified as being unable to ferment the disaccharide melibiose (α-1,6-linked D-galactose and D-glucose units), whereas bottom 'lager' yeasts can utilize it because they possess an appropriate α-galactosidase, melibiase (Fig. 12.5).

Yeast management

Stock cultures of yeasts are maintained at low temperature or stored in freeze-dried form. A stock sample is grown up through a specific propagation procedure to produce a suitable quantity for inoculation, otherwise known as '**pitching**', of the production fermenter. However, most brews are not 'pitched' with fresh yeast, but with yeast recovered from a previous fermentation. Yeast may be reused 5–10 times, sometimes more, before fresh inoculum is prepared.

Fermentations produce 4–5 times more yeast than is required for further fermentations. Surplus yeast may be used to supplement distiller's yeast in whisky fermentation, dried for animal feed, or processed into yeast extract and B vitamin supplements. The portion to be

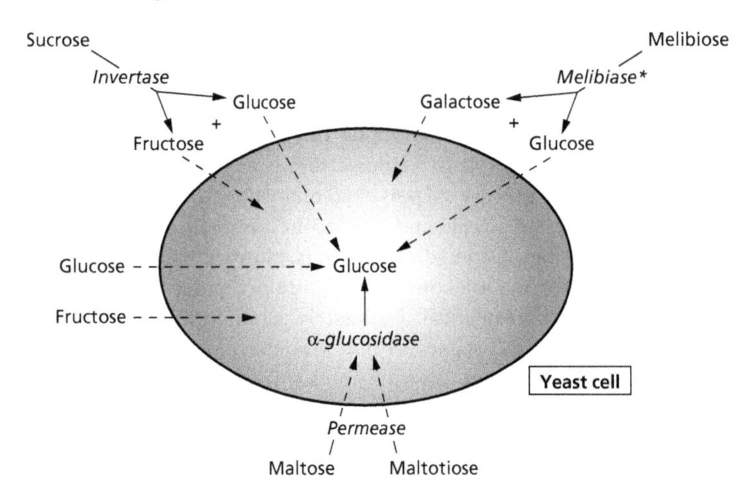

Fig. 12.5 Sugar uptake by *Saccharomyces cerevisiae* (*present in lager yeast strains).

reused is stored under refrigeration. Prior to repitching it may be subjected to acid-washing to remove any bacterial contaminants. This treatment is selective because yeasts are more tolerant of low pH than are bacteria. Acid treatment normally involves maintaining the yeast at pH 2.0–2.5 for up to 2 h in dilute phosphoric acid or acidified ammonium persulphate. Nisin, a natural food preservative (see Chapter 13), has been examined as a replacement for acid-washing. It kills the lactic acid bacteria that spoil beer, but has little or no effect on the yeast. The viability of yeast used for repitching should be at least 90–95%, otherwise subsequent fermentation rates are too slow.

Dried yeast preparations are easy to handle, store and transport. They are used extensively in bread and wine making, but have only recently been advocated for beer brewing.

Yeast nutrition

Brewing yeasts require the following nutrients, which are normally supplied by a well-balanced wort:

1 a carbon and energy source, provided by fermentable carbohydrates;

2 a nitrogen source, comprising amino acids and peptides;

3 a source of calcium, magnesium, phosphorus and sulphur, and traces of copper and zinc ions; and

4 growth factors, including biotin and pantothenate.

Yeast storage carbohydrates provide an endogenous source of carbon and energy during non-growth phases. For example, after pitching into wort there is a 4–5 h lag period, during which glycogen reserves are used, prior to the uptake of wort sugars. Consequently, the glycogen levels become depleted, but are replenished towards the end of the fermentation. These glycogen reserves are the main storage carbohydrate of anaerobically grown cells, whereas under aerobic conditions, trehalose is a major reserve.

Current brewer's yeast strains cannot directly ferment starch and higher dextrins, hence the necessity for malting and mashing processes to generate fermentable sugars. The resultant wort contains glucose, fructose, sucrose (only small quantities), maltose, maltotriose, and both linear and branched higher dextrins. They are utilized in that order, with some overlap up to maltotriose, but higher dextrins are not fermented. Usually, maltose and maltotriose are the main wort fermentable sugars (see Table 12.2), and it is essential that they are fermented efficiently. However, in some yeast strains, utilization of the two sugars is subject to catabolite repression by glucose (see Chapter 3, Microbial metabolism). Consequently, they cannot be metabolized until the wort glucose concentration falls below a certain level, which can be a major limiting influence on the overall fermentation rate.

Modes of uptake of the various sugars are somewhat different (Fig. 12.5). The monosaccharides enter via facilitated diffusion mechanisms, but sucrose is predominantly hydrolysed to glucose and fructose outside the cell by a wall-bound invertase. Maltose and maltotriose are transported intact by their respective permeases, which are inducible in some strains and constitutive in others. Once inside the cell, both are hydrolysed to glucose by intracellular α-glucosidases.

Free amino nitrogen (FAN) levels in wort are around

140 mg/L. The concentration and spectrum of wort amino acids are important for two reasons. First, they are required in protein synthesis and microbial growth, and second they have a major influence on beer flavour via their conversion to key flavour compounds (see p. 190). Amino acids are transported into the cell by a limited number of permeases and, like the sugars, they are taken from the wort in a specified order.

Although alcoholic fermentations are anaerobic, the initial oxygen concentration of wort is vitally important. The specific oxygen level required is strain dependent, but a minimum oxygen concentration of 10 mg/L is usually aimed for in wort prior to fermentation. If the oxygen level is inadequate, yeast growth and ethanol production are usually impaired. The oxygen is required for the synthesis of sterols; these unsaturated lipids are essential cell membrane components. Yeast that has been previously grown aerobically can grow satisfactorily in conditions of low initial oxygen concentration. However, the yeast in most beer fermentations is derived from a previous anaerobic fermentation and so ferments poorly under these conditions.

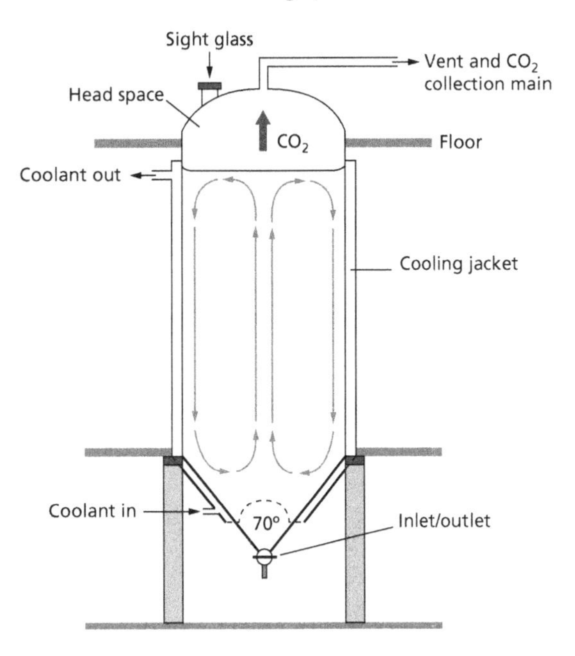

Fig. 12.6 Diagram of a cylindroconical fermentation vessel showing the flow of wort during fermentation.

The fermentation process

The main factors that affect the fermentation rate and influence beer quality are:
1 the amount of yeast used to inoculate or pitch the fermentation;
2 yeast cell viability and yeast quality;
3 the level of dissolved oxygen in wort at pitching;
4 wort soluble nitrogen concentration;
5 wort fermentable carbohydrate concentration; and
6 temperature.

British ales are traditionally produced in relatively shallow open circular or rectangular vessels, constructed of wood or stone. They use top-fermenting yeasts that rise to the surface and can be skimmed off. Traditional European lager fermentations use similar, albeit deeper, open vessels where the yeast ultimately collects at the bottom. Nowadays most modern brewery fermenters are closed cylindroconical vessels constructed of stainless steel, with a capacity of up to 200 000 L (Fig. 12.6). They normally have cooling jackets and are used to produce both ales and lagers. Advantages of cylindroconical vessels include:
1 improved fermentation rates;
2 greater flexibility, as they can be used to produce a range of beers;
3 relatively low construction costs, requiring relatively little civil engineering work and occupying less land area;
4 large fermentation capacity, giving benefits of economies of scale;
5 lower running costs;
6 easy CO_2 collection from the top of the fermenter and yeast removal from the conical base, where yeast collects at the end of the fermentation;
7 beer losses are reduced to a minimum;
8 efficient temperature control;
9 consistent product quality;
10 easy to clean using modern cleaning-in-place (CIP) systems; and
11 the same vessels may also be used for beer maturation following yeast removal.

The dimensions of these vessels aid rapid fermentation rates, due to greater mixing as the CO_2 bubbles rise through the deep fermentation (Fig. 12.6). Fermentation rates are primarily controlled by adjusting the temperature. However, care must be taken as higher temperatures may cause flavour defects. Ale fermentations in these vessels take 2–3 days at 12–24°C (Fig. 12.7), whereas in a traditional vessel they last for 4–7 days. Lager fermentations are usually performed at lower temperatures of 3–14°C. They last for 5–7 days in

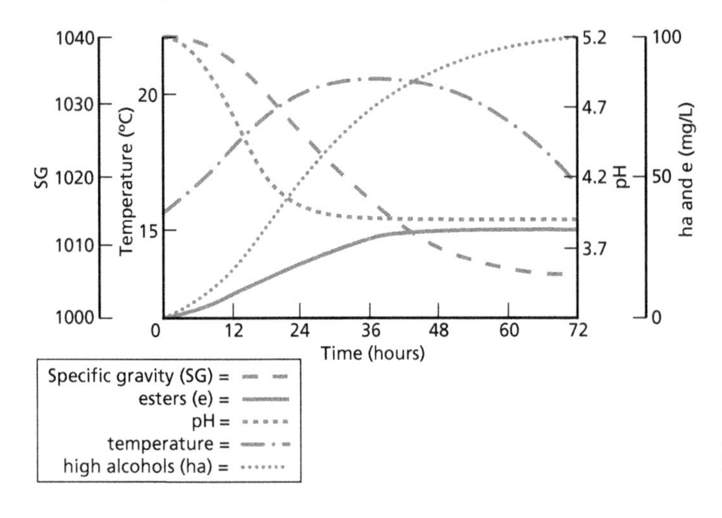

Fig. 12.7 An ale fermentation (after Lewis & Young (1995)).

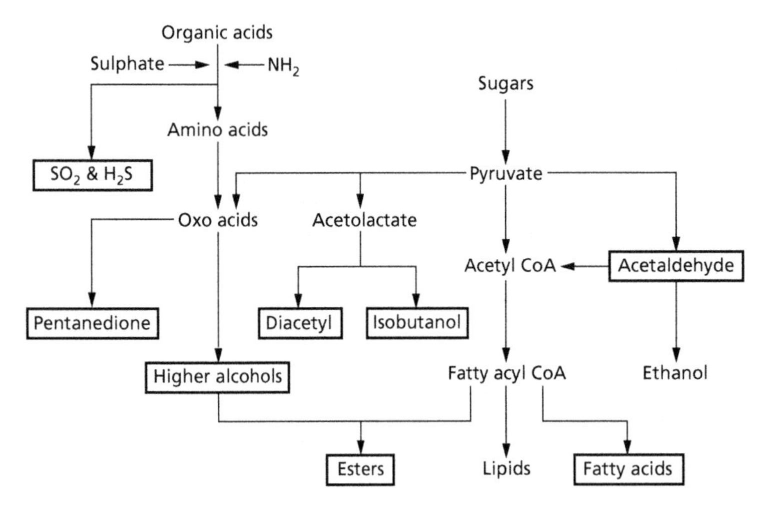

Fig. 12.8 The interrelationship between yeast metabolism and the production of organoleptic compounds (boxes).

these cylindroconical vessels, compared with 8–10 days for a traditional fermentation vessel.

Beer fermentations can be monitored by measurement of CO_2 evolution, ethanol production, the amount of heat generated or decline in wort specific gravity. The specific gravity of worts for ales is normally around 1.040 and falls to about 1.008 by the end of fermentation (Fig. 12.7). For lagers the initial gravity is often higher, at 1.050. Wort pH starts at around 5.2 and falls to about pH 4.0 in finished beer, whereas the number of yeast cells in suspension rises through the fermentation period and falls towards the end, as the yeasts flocculate.

Metabolic products of yeasts

Wort sugars are metabolized via the Embden–Meyerhof–Parnas (EMP) pathway to pyruvate, followed by its decarboxylation to acetaldehyde, which is then reduced to ethanol (see Chapter 3, Microbial metabolism). Apart from ethanol and CO_2, probably the next most abundant product is glycerol. This is formed in significant amounts, as a byproduct of the EMP pathway from dihydroxyacetone phosphate, via conversion to glycerol 3-phosphate. Acids, including acetic, lactic, succinic and caproic, are also excreted, causing the pH to fall to around 4.0 by the end of the fermentation. Other minor metabolic products that are important beer flavour compounds are shown in Fig. 12.8. They are as follows.

Higher alcohols (fusel oils). Higher alcohols include the aliphatic alcohols butanol, hexanol and propanol, and aromatic alcohols such as 2-phenylethanol. They can be synthesized from carbohydrates or formed from wort amino acids (Fig. 12.9), e.g.

valine → 2-oxo-3 methyl butanoate → 2 methyl propanol (isobutanol)

phenylalanine → phenylpyruvate → 2-phenylethanol

Greater amounts are generally produced at higher fermentation temperatures. Their production may be part of an alternate mechanism for the regeneration of NAD^+ (Fig. 12.9), if the main route via pyruvate conversion through to ethanol is inadequate. These compounds have flavour thresholds (the minimum concentration that can be detected on tasting) of 10–600 mg/L.

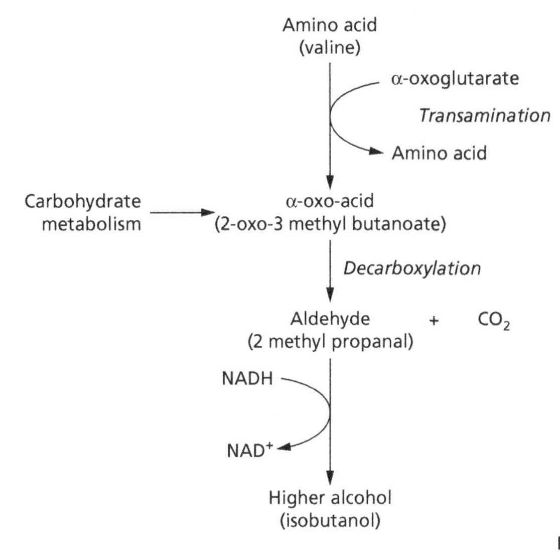

Fig. 12.9 Higher alcohol formation (example, isobutanol formation from valine).

Esters. Esters are synthesized by the reaction of acyl coenzyme A (CoA) esters with alcohols. For example, acetyl CoA plus ethanol gives ethyl acetate, which is the main beer ester, alongside isoamyl acetate, isobutyl acetate, ethyl caproate and phenylethyl acetate. They have flavour thresholds of around 2–25 mg/L.

Aldehydes. Acetaldehyde (ethanal), the most significant aldehyde, has a flavour threshold of 15 mg/L. It is formed by the decarboxylation of pyruvate. Excess acetaldehyde is produced if zinc ions are at low concentration in the wort, as they are required by the dehydrogenase responsible for the reduction of acetaldehyde to form ethanol.

Diketones. Diacetyl and pentane-2,3-dione are the main diketones of beer. Diacetyl has a sweet butterscotch aroma and flavour, with a flavour threshold of 0.1 mg/L. Overproduction may result from inadequate valine levels in the wort, which necessitates initiation of valine synthesis by the yeast, and consequent production of diacetyl as a byproduct (Fig. 12.10).

Sulphur compounds. Sulphur compounds have low flavour thresholds, often below 5 μg/L. Dimethyl sulphide is a characteristic flavour component of lagers and is primarily derived from malt, although yeasts can produce a proportion of it. Unwanted high levels of hydrogen sulphide are generated by some yeast strains if pantothenate or methionine are in short supply.

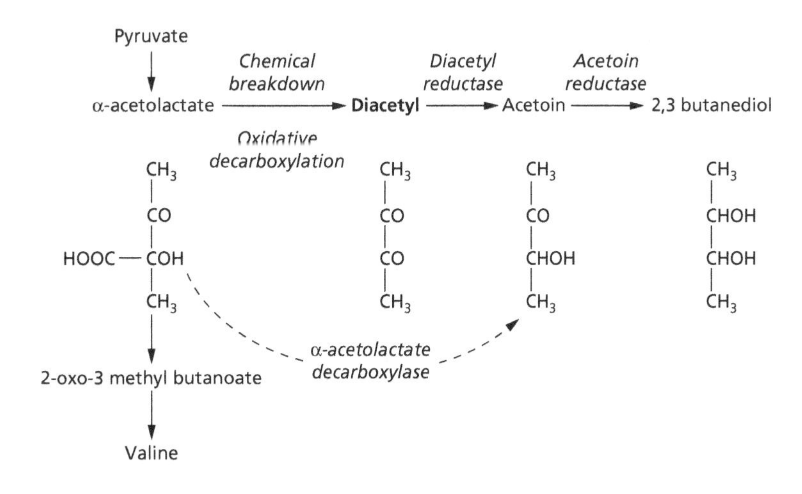

Fig. 12.10 The elimination of diacetyl from beer.

POST-FERMENTATION TREATMENTS AND MATURATION

Beer that comes from a primary fermentation in a cylindroconical vessel contains relatively little yeast, as the bulk is removed by sedimentation into the conical base. This young or 'green' beer then needs to undergo a period of maturation or conditioning before it is ready to be consumed. The exact nature of this maturation varies depending upon the beer type.

Cask and bottle conditioning of ales

This traditional method is performed by adding priming sugar to enable the remaining yeasts, approximately 2.0×10^6 cells/ml, to undergo a secondary fermentation to carbonate the beer within sealed bottles or casks. Other additions to the beer may include dry hops or hop products for flavour enhancement. Isinglass finings are also added, which are prepared from fish swim-bladder collagen. This material is positively charged and interacts electrostatically with negatively charged polysaccharide components of yeast cell walls, helping to sediment the yeast and thus clarify the product.

Krausening

Krausening is a traditional German process involving a secondary fermentation similar to cask conditioning, but it is performed in large tanks. Here the additional fermentable sugars and yeast are provided by adding 10% (v/v) of freshly fermenting wort, referred to as the krausen. This speeds removal of aldehydes and diketones, and later sealing of the vessel results in carbonation.

Lagering

In this treatment lager beer is held in a tank at 8°C for several weeks or even months. No priming sugars are added, but the remaining yeasts continue to slowly ferment residual wort sugars. The early stages of the secondary fermentation generate CO_2 that purges the beer to remove some unwanted volatile compounds, particularly aldehydes and diacetyl. Later the tanks are sealed to retain the CO_2 for carbonation of the beer.

Storage ageing

This is now the most commonly used bulk maturation procedure. No further fermentable sugars are added and the levels of yeast remaining after the primary fermentation are very low. The beer is stored at low temperature (–1 to 4°C) for 7–10 days, to encourage chill-haze particles to form (see below), which are later removed by filtration. Vessels may also be purged with CO_2 to eliminate some unwanted volatile compounds. At this stage, carbonation is usually unnecessary, as the beer is predominantly carbonated on packaging into bottles and cans, or when dispensed from bulk containers, such as kegs and tanks.

Protein haze control

The use of the plant protease papain to control protein hazes was first patented in the USA in 1911, but now higher purity microbial proteases from *Candida olea* and *Pichia pini* are often preferred. Great care is required in their application as some beer proteins are major contributors to foam stability and characteristics. Proteases are added to the fermentation or postfermentation to prevent **chill-haze**, which would otherwise appear on storage at 3°C. In contrast, **permanent haze** develops on prolonged storage at ambient temperatures and does not redissolve, whereas chill-haze redissolves on warming. Hazes are mostly formed from proteins and polyphenols (tannins), and may be influenced by the presence of certain metal ions. Stabilization to prevent chill-hazes is also commonly achieved by treatment with silica hydrogels, polyvinylpyrrolidone or tannic acid.

Prevention of oxidation

Oxygen remaining in beer can promote the activity of beer spoilage acetic acid bacteria. It is also responsible for the oxidation of lipids which can affect beer flavour stability and foam character. For example, linoleic acid may be oxidized to form carbonyl compounds such as trans-2-nonenal, a compound that has a very low flavour threshold and a stale cardboard-like flavour. Removal of oxygen may be achieved by using ascorbic acid or SO_2, but the latter can cause sulphury off-flavours. This can be avoided by using commercial glucose oxidase in combination with catalase, to scavenge for oxygen (see Chapter 9).

Speeding beer maturation

Maturation, particularly for lightly flavoured lagers and

ales, largely depends on the rate of removal of the diketones, **diacetyl** and **2,3 pentanedione**. They are derived from α-acetolactate and α-acetohydroxybutyrate, respectively, which are intermediates in the biosynthesis of the amino acids valine and iso-leucine. The α-acetolactate, for example, is secreted into the beer by yeast and certain spoilage microorganisms. Here it undergoes chemical breakdown to diacetyl and is slowly reduced to acetoin and 2,3 butanediol by yeast reductases (Fig. 12.10). A similar process occurs with 2,3 pentanedione, which is ultimately reduced to 2,3 pentanediol. The addition of extra diacetyl reducing enzymes has not proved successful in speeding diacetyl removal. However, addition of bacterial α-acetolactate decarboxylase before fermentation prevents diacetyl formation by direct conversion of its precursor, α-acetolactate, to acetoin.

Special finishing processes

Over recent years, low-alcohol beers have become increasingly important. The alcohol may be removed from conventionally produced beer by vacuum evaporation, dialysis or reverse osmosis. Alternatively, the fermentation may be performed in such a way as to restrict ethanol production. There are three categories of these products: low-alcohol, containing up to 1.2% (v/v) ethanol; dealcoholized, 0.05–0.5% (v/v) ethanol; and alcohol-free, with less than 0.05% (v/v) ethanol.

Other beers that have become popular since the mid-1990s and now constitute a significant proportion of the beer market include 'ice beer'. Ice beers are produced by cooling the beer to –2 to –4°C to form ice crystals. The beer is held under these conditions for several days and then filtered to remove the ice crystals. This treatment increases the relative amount of alcohol by about 15% and modifies the flavour, particularly removing components responsible for 'aftertaste'. This is achieved during ice crystal formation, as the crystals absorb the bitter tasting tannins, proteins and polyphenols.

Packaging

Most beer is clarified by filtration prior to final biological stabilization by sterile filtration or flash pasteurization, and eventual packaging into bottles, cans, kegs and tanks. Alternatively, small packaged products are often batch pasteurized after packaging. Carbonation of the small packaged products is performed before packaging. In some cases a mixture of CO_2 and nitrogen is now used for certain beers to improve foam stability.

MICROBIOLOGICAL PROBLEMS DURING BREWING

Microbial infections can produce off-flavours and aromas, slime and turbidity. They may arise from contaminated raw materials or poorly cleaned equipment and vessels. During mashing and sweet wort preparation, only thermophilic lactic acid bacteria are likely to cause problems. They are sensitive to hop constituents and generally do not survive in hopped wort. Cooled hopped wort is nutritionally rich, contains oxygen and is at a reasonable pH of 5.0–5.5. It is prone to infection by coliforms, acetic acid bacteria, lactic acid bacteria, *Obesumbacterium* (*Hafnia*) and wild yeasts. However, infections are relatively rare, unless the wort is stored for long periods prior to pitching or is pitched with contaminated yeast. As the fermentation progresses the conditions become less suitable for most potential spoilage organisms. Unlike wort, beer is a hostile environment for most microorganisms as it has:

1 a low pH of 3.5–4.5;
2 little or no oxygen;
3 hop constituents that have antimicrobial properties;
4 low levels of readily utilizable carbon sources; and
5 a relatively high ethanol concentration.

In finished beer, infections by acetic acid bacteria are now very rare, due to the ability to eliminate oxygen. *Zymomonas* infections of ales, particularly if cask conditioned, were once a problem, but are now relatively uncommon. Currently, the most problematic organisms are lactic acid bacteria, *Lactobacillus* species and *Pediococcus* species. In some regions the Gram-negative obligate anaerobes *Pectinatus* species and *Megasphaera* species are also found to infect lagers.

FUTURE DEVELOPMENTS IN BEER PRODUCTION

Continuous beer fermentation was first attempted over 100 years ago and further developed between the late 1950s and early 1970s. However, despite several potential advantages it failed to gain commercial acceptance, apart from in New Zealand. Its advantages include reduced labour costs, higher production due to a greater throughput from smaller volume plant, constant product quality, less down-time, lower cleaning costs, more efficient use of raw materials and greater overall profitability. Currently, established batch systems have spare capacity. Also, they require less monitoring, are more flexible (capable of producing a range of products) and microbial contamination is less potentially damaging. Nevertheless in the future, continuous systems, probably utilizing immobilized yeast, may be introduced. Some immobilized yeast reactors are already used for speeding maturation.

The mode of enzyme application in brewing may also move to immobilized systems, which have the advantages of not introducing foreign protein into the product and reuse. This may be particularly beneficial in treating finished beer, notably for reducing the levels of dextrins using amyloglucosidase, removing oxygen with glucose oxidase and catalase, and preventing diacetyl formation by α-acetolactate decarboxylase treatment. Attempts have been made to remove haze-forming proteins using immobilized pig pepsin, but it produced beer with poorer head-retention properties due to the hydrolysis of proteins needed for this phenomenon.

Now that techniques have been developed for the genetic engineering of polyploid and aneuploid brewer's yeast strains, attempts are being made to improve their properties. Major objectives are to reduce the need for commercial enzymes and to decrease the reliance on malt enzymes. Introduction of genes for appropriate enzymes and the ability to secrete them, where necessary, would reduce process costs and increase process efficiency. These targets for genetic engineering include:

1 amylolytic enzymes to facilitate starch and dextrin hydrolysis; one genetically engineered brewing yeast, containing the STA 2 gene, a glucoamylase, from *S. cerevisiae* var. *diastaticus*, has already been approved for use in the UK;
2 suitable proteases for haze control;
3 β-glucanases for reduction of hazes and filtration problems;
4 α-acetolactate decarboxylase, possibly from *Acetobacter* species, to speed maturation;
5 oxygen-scavenging systems;
6 enzymes allowing the fermentation of other carbohydrates, such as cellobiose, lactose and xylose;
7 high-gravity brewing characteristics, e.g. improved tolerance to ethanol, CO_2 and high osmotic pressure;
8 increased production of SO_2 to improve flavour stability and reduce staling;
9 improved alcohol acetyltransferase activity, producing more acetate esters to enhance flavour;
10 production and secretion of zymocins (proteinaceous yeast killer toxins, produced by some yeasts to kill other sensitive yeasts) to control potential wild yeast contaminants;
11 production and secretion of nisin from *Lactococcus lactis* to control contamination by lactic acid bacteria (see Chapter 13); and
12 derepressed strains to overcome catabolite repression and speed the fermentation.

There have also been several suggestions for the incorporation of genes for high-value heterologous proteins that have no brewing role but could be recovered from waste yeast.

Wine production

Wine can be made from any plant extract or fruit juice that contains sufficient levels of fermentable sugars. However, most wine is made from the berries of the grape vine, primarily cultivars of *Vitis vinifera*, which can be grown in most temperate regions of the world. These grape wines have become major fermentation products in Central and Southern Europe, North and South America, South Africa, Australia and New Zealand. Total world production is now approximately 10^{10} L/annum.

Grape juice is particularly successful for wine production because it contains high levels of fermentable sugars (mainly glucose and fructose; 160–240 g/L) and other nutrients that are necessary for successful yeast fermentation. The natural acidity (pH 2.8–3.8) inhibits potential spoilage organisms, as does the high level of ethanol that can be achieved during fermentation. Grapes also have very pleasant flavour and aroma components. Most wines are table wines that normally contain up to 14% (v/v) ethanol. However, some regions produce wines fortified with ethanol, in the form of added spirit or grape brandy, to give final ethanol levels up to 22% (v/v). The main fortified wines are sherry, port and madeira; along with vermouths, which contain various flavour supplements derived from herbs and spices.

There is an immense range of wines, which in the first instance may be classified by their colour, which ranges from white, through rose to dark red, and sweetness, which varies from very dry to very sweet. Other flavour components and quality factors are influenced by, or derived from (Fig. 12.11):

1 the specific grape variety or mixture used (e.g. Muscat grapes contain highly characteristic flavour and aroma components, the terpene alcohols linalool and geraniol);

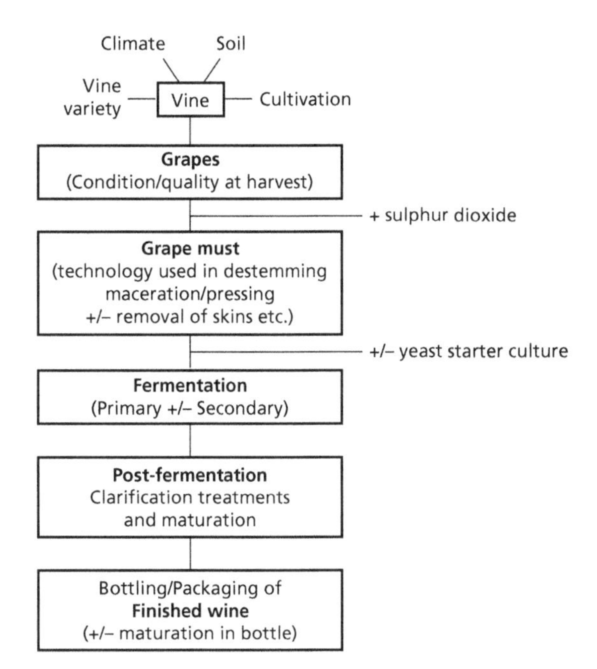

Fig. 12.11 Factors that influence the properties of a wine.

2 the soil, local climate and the mode of grape cultivation;
3 grape maturity and condition at harvest;
4 the methods of grape pressing and processing;
5 the primary fermentation, which generates variable quantities of ethanol, CO_2, higher alcohols, esters, aldehydes and ketones;
6 any secondary fermentation to produce wine carbonation, or for acidity reduction via a malo-lactic fermentation, whereby malic acid is converted to lactic acid; and
7 post-fermentation and maturation treatments.

PRODUCTION OF GRAPE MUST

Grapes may be picked by hand or mechanically harvested and then conveyed to the winery for processing. This involves mechanical destemming, crushing and pressing to form the **must**. Red wine must consists of the macerated fruit, whereas for white wines it is only the clarified juice that is fermented. Carbonic maceration, an alternative to pressing for red wine production, involves holding intact grapes for several days under a blanket of CO_2. This results in cell death and breakdown of cellular structure. For red wine, skins and seeds are not usually removed until the end of fermentation. This allows anthocyanins (red pigments that also add astringency) to be slowly extracted from the skins during the fermentation. Pectinolytic enzymes may be added during pressing to aid both juice and colour extraction.

White wines may be produced from either red or white grapes, provided that the source of colour, the skins, is removed. For white wines, irrespective of grapes used, early separation of the juice from skins, seeds and plant debris is essential, to reduce astringency and browning. Oxidative browning results from the oxidation of phenolic compounds, and may be prevented by maintaining anaerobic conditions, using a CO_2 blanket during and after crushing. Addition of SO_2 or centrifugation of the juice to remove membrane-bound phenol oxidases (browning enzymes) also helps to prevent browning.

In some regions addition of other fermentable sugar is allowed to supplement grape sugars and the acidity may be adjusted for low-acid musts.

WINE FERMENTATION

Alcoholic fermentation of wine is carried out by indigenous yeast or added starter yeast. During the fermentation the yeast must grow to a sufficiently high cell density to complete the fermentation. For dry wines this requires conversion of all sugar into alcohol. The fermentation rate is influenced by:
1 the starter yeast strain used or yeast strains that make up the natural microflora;
2 temperature;
3 pH;
4 initial sugar concentration; and
5 nutritional components of the batch of grape must and any supplements of thiamine, ammonium salts and sterols that may be added to ensure maintenance of yeast activity until the fermentation is complete.

Traditionally, open wooden fermentation vessels were used, but these have largely been replaced by stainless steel cylindroconical vessels with cooling coils or jackets. Fermentations require cooling, as for every 10 g/L sugar fermented the temperature rises by 1.3°C and it can rise to 38°C. White wine fermentations are often cooled to 10–20°C to control the development of certain flavour compounds and they continue for about 1–4 weeks. The temperature for red wine fermentation is normally maintained at 24–29°C. Consequently, the fermentation is shorter, lasting only 3–5 days. Yeast fermentation stops when all available sugar has been metabolized, leaving less than 1 g/L of residual sugar. However, the winemaker can terminate the fermentation earlier, if required, to retain sweetness. This is achieved by removing the yeast via filtration or inactivating it through cooling, heating or adding a metabolic inhibitor such as SO_2.

Traditional spontaneous wine fermentations

Once the juice is released from the grapes it will ferment spontaneously as it is immediately exposed to the natural indigenous microflora. These microorganisms enter the fermentation from the vineyard, winery environment, grape skins, plant debris, soil, equipment surfaces and the air. The natural microflora were traditionally, and still are in some wineries, the sole source of fermentation yeasts. Their proliferation is favoured by the low pH, high sugar concentration and anaerobic conditions. Such spontaneous fermentations result in a succession of wild yeast types, often initiated by *Kloeckera* species and other relatively alcohol-intolerant species. It culminates in the dominant activity of strains of the alcohol-tolerant *S. cerevisiae*. The specific succession varies depending upon the native microflora of the location and the temperature.

Traditional fermentations are comparatively slow due to the low levels of yeast that are initially present. They can also be prone to some flavour and aroma defects due to the overactivity of certain yeasts. To ensure complete fermentation, and avoid production of unwanted flavour and aroma compounds, winemakers encourage the growth and dominance of *S. cerevisiae*. This may be achieved by adding SO_2 to the grape must at a concentration of 30–50 mg/L, which inhibits non-*Saccharomyces* yeast and spoilage bacteria. As a result, *Saccharomyces* species gain an advantage and increase from less than 10^3 to 10^7–10^8 cells/ml, thereby dominating the fermentation. In addition, they are favoured by an initial must temperature of 16–20°C. Temperatures below 14°C favour other yeasts such as certain *Kloeckera* species that can produce large amounts of unwanted acetic acid and ethyl acetate. However, some strains of *Kloeckera* may be beneficial to wine flavour. In well-balanced fermentations, desirable complex flavour and aroma characters develop due to the variety of organisms involved, and can result in very fine complex wines.

Use of starter cultures

Generally, newer wine-producing regions have less well-developed natural microflora. Therefore, added starter cultures are normally used, which are often derived from dried *S. cerevisiae* preparations, originating from high-quality red, white, fortified and sparkling wines. These wine yeasts have generally been found to be homothallic diploids, whereas beer brewing strains are polyploid or aneuploid. The wine strains cannot be readily differentiated using classical methods, but suitable molecular techniques have now been devised.

Use of starter cultures allows specified quantities of selected strains of *S. cerevisiae*, with known viability and desired fermentation characteristics, to be added to the must to initiate fermentation. Also, they will be free from contamination by other yeasts and bacteria. The yeasts are usually added to the must to give an initial cell density of approximately 5×10^6 cells/ml. They grow to more than 10^7 cells/ml and dominate the fermentation. Any indigenous non-*Saccharomyces* yeasts are suppressed by adding 30–50 mg/L SO_2 and maintaining the fermentation temperature above 16°C. Use of starter cultures provides several advantages: it ensures completion of the fermentation and allows the fermentation rate to be controlled, to give a shorter lag period, and more even and rapid fermentation rates. Also, there is less opportunity for the production of off-flavours by wild yeasts and bacteria, giving more consistent flavour characteristics and quality.

The desirable characteristics that a starter yeast should possess include tolerance to: alcohol, high sugar musts, SO_2, low and high temperatures, and high pressure, especially for the production of sparkling wines. It is also important that they produce only small amounts of acetic acid, acetaldehyde, H_2S, mercaptans, diacetyl, SO_2 and higher alcohols. Other beneficial properties are low foam production, good flocculation properties to help clarify the young wine and suitability for drying, allowing long-term storage of starter cultures.

The construction of wine yeast strains that express additional beneficial characteristics is being pursued via recombinant DNA technology. In many cases, this will obviate the need for commercial enzymes. Targets include:

1 production of desirable organoleptic compounds;
2 metabolism of a large portion of malic acid in must;
3 low urea excretion;
4 improved fermentation efficiency;
5 suitable flocculation characteristics;

6 controllable ester formation;
7 protease production;
8 production of inhibitory substances against spoilage microorganisms, e.g. nisin to kill Gram-positive bacteria (see Chapter 13) and zymocins to kill wild yeasts;
9 pectinase production to improve wine filtration rates;
10 production of glucanases that release bound terpenes to enhance aroma; and
11 greater dihydroxyacetone phosphate reductase and glycerol phosphatase activities to increase glycerol levels.

Also, strains suitable for immobilization may be useful for some purposes. For example, their use in producing sparkling wine is currently being examined.

Continuous wine fermentation

Continuous fermentation has potential advantages over batch systems (see Beer brewing p. 193). However, it can be successfully operated only in regions where there is a large readily available supply of grapes for continuous maceration to feed into the fermenter. This has been attempted in southern France for the production of red wine using fermenters of 500 000 L capacity. Over 150 000 kg of grapes are processed each day and a corresponding amount of wine is generated. The average residence time of the wine in the fermenter is 3 days at 26–28°C.

Secondary wine fermentation

Following the primary alcoholic fermentation the winemaker can, for specific purposes, encourage the further activities of yeasts or other microorganisms. Such secondary fermentations include a yeast alcoholic fermentation in bottles or cask to create naturally carbonated sparkling wines, or the growth of aerobic surface/film yeasts to produce fino sherry; or lactic acid bacteria may be encouraged to perform a malo-lactic fermentation.

Malo-lactic fermentations can occur spontaneously in wine through the action of lactic acid bacteria (*Lactobacillus*, *Leuconostoc* and *Pediococcus* species), which are naturally present in the wine. This activity is undesirable for many wines and may be prevented by adding SO_2. However, it may be beneficial for certain wines, particularly for red wines from cool climates that often contain high levels of malic acid. The malo-lactic fermentation reduces the acidity through the decarboxylation of this dicarboxylic acid to lactic acid, a monocarboxylic acid (Fig. 12.12). Where malo-lactic fermentations are required, suitable lactic acid bacteria such as *Leuconostoc oenos* may be specifically inoculated into the wine at the end of the alcoholic fermentation. Not only is acidity reduced, but other useful flavour components are produced and the wine becomes more microbiologically stable. It is rare for other microorganisms to grow in wine after it has been subjected to a malo-lactic fermentation.

POST-FERMENTATION TREATMENTS

Following fermentation the wine must be protected against both microbiological and non-microbiological spoilage, particularly oxidation. It can be stabilized by cooling, maintenance of anaerobic conditions and preliminary clarification by settling and racking-off the wine from the sediment to remove microorganisms. Alternatively, the winemaker may choose to leave the wine on the yeast sediment (lees) for 2 weeks to 9 months, in order to release more yeast flavour compounds. A small amount of SO_2 may be added to inhibit further microbial activity. It also protects the wine against chemical oxidation and complexes with free acetaldehyde. Final SO_2 levels remaining in wine are normally 60–200 mg/L. Some wine is stored for a period of a few weeks to several years in wooden casks, where it acquires additional flavour characters. Ultimately the wine is filtered, pasteurized or sterile filtered, and filled into bottles or other packages. The wine components

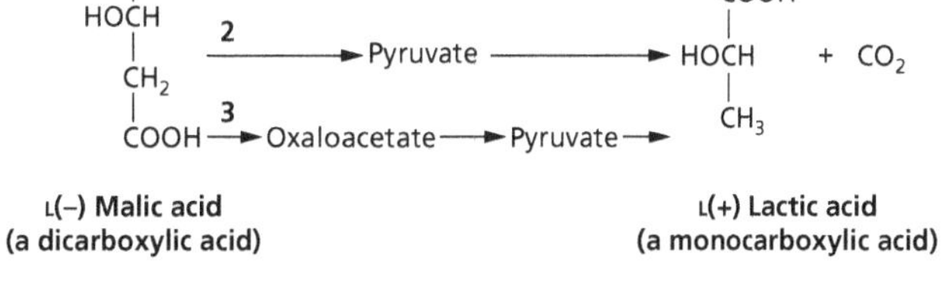

1 Direction conversion
2 → Pyruvate →
3 → Oxaloacetate → Pyruvate →
COOH–HOCH–CH_2–COOH
COOH–HOCH–CH_3 + CO_2
L(–) Malic acid (a dicarboxylic acid)
L(+) Lactic acid (a monocarboxylic acid)

Fig. 12.12 The conversion of L (–) malolactate to L (+) lactic acid (three enzymic pathways for conversion of malic acid to lactic acid, resulting in loss of a carboxyl/acid group, hence reduction in wine acidity).

continue to interact during further storage, slowly modifying colour, flavour and aroma.

Cider production

Cider is an alcoholic beverage prepared from apple juice. It has been produced for over 2000 years and is now made almost everywhere that apples (*Malus pumila*) are grown. In the UK, for example, the annual production is over 4.5 million litres. The cider making process is very similar to wine production and results in products that range from sweet to very dry, usually containing 2–8% (v/v) alcohol. Traditional cider is often uncarbonated and may not be clarified. Some may be served as naturally conditioned cask cider, analogous to cask-conditioned ale. However, most ciders are clear and sterile filtered or flash pasteurized products. They are artificially carbonated and analogous to bottled and keg beer. In France, carbonation may be achieved via the Charmat process. This technique, also used to produce some sparkling wines, involves a bulk secondary fermentation under pressure.

APPLE JUICE EXTRACTION

Cider making starts with the picking of the apples, usually specific cider varieties that contain high levels of phenolic compounds. Single apple cultivars can be used, but more often cider is prepared from blends. The cleaned fruit is coarsely milled and the resulting apple pulp is pressed to extract the juice using either a large cylindrical press or a continuous belt press. Pressed pomace, consisting of skin, pips and core, can be used for cattle feed or pectin may be extracted for use in jam and confectionery manufacture. Freshly pressed juice may be fermented immediately, or concentrated up to eightfold and stored for later fermentation, following pasteurization and pectin removal. Care has to be taken in the production of concentrate as furfurals (sugar degradation product) may be formed, which could impede subsequent fermentation.

COMPOSITION OF APPLE JUICE

Apple juice composition varies depending upon the apple varieties used. Compared with brewer's wort, it has a lower pH of 3.0–4.0, contains 6–8 times less soluble nitrogen, and mono- and disaccharides are virtually the only sugars present (Table 12.5). A typical sugar composition is 75% fructose, 15% sucrose and 10% glucose, with only small quantities of other simple sugars, oligosaccharides and starch. Apple juice also contains soluble pectin, which consists of polymers of galacturonic acid esterified with methanol. The major acid is L(−) malic acid, along with variable quantities of the phenolic acids: quinic, chlorogenic, shikimic and *p*-coumarylquinic. Polyphenolic constituents are tannins, primarily epi-catechin, and dimeric and trimeric pro-anthocyanidin. Other minor components include ascorbic acid, minerals and esters such as ethyl-methyl butyrate.

Table 12.5 Apple juice composition

Total sugars	100–130 g/L
fructose	45–65%
sucrose	14–45%
glucose	5–25%
sorbitol	1–10%
Amino nitrogen (FAN)	10–110 mg/L
Acidity	20–180 mmol/L (pH 3.2–4.2)
Tannins	0.6–4.6 g/L
Specific gravity	1.050–1.060

CIDER FERMENTATION

Traditional wooden fermentation casks are used without temperature control, whereas modern commercial vessels are usually temperature-controlled vessels of 2000–9000 L capacity, constructed of lined concrete, lined mild steel or stainless steel. Fermenter depths in excess of 14.5 m, which produce hydrostatic pressures of 1.5 atm, have been shown to impair cider yeast performance.

Fresh juice or reconstituted concentrate may be supplemented with fermentable sugars before fermentation. Cane sugar is the usual source of additional fermentables, without which alcohol levels rarely exceed 6.0% (v/v). Traditional fermentations are performed by indigenous yeasts that originate from the skins of the fruit, the environment and processing equipment. Normally, sound ripe apples have less than 500 yeast cells per gram of fruit. These are primarily *Kloeckera apiculata*, *Aureobasidium pullulans* and species of *Rhodotorula*, *Torulopsis*, *Candida* and *Metschnikowia*, whereas *Saccharomyces* species and other sporulating yeasts are rarely found. Acid-tolerant bacteria such as *Gluconobacter* species are usually present but lactic acid bacteria are rare. However, levels of microorganisms rise if the fruit is allowed to fall to the ground and especially when the skin is damaged. The traditional-type fermentations are usually performed at ambient temperature. They are slow to start, usually beginning within 1–2 days, and continue for several weeks.

In modern commercial operations the juice is treated to remove microbial contaminants by high-speed centrifugation, sterile filtration or pasteurization. If the apple juice is heat-treated constituent pectinases are denatured. As a result, cider does not clear as pectins remain undegraded unless additional pectinases are added. Apple juice is commonly treated with SO_2, which has both antioxidant and antimicrobial function. The quantity of SO_2 required to prevent oxidative browning and control unwanted microorganisms is dependent upon the pH of the juice: the lower the pH, the less SO_2 is required. In many countries the maximum legal limit for sulphur dioxide is around 200 mg/L, which adequately retards aerobic yeasts, and both lactic and acetic acid bacteria.

The treated juice is fermented with an added starter culture of *S. cerevisiae*. These yeast preparations originate from cider fermentations or may be selected winemaking strains. Lower soluble nitrogen levels, compared with brewer's wort, lead to relatively slow fermentation rates, which is exacerbated by the lower concentrations of yeasts used for pitching. This is often 5–15 times lower than in beer production, starting at only 10^6 cells/ml, rising to 5×10^7 cells/ml by the end of the fermentation. Supplementing the juice with ammonium salts does not increase the fermentation rate, but makes its progress more predictable. These cider fermentations are normally performed at 20–25°C and last for 1–4 weeks. Ideally, the cider yeasts should exhibit the following properties:

1 rapid initiation of fermentation;
2 resistance to SO_2 and high ethanol concentrations;
3 low requirement for growth factors;
4 ability to complete attenuation by fermenting all sugars;
5 suitable flocculation characteristics;
6 development of a sound organoleptic profile; and
7 production of polygalacturonase to degrade soluble pectin.

During the yeast fermentation there is a decrease in pH due primarily to the formation of L(−) malic acid by the yeast. Also, minor quantities of a range of other acids are produced, including gluconic, lactic and succinic acids, and mono-, di- and trigalacturonides, derived from the enzymic degradation of pectin. Higher alcohols are important organoleptic components, but their actual levels depend on the apple variety used, juice treatment, the yeast strain employed, and both the fermentation and storage conditions. Generally, low pH and low nitrogen levels result in ciders with increased higher alcohol levels, whereas use of clarified apple juice and sulphur dioxide produces lower concentrations.

Towards the end of the fermentation, the yeasts release nitrogenous compounds into the cider, including amino acids and peptides, along with pantothenic acid, riboflavin and some phosphorus compounds. This release of nutrients is important for any subsequent bacterial malo-lactic fermentation. After the primary fermentation is completed, the yeast is allowed to settle to the bottom of the fermentation vessel. Cider is either racked (decanted from the sediment) or partially clarified by centrifugation before storage under an inert gas blanket for several months.

CIDER MATURATION

Naturally carbonated cider is a relatively small proportion of the market and is produced by carrying out a secondary yeast fermentation. However, a more common secondary fermentation is malo-lactic fermentation (see Wine production, p. 197), which can sometimes occur earlier, during the primary yeast fermentation. It is performed by lactic acid bacteria, usually non-slime-forming strains of *Leuconostoc mesenteroides*, *Lactobacillus collinoides* and occasionally *Pediococcus cerevisiae*. In addition to the conversion of malic to lactic acid, thereby reducing acidity, it brings about the conversion of quinic acid to dihydroshikimic acid and generates other flavour compounds.

At the end of maturation the cider can be blended with new and old ciders to moderate the flavour, thus maintaining a consistent flavour profile for the product. Those ciders produced with added sugar reach 9–11% (v/v) alcohol and may be diluted with water before further processing. Cider blends are normally clarified by centrifugation or filtration, followed by sterile filtration or flash pasteurization at 85°C for 20 s. These products are artificially carbonated during packaging and to maintain stability, chemical preservatives may be added, e.g. SO_2, sodium benzoate and potassium sorbate.

Distilled beverages

Distilled beverages are prepared by the fermentation of sugars derived directly from plant extracts and fruit juices, or indirectly from hydrolysed grain and root starch. The yeast fermentation normally produces an intermediate product containing a maximum of 14% (v/v) ethanol. Many of these are akin to products that are desirable alcoholic beverages in their own right, such as beer, wine and cider (Table 12.1), and whose production has been discussed earlier. Distillation of these intermediate products results in whisky from unhopped beer, and brandies from wine and cider. In some instances the intermediates are not related to a product of direct beverage use and it is only the final distilled product that is of commercial interest. Several of these, including gin and vodka, can be prepared from virtually any source of ethanol. They are colourless neutral spirits with little innate flavour, but may be later coloured and flavoured with plant materials, such as herbs and spices. Many distilled beverages are self-flavoured, notably grape and other fruit brandies, whiskies prepared from malt and various starch sources, and rum derived from fermented sugar cane extract or molasses.

WHISKY PRODUCTION

Whiskies are prepared from malted barley, exclusively in the case of Scottish and Irish malt whiskies, or with various proportions of added cereal starch for other whiskies, such as bourbon, grain and rye whiskey. Blended Irish and Scottish whiskies contain a mixture of grain and malt whiskies. The malt, and other cereals where used, are extracted and saccharified to form a wort, essentially as for beer brewing, except that no hops are added (see Beer brewing, p. 179). However, the wort is often not boiled prior to fermentation, thus retaining the activity of mash enzymes. In some cases, the whole mash is fermented rather than separating the spent grains from the wort and is termed 'in grains' fermentation. Certain fermentations, referred to as 'sour mash', also have a preliminary fermentation with lactic acid bacteria, usually *Lactobacillus delbrueckii*. They are allowed to act for 4–6 h to produce 1.5% (w/v) lactic acid before being killed by boiling. The yeast fermentation is then performed by specific distilling strains of *S. cerevisiae*, which are sometimes supplemented with surplus beer brewing yeast. Some whisky yeasts are able to ferment maltodextrins because they are hybrids between *S. cerevisiae* and *S. cerevisiae* var. *diastaticus* which possess amyloglucosidase (glucoamylase) activity. The fermentation is pitched to give an initial yeast concentration of 2×10^7 cells/ml that increases to 2×10^8 cells/ml by the end of the fermentation. Fermentations last for about 30–40 h and the emphasis is on ethanol yield. The resultant beer is then distilled, often without separating the yeast. Distillation residues may be used as an animal feed supplement.

Scottish malt whisky and Irish whiskey are distilled in pot stills, twice and three times, respectively. Others such as Scottish grain, bourbon and rye whiskies are distilled in a continuous still, usually a Coffey-type still (see Chapter 7). Products from pot stills are about 110° proof (62.9%, v/v ethanol), whereas those from a Coffey still are at 160° proof (91.4%, v/v ethanol) and are usually diluted to around 110° proof prior to maturation (note: 100° proof is the minimum concentration of ethanol in water that still allows ignition when mixed with a standard recipe of gunpowder; pure ethanol is 175° proof).

The distillate is always colourless; any colour in the final product is derived from the wooden barrels in which it is matured or from added caramel colouring. Maturation is usually performed in oak casks for several years, the minimum period depending upon the specific local legislation. Casks can be made of new wood, often internally charred to make the wood more 'reactive'. Used casks, which have previously held sherry or bourbon, can also be used to impart particular characteristics to the whisky. Wood components, referred to as 'wood congeners', are extracted into the whisky during maturation, giving colour and flavour. These compounds include aromatic aldehydes, lactones, furfurals and tannins. At the end of maturation the whisky may be blended and finally diluted, usually to 70° proof (40%, v/v ethanol) before bottling.

Vinegar production

The normal course of events when a natural solution of fermentable sugars (fruit juice or other plant extracts) is exposed to indigenous microbial activities is an alcoholic fermentation by yeasts, followed by bacterial oxidation of the ethanol to acetic acid (acetification). If sufficient acetic acid is produced it may be classified as vinegar. This term is derived from the French words *vin* (wine) and *aigre* (sour). Generally, vinegar is classified as a condiment that contains a minimum of 4% (w/v) acetic acid.

Vinegars have been produced for several thousand years from local fermentable substrates and alcoholic beverages. They are mainly produced from fruit juices and sugar syrups, including maple syrup, molasses and honey, but may also be manufactured from saccharified cereal or root starch. Where alcohol sources are used directly, including wine, cider and spirit alcohol, they require only an acetification step. If spirit alcohol is the starting material, additional nutrients must be added prior to acetification.

Sources of table and preserving vinegar vary around the world. In the USA, cider vinegar is the major source, whereas red and white wine vinegars predominate in central and southern Europe, and malt vinegar is commonly used in the UK. Small quantities of other vinegars, such as various fruit and balsamic vinegars, are also produced. Today, the worldwide annual production of vinegars is around 2000 million litres.

Vinegar fermentations

ALCOHOLIC FERMENTATION

This is an anaerobic process performed by specific vinegar-making strains of yeast, usually *S. cerevisiae* var. *ellipsoideus*. In some traditional processes endogenous yeasts may be employed rather than a specific starter culture. For malt vinegar production, the preparation of the fermentation medium, the wort, and the yeast fermentation are essentially the same as for beer brewing. Fermentation products are ethanol, CO_2, and small amounts of glycerol, acetic acid and some higher alcohols.

ACETIC ACID FERMENTATION

The acetic acid fermentation is a highly aerobic process, essentially a biotransformation by acetic acid bacteria, involving incomplete oxidation of ethanol to acetic acid (Fig. 12.13). Other minor products include acetaldehyde, ethyl acetate and acetoin. Industrial acetic acid bacteria are members of the genera *Acetobacter* and *Gluconobacter*, mostly *A. aceti*, *A. europaeus*, *A. hansenii*, *A. rancens*, *A. xylinum* and *G. oxydans*. In some instances, mixed cultures are more efficient than a single organism. These bacteria may be divided into two main groups by their action on acetic acid. Members of the peroxidans group, including *A. aceti* and *A. pasteurianum*, can metabolize acetic acid to CO_2 and H_2O, whereas the suboxydans group, such as *Gluconobacter* species, cannot perform this so-called over-oxidation. Some *Acetobacter* species are of an intermediate type that carry out some over-oxidation at lower acetic acid concentrations.

These aerobic bacteria are Gram-negative rods that are naturally found on plant material, often in association with yeasts. They resemble pseudomonads, but may be distinguished by their high acid tolerance, low peptolytic activity, limited motility and lack of pigments. The group all share the ability to form acids by the incomplete oxidation of sugars and alcohols. This is facilitated by the possession of a family of dehydrogenases having alcohols, glucose or polyols as substrates. All possess an unusual prosthetic group (non-protein chemical group bound to the enzyme) called pyrrolquinoline quinone (PQQ), which is an electron mediator that delivers electrons to the dehydrogenase. These dehydrogenases are localized on the outer surface of their cytoplasmic membrane.

Methods of acetification in vinegar manufacture

Conversion of alcohol to acetic acid is largely a problem of aeration technology, which must bring into intimate contact the liquid to be oxidized with acetic acid bacteria and oxygen. There are three main techniques that can be used to achieve this.

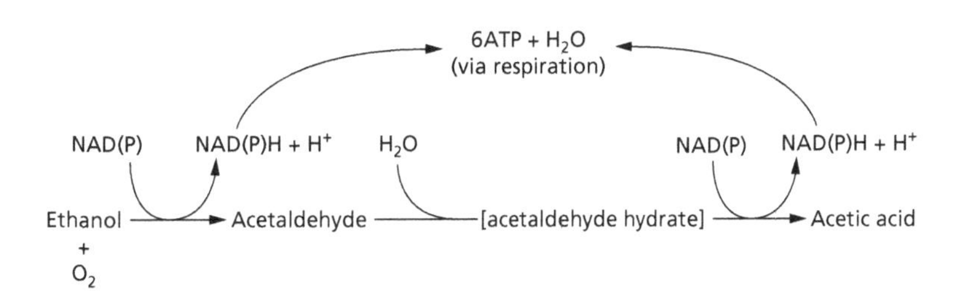

Fig. 12.13 The oxidation of ethanol to acetic acid.

TRADITIONAL SURFACE METHODS

Surface processes involve natural microflora that first carry out an alcoholic fermentation of sugars derived from plant juices or saccharified starch. The resultant alcoholic solution, contained within part-filled barrels, then undergoes a spontaneous natural fermentation to convert ethanol to acetic acid. This is reliant upon appropriate acetic acid bacteria establishing themselves on the liquid surface where ethanol oxidation occurs. These traditional batch methods are generally slow and unpredictable. However, the Orleans or French method, initially developed for wine acetification at the end of the 14th century, was an early attempt at semicontinuous production. It involved charging fresh wine with some raw vinegar from a previous run to speed establishment of acetifying microflora. Some surface methods are still used to produce small quantities of special vinegars.

TRICKLING GENERATORS

These methods involve the movement of alcoholic liquid over surfaces on which films of acetic acid bacteria have become attached (a supported film system), and a good air supply is provided to facilitate ethanol oxidation.

A widely used system based on this technique is the vinegar generator (Fig. 12.14). This process was first developed by Schutzenbach (1823) for the commercial production of vinegar in Germany. Traditional generators consist of a cylindrical wooden vessel of up to 60 m^3, divided horizontally into three sections. The upper section receives the alcoholic solution, which then trickles down through the main middle section that is loosely packed with inert materials, such as beech wood or rattan shavings, corn cobs and charcoal. These materials provide a very large surface area on which acetic acid bacteria can become established and over which forced air is passed from below. The bottom section serves to collect the vinegar, from where it may be returned to the top of the same generator for recirculation or to a second generator in series.

Oxidation of ethanol produces heat: the top of the generator is often at 29°C, increasing to 35°C at the bottom. Therefore, cooling coils or jackets are normally provided to prevent overheating. The process time is about 3 days, a considerable improvement over surface methods, resulting in final acetic acid concentrations of 10–12% (v/v). Productivity of these generators is in the order of 10–15 kg/m^3 of reactor volume per day, with a yield of 85%, based on ethanol added.

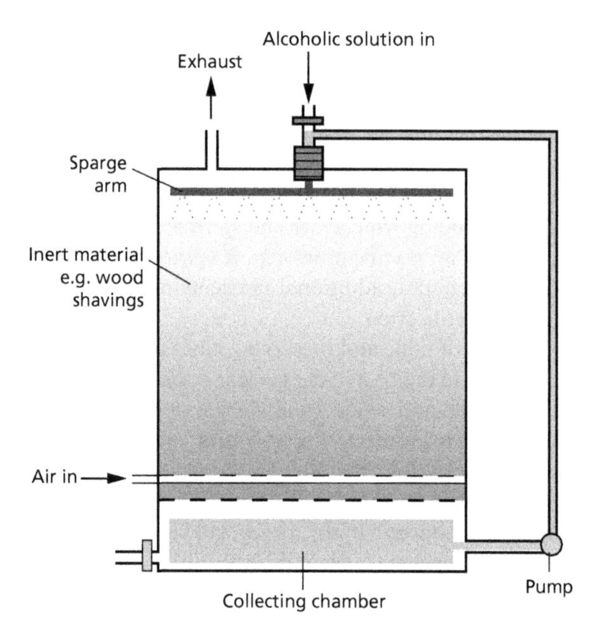

Fig. 12.14 Diagram of a vinegar generator.

SUBMERGED METHODS

Submerged methods were first devised in the late 1940s. They are five times faster than the trickling methods and require less capital investment. Since their introduction there have been several modifications that have improved aeration efficiency. The process frequently used, often based on Frings acetators, consists of a highly aerated stainless steel stirred tank reactor. This vessel is continuously stirred at 1450–1750 r.p.m. and is normally maintained at 24–30°C, although temperatures of up to 40°C have been reported. Specially selected strains of acetic acid bacteria are used that function well in suspension cultures. They grow in a suspension of very fine air bubbles within the fermenting liquid. Initial cell concentrations of around 1.5×10^8 cells/ml at the start of a cycle rise to 2.25×10^8 cells/ml, a 1.5-fold increase in population. Submerged processes are usually operated automatically on a semicontinuous basis. Starting medium for each cycle contains 7–10% (v/v) acetic acid and 5% (v/v) ethanol, higher ethanol concentrations being inhibitory. When the ethanol concentration falls to 0.05–0.3% (v/v), a portion of medium,

40–50%, is removed and replaced with fresh 'starting' medium. The production bacteria are very sensitive to both low oxygen and ethanol concentrations. They die rapidly if the oxygen supply is halted or if the ethanol becomes depleted. Final acetic acid concentrations of up to 14% (v/v) can be achieved using these methods. Their productivity is 50–60 kg/m^3 per day with a yield of over 90%.

Finishing processes

Vinegars produced by slow surface methods are less harsh and require shorter maturation. Those produced by the quicker methods benefit from a period of maturation to improve flavour and body. The vinegars are then diluted and clarified, which is accomplished by fining with bentonite and filtration. Some vinegars may be decolorized using potassium ferrocyanide. Finally, the liquor is pasteurized at 75–80°C for 30–40 s prior to bottling, during which SO_2 may be added at levels of 50–100 mg/L.

Defects and diseases of vinegar include production of precipitates or cloudiness caused by certain metal ions. Microbial infections by lactic acid bacteria may result in slime production and some acetic acid bacteria can cause overoxidation. Certain wild yeasts and fungi also oxidize acetic acid.

Recovery of acetic acid from low-quality or waste vinegars, for use as a chemical feedstock, can be achieved by methods such as fractional distillation and solvent extraction.

Process developments

Improved production efficiency of acetic acid could be achieved by modifying the producer organisms or the process. Targets for genetic modification of the acetic acid bacteria include: the introduction of alcohol dehydrogenases (ADH) from other organisms that possess better temperature, pH and kinetic characteristics; the amplification of ADH activity; and improved thermotolerance. Greater phage resistance is a further target, as bacteriophage problems have been encountered in both trickling generator and submerged methods of vinegar production. Regarding the process, acetic acid productivity could be increased by using cell immobilization within hollow fibres, ceramics or gel beads. However, the high oxygen requirement may restrict viable cells to the surface of the support medium. A more attractive approach may be through the use of membrane bioreactors, to facilitate cell recycle that maintains high cell concentrations. Also, the possible use of cyclone bioreactors is currently being evaluated.

Dairy fermentations

Milk from any mammal, but particularly from cow, horse, sheep, goat and buffalo, may be used to produce fermented dairy products such as cheese, butter, sour cream, kefir and yoghurt. Their production has traditionally been a means of preservation, which is accomplished through changes in water content, and acid and bacteriocin formation. The final products also have enhanced texture, flavour and aroma. More recently popularized dairy products include fermented 'health' drinks and foods, referred to as probiotics. These products contain live bacteria such as *Lactobacillus acidophilus* and *Bifidobacterium* species, which are alleged to improve the functioning of the gut and stabilize its microflora.

The microorganisms used in the production of fermented dairy products are primarily lactic acid bacteria, which are naturally present in milk. Other organisms, notably filamentous fungi, may be involved in later maturation processes of some cheeses. Starter cultures, often containing specific mixtures of lactic acid bacteria, are now predominantly used to inoculate pasteurized, rather than raw milk. Use of pasteurized milk not only reduces the possibility of growth of potential pathogens, but also lowers the activity of certain milk enzymes that can affect some dairy fermentations.

Basic manufacturing processes are very similar for many of these dairy products. Fermentation generates lactic acid, which modifies milk proteins, and forms flavour and aroma compounds, notably diacetyl, which imparts a buttery flavour. The lactic acid bacteria used as starter cultures are somewhat prone to infection by destructive bacteriophages. Attempts to overcome these problems involve the use, wherever possible, of aseptic techniques and the introduction of phage-resistant bacterial strains.

Butter production

There are two main types of butter, sweet cream butter and cultured butter, but only the latter involve the use of microorganisms. Cultured butter is usually prepared from pasteurized cream ripened with bacteria for 24–48 h prior to churning. *Lactococcus lactis* ssp. *diacetylactis* and *Leuconostoc citrovorum* are often used, which produce acid and flavour compounds, particularly diacetyl.

Yoghurt production

Yoghurts are traditional sour milk preparations that have become major dairy products. There are several variants, but commercial forms are usually prepared from whole milk which may be supplemented with protein such as skim milk powder. This helps to form the protein-gel structure of the product. These raw materials are heated and then cooled prior to inoculation; the heating is necessary, otherwise later protein coagulation does not produce a smooth gel. Inoculation involves a mixed starter culture containing thermophilic strains of *Streptococcus thermophilus* and *Lactobacillus delbrueckii* ssp. *bulgaricus* in a ratio of 1:1. The former produces mainly acid, whereas the latter generates more organoleptic compounds, particularly acetaldehyde. Their proteolytic enzymes and extracellular polymers also aid protein-gel formation. Yoghurt can be pasteurized to improve storage-life or remain 'live', the latter reputedly having probiotic qualities.

Cheese production

The bulk of world cheese production, which is now over 10^{10} kg/annum, is from 5×10^{10} L of cow's milk. Cheese making essentially involves concentration of milk fat and protein by removing water. However, there are numerous different types of cheese with widely varying texture and flavour. Textures range from soft through to very hard, and flavours vary from mild to very strong (Table 12.6). The vast range is due to differences in:

1 the type of milk used;
2 the diet of the milk producer, which in turn is influenced by the local soil, vegetation and climatic factors;
3 the particular strain of microorganisms used for inoculation at each phase of production, including starter and ripening stages, which contributes different organoleptic compounds and modifies the texture;
4 the processing methods employed; and
5 the environmental conditions during ripening, particularly temperature and humidity.

There is no standard method of cheese making, as limitless variations exist for all stages of the process and a vast quantity is still produced with little or no reference to the underlying science. However, for most cheeses, the process basically involves (Fig. 12.15):

1 pretreatment of raw milk;
2 formation of solid curd;
3 removal of the liquid whey from the curd;
4 curd processing; and
5 ripening and ageing.

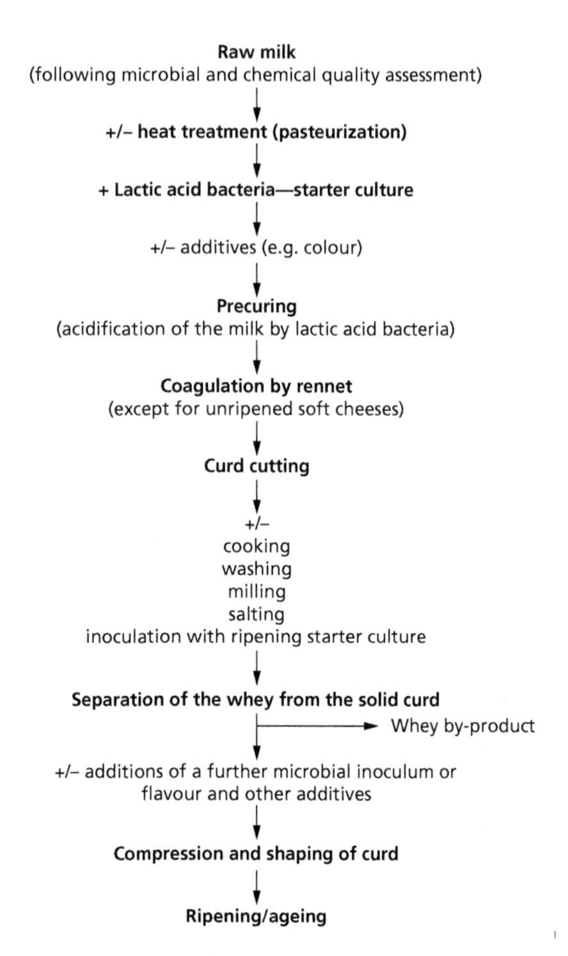

Fig. 12.15 Stages involved in the manufacture of cheese.

First, raw milk is checked for various chemical and microbiological quality parameters and then pasteurized. The production of coagulated milk proteins or curd is then achieved by the activities of lactic acid bacteria, such as *Lactococcus lactis*, *L. cremoris* and *Streptococcus thermophilus*. These bacteria have the ability to lower the pH through the fermentation of lactose to lactic acid, which facilitates protein coagulation. They also influence the flavour of the final product by producing specific flavour and aroma compounds, and perform essential proteolysis and lipolysis in later maturation.

A mixed starter culture is often used, consisting of

Table 12.6 Examples of cheeses with different texture and the microorganisms involved in their manufacture

Cheese texture	Microorganisms involved in manufacture
Soft (unripened)	
Cottage	*Lactococcus lactis*
	Leuconostoc citrovorum
Cream	*Lactococcus cremoris*
Mozzarella	*Lactobacillus bulgaricus*
	Streptococcus thermophilus
Soft (ripened, 1–5 months)	
Brie and Camembert	*Lactococcus lactis*
	Lactococcus cremoris
	*Penicillium camemberti**
	*Penicillium candidium**
	*Brevibacterium linens**
Semi-soft (ripened, 1–12 months)	
Gorgonzola and Roquefort	*Lactococcus lactis*
	Lactococcus cremoris
	*Penicillium glaucum**
	*Penicillium roqueforti**
Hard (ripened, 3–12 months)	
Cheddar	*Lactococcus lactis*
	*Lactobacillus casei**
	Lactococcus cremoris
	Streptococcus durans
Edam and Gouda	*Lactococcus lactis*
	Lactococcus cremoris
Gruyère	*Lactococcus lactis*
	Lactobacillus helveticus
	Streptococcus thermophilus
	Propionibacterium freudenreichii ssp. *shermanii**
Very hard (ripened, 12–18 months)	
Parmesan	*Lactococcus lactis*
	*Lactobacillus bulgaricus**
	Lactococcus cremoris
	Streptococcus thermophilus

* Involved in later stages of the process.

several strains of these mesophilic or thermophilic streptococci and lactobacilli, which may be prepared in heat-treated milk or whey-based media. Use of defined starter cultures reduces batch-to-batch variations in both production time and levels of acid generated. An inoculum of 0.5–2.0% (v/v) is added and the fermentation is performed at around 32°C for 10–75 min. For some cheeses this may be further controlled by heat treatment at 55°C, which inhibits mesophiles and promotes the action of thermophiles. Thus, the initial selection of suitable microbial strains, the amount of starter culture used, the length of preripening and the incubation temperature are important in creating many subtle differences in the final colour, flavour and aroma.

Curd formation may be promoted, in all but soft, unripened cheeses such as cottage and cream cheeses, by the addition of specific proteolytic enzymes. Traditionally, rennin (chymosin, aspartic protease EC 3.4.23.4) is used, which is prepared in a crude form from the stomach, abomasum, of veal calves and is referred to as **rennet**. Due to a shortage of available calf chymosin and the requirement for so-called 'vegetarian' cheeses, fungal proteases with similar properties to calf chymosin are now also employed. In addition, the calf chymosin gene has been introduced into several microorganisms for the commercial production of recombinant enzyme (see Chapter 9).

The milk component primarily involved in curd formation is the protein casein, which is a mixture of α-1, α-2, β and κ caseins. κ casein is important in maintaining the colloidal stability of milk proteins. Addition of rennin results in the removal of surface glycopeptide from the κ casein. Consequently, the casein becomes unstable and aggregates in the presence of calcium ions to form a gel. As the gel forms it entraps fats and ultimately forms white creamy lumps, referred to as curd. Precipitated curd is soft but can be readily separated from the liquid whey by holding the mixture in cheese cloth. Semi-dried curd that remains in the cheese cloth is usually salted and other ingredients may be added, such as colouring agents, herbs, or a further microbial inoculum. It is then pressed and placed into a shaping mould or cut into blocks. For many cheeses this is followed by a period of ripening or ageing to develop the final flavour and texture.

Ripening involves the modification of proteins and fats by microbial and milk proteases and lipases that remain in the young cheese. Some countries allow acceleration of flavour development through the addition of commercial enzyme preparations. Lysozyme may also be added in the manufacture of hard-cooked Emmental, Gouda and Gruyère-type cheeses to prevent growth of the spoilage organism, *Clostridium tyrobutyricum*, which otherwise would be troublesome in the later stages of ripening. In addition, the curd may have been inoculated with a bacterial or fungal culture before ripening. For example, *Propionibacterium freundenreichii* ssp. *shermanii* is used for the production of Swiss-type cheeses, e.g. Emmental and Gruyère. As these bacteria grow they modify the flavour and generate gas bubbles that result in holes or eyes within the cheese.

Internally mould-ripened blue-veined cheeses, including Danish Blue, Gorgonzola, Roquefort and Stilton, primarily use *Penicillium roqueforti*. Traditional manufacture relies on the natural development of the mould that originates from spore populations that become established in the local cheese-making environment. Large-scale production now uses spore inocula, which are mixed into the curd before the cheeses are pressed and formed. Young cheeses are usually punctured with stainless steel rods to promote fungal growth by increasing oxygen levels within the cheese and then stored under controlled humidity at around 9°C. Cheeses are ripened for up to a year during which they develop the characteristic blue veins, and the aroma and flavour that is due to methyl ketones such as 2-heptanone.

Camembert-type and other surface-ripened cheeses use *Penicillium camemberti*, which originates naturally from the environment, or the surface of the cheese may be sprayed with a spore inoculum. The mould grows on the surface of the cheese for 1–6 months to produce the characteristic white crust or rind. At the same time, its hydrolytic enzymes are secreted into the cheese where they modify the flavour and texture.

DEVELOPMENTS IN CHEESE MAKING

Bacteriophage infections of starter cultures have been a major problem in cheese and other dairy fermentations. The use of phage inhibitory media, defined strains, the rotation of strains and phage-insensitive bacterial strains has improved the situation. However, phage infections continue to cause problems within the industry. Attempts are being made to harness the natural defences of the bacteria, such as inhibition of bacteriophage adsorption to the bacterium and blocking of phage nucleic acid penetration. The genes for these phage resistance factors appear to be plasmid encoded and resistance plasmids may be exploited to improve the ability of industrial strains to resist phage attack.

In lactic acid bacteria, the genes for many of the key enzymes involved in lactose and casein metabolism are not chromosomal, but plasmid-borne and prone to being lost. The mapping of the plasmids and subsequent isolation of these genes has allowed them to be transferred to the chromosome, providing strains with greater stability. Such techniques may also lead to the introduction of genes from other lactic acid bacteria and heterologous genes. These may provide additional properties, especially the ability to produce extra organoleptic compounds, remove off-flavours, speed ripening and destroy microbial contaminants.

Alternate methods of introducing additional enzymes, and antimicrobial agents to prevent spoilage include the use of encapsulation technology. The enzyme or agent can be contained within liposomes (microscopic phospholipid spheres) and added, probably most conveniently during curd processing.

Probiotics

Probiotics are organisms and substances that promote the development and maintenance of balanced intestinal microflora. Lactic acid bacteria, particularly strains of *Lactobacillus acidophilus*, *L. casei* and *Bifidobac-*

terium bifidum, are considered capable of helping to restore this balance and in so doing, improve health. Several fermented yoghurt-like dairy products and milk-based carriers containing these specially selected live probiotic microorganisms are now available. However, the actual mechanism underlying these activities has yet to be elucidated. Their effects appear to include the ability to stimulate the immune response, often seen as an increase in the production of immunoglobulin A (IgA). Within the gut these bacteria promote colonization, and modify both gut bacterial numbers and their metabolic activity. They may improve intestinal integrity and reduce the mutagenic effect of certain intestinal metabolites. Their ability to aid re-establishment of gut microflora could be particularly useful in animals and humans following the application of oral antibiotics. Also, lyophilized preparations of viable *Saccharomyces boulardii* cells (e.g. Perenterol 250) are used in the treatment of diarrhoea, particularly when associated with oral antibiotic treatment.

Probiotics may also have beneficial effects when given to farm animals. For example, when food and water inoculated with *Lactobacillus salivarius* is fed to 1-day-old chickens infected with *Salmonella enteritidis* (a cause of salmonellosis) the salmonellae are prevented from colonizing their gastrointestinal tracts.

Other traditional fermented foods

Fermented meat and fish

Gram-positive micrococci are particularly important in the production of fermented meat and fish products. Their fermentative activities are used as a means of preservation for a wide range of dry (pepperoni, salami, etc.), semidry (cervelat, mortadella, etc.) and some undried (Braunschweiger, teewurst, etc.) meat sausages. The biological stabilization achieved is usually through the addition of salts and the generation of lactic acid by these bacteria, which are tolerant to both salt and low water activity. Bacteria involved are usually *Pediococcus cerevisiae*, *Lactobacillus plantarum* and *Staphylococcus carnosus*. These fermentations may now be initiated by adding starter cultures rather than merely relying upon the natural indigenous microflora. Over several days fermentation and then weeks of ripening, the pH is reduced to around 5.0, and there are also changes in texture, flavour and colour. Control of *Clostridium botulinum* is a major objective, which may be promoted by the presence of nitrites. Reduction of added curing agent, nitrate, to nitrite may be enhanced by the nitrate reductase activity of certain strains of these micrococci.

Fermented fish products, primarily sauces and pastes, are used as food flavouring agents. They are mainly produced in south-east Asia, where they are major commodities with an annual production of 250 000 tonnes. These products are prepared from whole, chopped or mashed fish and shrimps which are salted and sealed into vessels for several months. Some preparations incorporate other ingredients such as rice and flavourings. The fermentation is performed by the natural microflora that develops under these specific conditions, which is mostly strains of *S. carnosus* and *Staphylococcus piscifermentans*.

Fermented plant products

Fruits, cereals, leaf and root vegetables, legumes and oilseeds are all used for the production of fermented foods. The fermentation is generally a means of preservation, but can also improve digestibility, nutritional value, texture and flavour.

BREAD (see also Chapter 14, Baker's yeast production)

Use of yeast to leaven cereal products dates back over 4000 years. Their primary function is to generate carbon dioxide, which can be controlled by the quantity of yeast added, level of fermentable sugars present and the temperature. The rising of bread relies on the fact that wheat and several related cereal grains contain gluten proteins. Glutens contribute to the final flavour of bread and have the characteristic of forming long molecular strings when they are 'kneaded'. These bind the bread together and have important elastic properties, allowing the formation of a dough. Bread dough traps the carbon dioxide generated by the baker's yeast (normally specific strains of *S. cerevisiae*, see p. 219) and rises due to the pressure of the carbon dioxide build-up. The yeast also add flavours, alcohol and acids. Importantly, it helps mature or chemically modify the gluten to promote even expansion of the dough and gas retention during baking. On baking, the bread protein is denatured and, along with the starch, forms the typical open crumb texture.

Bread dough containing more than 20% rye flour must be acidified to produce acceptable bread. This is due to its lack of gluten, which requires other components to bind the necessary water (e.g. pentosans). They do this much better at lower pH and the acidic conditions also inhibit unwanted amylase activity. Acidification may be achieved by adding citric acid or by sour-dough fermentation; the latter also contributes additional flavour. The dough may be fermented by natural microflora of the dough ingredients. Initially, Gram-negative enteric bacteria are involved, followed by lactic acid bacteria that reduce the pH to 3.6–3.9. Alternatively, a portion of dough from a previous batch, predominantly containing a mixture of lactic acid bacteria, may be used as an inoculum. More recently, single strain commercial starter cultures of heterofermentative lactic acid bacteria have become available for this purpose.

SAUERKRAUT PRODUCTION

This product is prepared from wilted shredded cabbage mixed with 2–3% (w/w) salt to reduce water activity and inhibit the growth of Gram-negative bacteria. Modern commercial fermentations are conducted in fibreglass or concrete tanks with acid-resistant liners of up to 100 tonne capacity. In traditional processes natural microflora are used to perform the fermentation. The prevailing conditions favour lactic acid bacteria, especially *Lactobacillus plantarum* and *Lactobacillus brevis*, along with *Leuconostoc mesenteroides*, which tends to dominate the early fermentation until the acid level reaches 0.7–1.0% (v/v). These organisms are also involved in the production of fermented gherkins and olives. Alternatively, a starter culture of suitable lactic acid bacteria can be added to perform the fermentation. Where potential spoilage organisms are controlled by the addition of nisin (see Chapter 13), nisin-resistant starter cultures are obviously required. Sauerkraut fermentation lasts for 20–30 days at 18–20°C and final lactic acid levels reach 1.5% (v/v).

SOYA BEAN FERMENTATIONS

Fermented soya bean products are a major part of the diet in south-east Asia, where they are largely substitutes for fermented dairy products. Some of these foods are also becoming popular in the west. There are several different solid, paste and liquid products such as tempeh, miso and soy sauces, respectively. Tempeh production involves fermentation of cooked whole or dehulled soya beans by *Rhizopus* species, *R. oigosporus* often being the preferred organism in commercial processes. The resulting cake-like product can be cut into cubes and fried or cooked with other ingredients.

Soy sauces are used as condiments, or colouring and flavouring agents. There are many different types, but their manufacture involves three basic stages. The first stage, referred to as **koji**, is a solid substrate aerobic fermentation of cooked soya beans or steamed defatted soya flakes, along with wheat flour or rice. Normally, a mixture of strains of *Aspergillus oryzae* is used and at 25–30°C the hydrolysis of constituent starch, protein and pectins is accomplished in 2–3 days. **Moromi** is the second stage and involves an anaerobic fermentation of the liquid slurry that results from the addition of brine (24% (w/v) sodium chloride solution) to the koji. During this phase, the activities of *Pediococcus halophilus* initially acidify and prevent spoilage. Subsequently, *Candida* species or *Zygosaccharomyces rouxii* perform an alcoholic fermentation and produce additional flavour compounds such as furanones. The product is traditionally matured for 6–9 months to develop the full flavour before the liquid is filtered, pasteurized and finally bottled.

COFFEE, COCOA AND TEA FERMENTATIONS

Berries of the coffee plant (*Coffea arabica* and *Coffea canephora*) are processed to obtain the bean from within by either a dry or wet process. The wet process, which generally produces better quality beans, involves the activities of indigenous bacteria, filamentous fungi and yeasts. The pulpy mucilaginous material that surrounds the beans is removed by the action of endogenous plant enzymes, pectinolytic bacteria and fungi. This is followed by an acid fermentation by lactic acid bacteria, primarily *Leuconostoc mesenteroides* and *Lactobacillus brevis*. The residual pulp is washed from the beans, which are then dried, hulled and roasted.

Cocoa production involves similar microbial activities, but these are rather more important than for coffee. The pods of *Theobroma cacao*, which contain the beans, are opened and the mucilaginous material is removed by fermentation. Initially there is an alcoholic fermentation performed by a mixture of yeasts, notably *Candida*, *Hansenula*, *Pichia* and *Saccharomyces*. This is followed by an increase in the activity of lactic acid bacteria, particularly *Lactobacillus plantarum* and *Lactobacillus fermentum*. Finally, acetic acid bacteria from the genera *Acetobacter and Gluconobacter* predominate and oxidize the ethanol to acetic acid.

Tea is prepared from the leaves of *Camellia sinensis*. Although their preparation is referred to as a fermentation, production of fermented black tea and partially fermented teas does not involve microorganisms. Endogenous plant enzyme activities are responsible for the changes within the leaves following loss of moisture, withering and leaf maceration.

Further reading

Papers and reviews

Cogan, T. M. (1995) Flavour production by dairy starter cultures. *Journal of Applied Bacteriology* Supplement **79**, 49S–64S.

Hammes, W. P., Bosch, I. & Wolf, G. (1995) Contribution of *Staphylococcus carnosus* and *Staphylococcus piscifermentans* to the fermentation of protein foods. *Journal of Applied Bacteriology* Supplement **79**, 76S–83S.

Hammond, J. R. M. (1993) Brewer's yeasts. In *The Yeasts*, 2nd edition, Vol. 5: *Yeast Technology* (eds A. H. Rose & J. S. Harrison), pp. 7–67. Academic Press, London.

Hammond, J. R. M. (1995) Genetically modified brewing yeast for the 21st century—progress to date. *Yeast* **11**, 1613–1627.

Henick-Kling, T. (1995a) Control of malo-lactic fermentation in wine: energetics, flavour modification and methods of starter culture preparation. *Journal of Applied Bacteriology* Supplement **79**, 29S–38S.

Henick-Kling, T. (1995b) Removal of alcohol from beverages. *Journal of Applied Bacteriology* Supplement **79**, 19S–28S.

Huis In't Veld, J. H. J., Bosschaert, M. A. R. & Shortt, C. (1998) Health aspects of probiotics. *Food Science and Technology Today* **12**, 46–49.

Martini, A. (1993) Origin and domestication of the wine yeast *Saccharomyces cerevisiae*. *Journal of Wine Research* **4**, 165–176.

Quero, A. & Ramon, D. (1996) Review: The application of molecular techniques in wine microbiology. *Trends in Food Science and Technology* **7**, 73–78.

Randez-Gil, F., Sanz, P. & Prieto, J. A. (1999) Engineering baker's yeast: room for improvement. *Trends in Biotechnology* **17**, 237–244.

Rocken, W. & Voysey, P. A. (1995) Sour-dough fermentation in bread. *Journal of Applied Bacteriology* Supplement **79**, 38S–48S.

Rombouts, F. M. & Nout, M. J. R. (1995) Microbial fermentation in the production of plant foods. *Journal of Applied Bacteriology* Supplement **79**, 108S–117S.

Sievers, M. & Teuber, M. (1995) The microbiology and taxonomy of *Acetobacter europaeus* in commercial vinegar production. *Journal of Applied Bacteriology* Supplement **79**, 84S–95S.

Vaughan, E. E., Mollet, B. & deVos, W. M. (1999) Functionality of probiotics and intestinal lactobacilli: light in the intestinal tract tunnel. *Current Opinion in Biotechnology* **10**, 505–510.

Books

Angold, R., Beech, G. & Taggart, J. (1989) *Food Biotechnology: Cambridge Studies in Biotechnology* 7. Cambridge University Press, Cambridge.

Bamforth, C. W. (1998) *Beer: Tap into the Art and Science of Brewing*. Plenum, London.

Boulton, R. B., Singleton, V. L., Bisson, L. F. & Kunkee, R. E. (1996) *Principles and Practice of Winemaking*. Chapman & Hall, New York.

Campbell-Platt, C. H. & Cook, P. E. (1994) *Fermented Meats*. Blackie Academic and Professional, London.

Coultate, T. P. (1996) *Food—The Chemistry of its Components*, 3rd edition. Royal Society of Chemistry, Cambridge.

Demain, A. L., Davies, J. E. & Atlas, R. M. (1999) *Manual of Industrial Microbiology and Biotechnology*. American Society for Microbiology, Washington DC.

Fleet, H. (ed.) (1993) *Wine Microbiology and Biotechnology*. Harwood, Chur.

Fox, P. F. (ed.) (1993) *Cheese: Chemistry, Physics and Microbiology*, 2nd edition. Chapman & Hall, London.

Hough, J. S. (1985) *Biotechnology of Malting and Brewing*. Chapman & Hall, Cambridge.

Jackson, R. S. (1994) *Wine Science: Principles and Applications*. Academic Press, London.

King, R. D. & Cheeltham, P. S. J. (eds) (1988) *Food Biotechnology* 1. Elsevier Applied Science, London.

Lewis, M. J. & Young, T. W. (1995) *Brewing*. Chapman & Hall, London.

Priest, F. G. & Campbell, I. (eds) (1996) *Brewing Microbiology*, 2nd edition. Chapman & Hall, London.

Rose, A. H. (ed.) (1977) *Alcoholic Beverages, Economic Microbiology*, Vol. 1. Academic Press, London.

Rose, A. H. (ed.) (1982) *Fermented Foods*. Academic Press, London.

Varnam, A. H. & Sutherland, J. P. (1994) *Beverages: Technology, Chemistry and Microbiology*. Chapman & Hall, London.

14 Microbial biomass production

In most of the fermentation processes already discussed, the conversion of a proportion of the substrate to biomass is somewhat incidental, or for some purposes biomass formation may be actively suppressed. The main aim has been the conversion of substrate into a useful primary or secondary metabolic product, such as antibiotics, ethanol and organic acids. In such cases, once the optimal amount of target product has been achieved, the organisms produced are often merely waste materials that have to be disposed of safely and at a cost, or are simply used as a cheap source of animal feed. However, in dedicated biomass production, the cells produced during the fermentation process are the products. Consequently, the fermentation is optimized for the production of a maximum concentration of microbial cells. Microbial biomass is broadly used for three purposes:

1 viable microbial cells are prepared as fermentation starter cultures and inocula for food and beverage fermentations (Chapter 12), waste treatment processes, silage production, agricultural inoculants, mineral leaching and as biopesticides (Chapter 15);

2 as a source of protein for human food, because it is often odourless and tasteless, and can therefore be formulated into a wide range of food items; and

3 animal fodder.

Manufacture of baker's yeast

In human diets, microbial protein constitutes a relatively minor portion of the total protein consumed. This primarily comes from edible macrofungi (mushrooms, truffles, etc.) and a small contribution in the form of yeast in bread, estimated at 2 g protein per person per week in the western European diet (see Chapter 12, Bread).

A major fermentation industry has developed to manufacture the vast quantity of baker's yeast required for making bread and associated bakery products. The 'skimming method' was one of the first procedures employed for the commercial production of baking strains of *Saccharomyces cerevisiae*. This method used media derived from cereal grains, and was similar to brewing and distilling fermentation processes. Here the yeast floated to the top of the fermentation and was skimmed off, washed and press-dried. However, during World War I, due to shortages of cereal grains, the yeast industry sought alternate fermentation materials. In Germany a process was devised whereby molasses, ammonia and ammonium salts were used in place of cereal-based media. This remains, with certain refinements and automation, as the basis of current methods of manufacture that now generate 2×10^9 kg/annum worldwide.

Baker's yeast production commences with propagation of a starter culture, which originates from a pure freeze-dried sample or agar-medium culture. Yeast cells are initially transferred to small liquid culture flasks, then on to larger intermediate vessels before being finally used to inoculate the large production fermenters of 50–350 m^3 capacity. Overall, this may involve up to eight scale-up stages to produce the necessary final inoculum volume.

Medium for the production fermentation normally contains molasses as the carbon and energy source, which may be pretreated with acid to remove sulphides and heated to precipitate proteins. Molasses is often deficient in certain amino acids, and supplements of biotin and pantothenic acid are usually necessary. Further nitrogen sources (ammonium salts or urea) may be added, along with orthophosphate and other mineral ions, and the pH is adjusted to 4.0–4.4.

The main objective of the process is to generate a high yield of biomass that exhibits an optimal balance of properties, including a high fermenting activity and good storage properties. Aerobic fermentation favours a high biomass yield, as approximately 50% of the available carbon can be potentially converted to biomass. The maximum theoretical growth yield (Y_s) is 0.54 g/g, whereas under anaerobic conditions this value is reduced to around 0.12 g/g. Therefore, the aim is to maintain the cells under aerobic conditions. This is partially achieved by strong aeration of the fermentation broth, which is usually further increased as the fermentation progresses, via an aeration system that must achieve an oxygen transfer rate of 150 mmol/L/h. In addition, the fermentation is operated as a fed-batch process. The nutrients are added at a specific rate to prevent the operation of the Crabtree effect, a phenomenon where substrate suppresses aerobic respiration (see Chapter 3). This feeding regime limits anaerobic metabolism and ethanol production, which would otherwise result in lower biomass yields. By maintaining a specific growth rate (μ) of 0.2–0.25/h at 28–30°C, cell concentrations of up to 60 g/L can be achieved.

At the end of the growth phase, when all the nutrients are depleted, the aeration is continued for a further 30 min to 'ripen' the yeast. This encourages production of the storage carbohydrate trehalose, reduces protein and RNA synthesis, and stabilizes the cells to give a longer storage life.

Centrifugal separators, run at a minimum speed of 5000 rev/min, are used to remove the yeast from the fermentation broth. Harvested cells are then washed several times with water, chilled to 2–4°C and finally dried to around 70–75% (w/w) moisture using vacuum filter dehydrators. The yeast is usually packed in 1 kg blocks and kept under refrigeration. Alternatively, the yeast may be dried further to 7–10% (w/w) moisture to form dried yeast, which can then be stored for long periods without refrigeration. Overall, the cycle of operations, from the initial pure yeast sample to the yeast blocks ready for sale, takes about 2 weeks, passing through 60 generations.

Desirable features of baker's yeast and areas where improvements are being sought via genetic engineering include:

1 high glycolytic activity, a key feature being CO_2 generation rates or 'gassing power';
2 rapid utilization of maltose, which is the main sugar of bread dough;
3 osmotolerance, i.e. the ability to function in the presence of high levels of sugars and salt within dough;
4 good storage characteristics, including cryotolerance (the suitability for frozen storage); and
5 high growth rates.

Single cell protein production

During World Wars I and II, interest in microbial protein as human food and animal feed increased as conventional protein sources were in short supply. Attempts were made in some countries to use yeasts, particularly strains of *S. cerevisiae* and *Torula* yeast (*Candida utilis*), to supplement the shortfall of protein. In more recent times, use of microbial protein has been considered as a potential means of fulfilling the urgent need for low-cost protein in certain parts of the world, which the agriculture in those regions arguably cannot provide. The increased demand is a result of the ever-increasing populations in developing countries. This objective has not been achieved, despite the obvious need and advantages that microbial protein provides over conventional protein sources. More effort has been directed towards producing either premium products, as meat substitutes for the western diet, or animal fodder. Interest in microbial protein for animal fodder largely depends on production costs in relation to the prevailing price of the main market competitors, particularly soya protein and fish meal. The reason that more microbial protein is not currently produced for fodder is due to the present low price of these conventional protein sources. However, this may change as there have been forecasts of future shortages of soya and fish meal.

Rapid developments in microbial protein production occurred during the 1960s and 1970s. Extensive research was conducted on a wide range of microorganisms as possible alternate protein sources, motivated by large increases in the price of conventional animal feed. It was during this period that the term single cell protein (SCP) was first coined at the Massachusetts Institute of Technology. SCP is not pure protein (Table 14.1), but refers to the whole cells of bacteria, yeasts, filamentous fungi or algae, and also contains carbohydrates, lipids, nucleic acids, mineral salts and vitamins. It has several advantages over conventional plant and animal protein sources, which include:

1 rapid growth rate and high productivity;
2 high protein content, 30–80% on a dry weight basis;

Table 14.1 Protein and nucleic acid content of microorganisms

Microbe	Protein percentage	Nucleic acid
Bacteria	50–85	10–16%
Yeast	45–55	5–12%
Filamentous fungi	30–55	3–10%
Algae	45–65	4–6%

Note: soya beans contain approximately 40% protein.

3 the ability to utilize a wide range of low cost carbon sources, including waste materials;
4 strain selection and further development are relatively straightforward, as these organisms are amenable to genetic modification;
5 the processes occupy little land area;
6 production is independent of seasonal and climatic variations; and
7 consistent product quality.

The protein content and quality is largely dependent on the specific microorganism utilized and the fermentation process (Table 14.2). Fast-growing aerobic microorganisms are primarily used due to their high yields and high productivity. Bacteria generally have faster growth rates and can grow at higher temperatures than yeasts or filamentous fungi, and normally contain more protein. Yeasts grow relatively rapidly and, like bacteria, their unicellular character gives somewhat fewer fermentation problems than do filamentous organisms. However, many filamentous fungi have a capacity to degrade a wide range of materials and, like yeasts, can tolerate a low pH, which reduces the risk of microbial contamination. They are also more easily harvested at the end of fermentation than yeasts or bacteria. Selection of a suitable microbial strain for SCP production must take several characteristics into account, including:
1 performance (growth rate, productivity and yields) on the specific, preferably low-cost, substrates to be used;
2 temperature and pH tolerance;
3 oxygen requirements, heat generation during fermentation and foaming characteristics;
4 growth morphology and genetic stability in the fermentation;
5 ease of recovery of SCP and requirements for further downstream processing; and
6 structure and composition of the final product, in terms of protein content, amino acid profile, RNA level, flavour, aroma, colour and texture.

Other major factors are safety and acceptability. Most SCP products are currently used as animal feed and not for human consumption. Nevertheless, these products must meet stringent safety requirements. Obtaining regulatory approval for the production of proteins for human consumption is an even lengthier and more expensive process, and obviously influences the choice of production organism. A safety aspect that must be considered for all SCP products is nucleic acid content. Many microorganisms have naturally high levels and the problem is further exacerbated because fermentation conditions favouring rapid growth rates and high protein content also promote elevated RNA levels. This can be problematic as the digestion of nucleic acids by humans and animals leads to the generation of purine compounds. Their further metabolism results in elevated plasma levels of uric acid, which may crystallize in the joints to give gout-like symptoms or forms kidney stones. Slow digestion or indigestion of some microbial cells within the gut and any sensitivity or allergic reactions to the microbial protein must also be examined. For filamentous fungi, the possibility of aflatoxin production must be eliminated. An additional concern is the absorption of toxic or carcinogenic substances, such as polycyclic aromatic compounds, which may be derived from certain growth substrates.

Carbon substrates

The overall cost of SCP is dependent on several factors, including the yield of biomass per unit of substrate utilized, productivity (kilograms of biomass produced per unit volume of reactor per hour), and the costs of substrates, process operation and any additional process-

Table 14.2 Percentage composition (dry weight basis) of single cell proteins and soya meal

	Quorn protein *Fusarium venenatum*	Pekilo protein *Paecilomyces variotii*	Pruteen *Methylophilus methylotrophus*	Pronin *Methylococcus capsulatus*	Brewer's yeast *Saccharomyces cerevisiae*	Spirulina protein *Spirulina maxima*	Soya meal *Glycine max*
Crude protein	58	59	72	70	49	53	48
Nucleic acids	10*	10.6	16	–	12	–	–
Lipid	1.0	1.4	5–6	10	3	3	1–2
Ash	6.4	6.4	10–12	7	7	9	5–6

* Reduced to 2% after processing.

ing, labour and capital investment. A major consideration in commercial biomass production is the cost of the carbon substrate, which may be up to 50–60% of the total production expenditure. Many carbon and energy sources have been considered, including carbohydrates, hydrocarbons, alcohols and even carbon dioxide. In the case of carbon dioxide, it may be photosynthetically fixed by cyanobacteria and green algae, such as *Spirulina* and *Chlorella*, respectively, grown in surface culture using light energy. However, most processes involve chemoheterotrophs grown on readily available, low-cost, bulk substrates. Renewable resources are desirable substrates, particularly waste streams that include agricultural, dairy and wood processing wastes. Production of SCP from liquid waste materials achieves two objectives: it removes pollution (biological oxygen demand, BOD; see Chapter 15) and can generate good quality SCP.

The major substrates that have been used in commercial SCP production are alcohols, *n*-alkanes, molasses, sulphite liquor and whey (Table 14.3). However, in choosing a substrate, consideration must be given to:
1 the cost of the substrate;
2 biomass yield;
3 oxygen requirement during fermentation (Table 14.4);
4 heat produced and the level of fermenter cooling required; and
5 downstream processing costs, including the removal of possible toxic components.

The biomass yield from most carbon substrates, which also function as an energy source, can be predicted if specific information is available regarding the metabolism of the microorganism; this includes:
1 the dissimilation pathway of the carbon source, i.e. an estimate of adenosine triphosphate (ATP) produced per unit of substrate oxidized;
2 the assimilation pathway, i.e. an estimate of ATP required for the synthesis of a unit of cell biomass;
3 the maintenance requirement for the organism, i.e. the quantity of ATP required in order to maintain cell integrity; and
4 the extent of uncoupling, i.e. non-productive uses of ATP.

Overall, this information indicates the amount of ATP available for biomass synthesis (Fig. 14.1), but in

Table 14.3 Microorganisms used for SCP production using various carbon sources

Carbon substrate	Microorganism
Carbon dioxide	*Spirulina* species *Chlorella* species
Liquid hydrocarbons (*n*-alkanes)	*Saccharomycopsis lipolytica* *Candida tropicalis*
Methane	*Methylomonas methanica* *Methylococcus capsulatus*
Methanol	*Methylophilus methylotrophus* *Hyphomicrobium* species *Candida boidinii* *Pichia angusta*
Ethanol	*Candida utilis*
Glucose (hydrolysed starch)	*Fusarium venenatum*
Inulin (a polyfructan)	*Candida* species *Kluyveromyces* species
Molasses	*Candida utilis* *Saccharomyces cerevisiae*
Spent sulphite waste liquor	*Paecilomyces variotii*
Whey	*Kluyveromyces marxianus* *Kluyveromyces lactis* *Penicillium cyclopium*
Lignocellulosic wastes (solid substrate)	*Chaetomium* species *Agaricus bisporus* *Cellulomonas* species

Table 14.4 SCP yields and oxygen requirements for yeasts in continuous culture on various substrates

Substrate	Microorganism	Yield (g cells/g substrate)	Oxygen requirement (g/g dry weight)
n-Alkanes	*Candida tropicalis*	1.1	–
Ethanol	*Saccharomyces cerevisiae*	0.63	2.0
Glucose	*Candida utilis*	0.54	0.4
Lactose	*Kluyveromyces marxianus*	0.55	–
Methanol	*Pichia angusta*	0.36	2.6
Sucrose	*Saccharomyces cerevisiae*	0.5	–

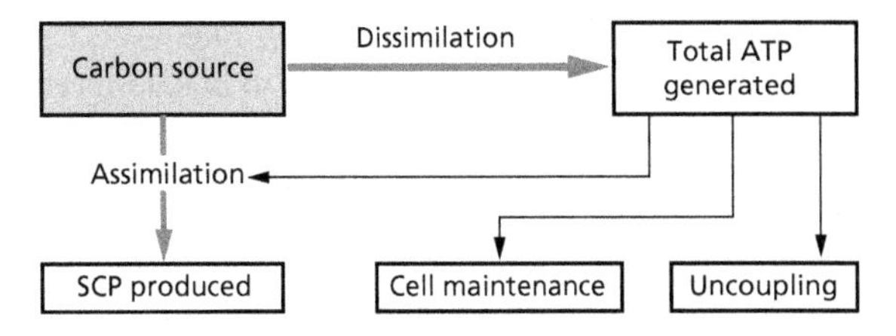

Fig. 14.1 Prediction of biomass yield.

some instances it is also important to know the supply of reduced nicotinamide adenine dinucleotide (phosphate) (NAD(P)H).

Single cell protein production processes

Many pilot plants have been developed over the last 30 years that utilize a range of substrates and microorganisms. However, relatively few have operated commercially, due to obstacles encountered on scale-up or for economic reasons. The physiological problems that are often encountered on scale-up include difficulties with:

1 oxygen requirements and oxygen transfer rates;
2 nutrient and temperature gradients;
3 effects of CO_2, as high levels may inhibit respiration in certain microorganisms; and
4 hydraulic pressure in deep fermenters.

In some instances, the economics of production can be improved by either increasing the value of the product or reducing the production costs through:

1 use of cheaper substrates;
2 improvements in the efficiency of the organism;
3 enhanced nutritional value/composition of the microbial protein;
4 marketing the protein as a premium product for human rather than animal consumption;
5 production of other valuable byproducts, i.e. development of a multiproduct process; and
6 lowering downstream processing costs, e.g. by reducing endogenous RNA levels.

The SCP production processes essentially contain the same basic stages irrespective of the carbon substrate or microorganism used.

1 Medium preparation. The main carbon source may require physical or chemical pretreatment prior to use. Polymeric substrates are often hydrolysed before being incorporated with sources of nitrogen, phosphorus and other essential nutrients.
2 Fermentation. The fermentation may be aseptic or run as a 'clean' operation depending upon the particular objectives. Continuous fermentations are generally used, which are operated at close to the organism's maximum growth rate (μ_{max}), to fully exploit the superior productivity of continuous culture.
3 Separation and downstream processing. The cells are separated from the spent medium by filtration or centrifugation and may be processed in order to reduce the level of nucleic acids. This often involves a thermal shock to inactivate cellular proteases. RNase activity is retained and degrades RNA to nucleotides that diffuse out of the cells. Depending upon the growth medium used, further purification may be required, such as a solvent wash, prior to pasteurization, dehydration and packaging.

The various processes described below have been relatively successful in commercial terms, and/or involve notable technological developments.

THE BEL PROCESS

The worldwide dairy industry generates over 80 million tonnes of whey each year. This byproduct of cheese manufacture has a high pollution load with a chemical oxygen demand (COD, see Chapter 15) of 60 g oxygen per litre. Consequently, it usually has to be disposed of at a high capital cost to the dairy industry. Whey contains approximately 45 g/L lactose and 10 g/L protein. It is particularly suitable for the production of SCP using lactose-utilizing yeast, although attempts have also been made to grow other organisms, including *Penicillium cyclopium*. Several processes have been developed for the utilization of lactose in milk whey. Some of the more successful have been those operated by Bel Industries in France. The Bel process was developed with the aim of reducing the pollution load of dairy industry waste, while simultaneously producing a marketable protein product. A number of plants are operated using *Kluyveromyces lactis* or *K. marxianus* (formerly *K. fragilis*) to produce a protein, Protibel, which is used for both human and animal consumption.

These processes initially involve whey pasteurization, during which 75% of whey proteins are precipitated. The lactose concentration is adjusted to 34 g/L and mineral salts are also added. This supplemented whey is introduced into a 22 m^3 continuous fermenter, maintained at 38°C and pH 3.5, with an aeration rate of 1700 m^3/h. The yeasts utilize the lactose and attain biomass concentrations of 25 g/L, with a biomass yield of 0.45–0.55 g/g lactose. Yeast cells are recovered by centrifugation, then resuspended in water, recentrifuged and finally roller-dried to 95% solids. Levels of residual sugar remaining in the spent medium are less than 1 g/L.

THE SYMBA PROCESS

The Symba process was developed in Sweden to produce SCP for animal feed from potato processing wastes. It is not economically attractive as a stand-alone operation. However, alternative routes for the purification of these waste-waters are difficult and expensive, as they contain up to 3% solids and have COD values of over 20 g oxygen per litre. A high proportion of the available substrate is starch, which many microbes cannot directly utilize. To overcome this problem the process was developed with two microorganisms that grow in a symbiotic association. They are the yeasts *Saccharomycopsis fibuligera*, which produces the hydrolytic enzymes necessary for starch degradation, and *Candida utilis*. The process is operated in two stages (Fig. 14.2). In the first stage, *S. fibuligera* is grown in a small reactor on the sterilized waste, supplemented with a nitrogen source and phosphate. At this point, the starch is hydrolysed, which is the rate-limiting step of the whole process. The resulting broth is then pumped into a second larger fermenter of 300 m^3 capacity where both organisms are present. However, *C. utilis* comes to dominate the second stage and constitutes up to 90% of the final product. The Symba process operates continuously and after 10 days the pollution load of the waste is reduced by 90%. Resultant protein-rich biomass (45% protein) is concentrated by centrifugation and finally spray or drum dried.

THE PEKILO PROCESS

This process began operating in 1975 and was the first commercial continuously operating process for the production of a filamentous fungus. It had to overcome the special problems caused by the pseudoplastic rheological behaviour of submerged cultures of fungal mycelium, which particularly affect oxygen transfer rates (see Chapter 6). The process was developed in Finland for the utilization of spent sulphite liquor, derived from wood processing, that contains monosaccharides and acetic acid. Supplements of other carbon sources, usually molasses, whey and hydrolysed plant wastes, may also be added prior to inoculation with *Paecilomyces variotii*. This continuous process is operated aseptically and produces over 10 000 tonnes of SCP a year from two 360 m^3 fermenters. Resulting dried Pekilo protein (Table 14.2), containing up to 59% crude protein, is used in the preparation of compounded animal feed.

THE BIOPROTEIN PROCESS

There has been a considerable amount of research into the production of SCP using alkanes as carbon sources, notably methane and liquid straight chain hydrocar-

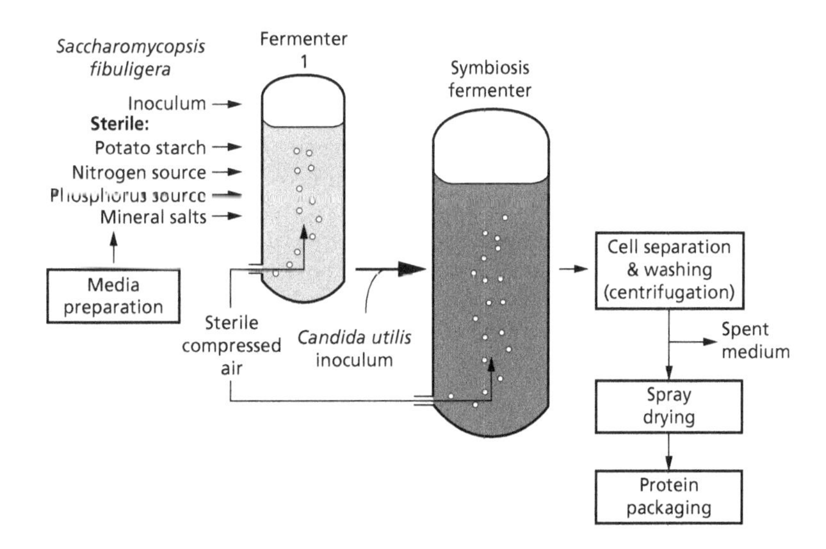

Fig. 14.2 The Symba process.

bons from C_{12} to C_{20}. Much of the work was conducted by several oil companies during the late 1960s and 1970s, in the period when conventional feed protein prices were high and oil prices were low. However, these compounds present certain technical problems. They are not miscible with water, and methane, in particular, is explosive when mixed with oxygen. Some of these substrates also require purification, or the protein product derived from them needs to be treated to remove adsorbed toxic compounds. In addition, these fermentations present cooling and aeration problems, as these highly exothermic processes require substantially more oxygen than when carbohydrates are used.

Most processes based on these substrates never went beyond the pilot stages. Nevertheless, at least one similar process, developed more recently, has proved to be relatively successful. The Bioprotein process, developed in the 1990s by Norferm, uses methane-rich natural gas as a sole carbon and energy source for the growth of *Methylococcus capsulatus*. A mixture of heterotrophic bacteria is also present, which helps to stabilize the process. The manufacturing plant, with an annual capacity of 10 000 tonnes, is at Tjedbergodden in Norway and was completed in 1998. This highly aerobic continuous fermentation is performed in a loop-fermenter, with medium containing ammonia, minerals and methane, obtained from North Sea gas fields. Biomass is continuously harvested by centrifugation and ultrafiltration, prior to heat inactivation and spray drying. The final product contains 70% protein and is currently marketed as Pronin. It is approved in the EU for use as a fish and animal feed, but may be used in human foods in the future.

THE PRUTEEN PROCESS

Methanol has several advantages over methane and many other carbon sources, particularly as it is completely miscible with water and is available in a very pure form. Consequently, the resultant protein does not have to undergo purification. As methane is readily converted to methanol, several oil and gas companies developed processes based on this attractive carbon source in the 1970s. However, there are some problems associated with methanol as a substrate. Only relatively low concentrations, 0.1–1.0% (v/v), are tolerated by the microorganisms that utilize it, and some methylotrophic yeasts form pseudo-mycelium while growing on methanol. During the fermentation, the oxygen requirement is high, as is the heat of fermentation. Nevertheless, the oxygen demands are somewhat lower than when using methane or other hydrocarbons.

Attempts to develop methanol-based processes were made in Europe, the former Soviet Union, Japan and the USA. They involved bacteria (*Hyphomicrobium* species, *Methylococcus* species and *Methylophilus methylotrophus*), yeasts (*Candida boidinii*, *Pichia angusta* and *Pichia pastoris*) and even filamentous fungi (*Gliocladium deliquescens*, *Paecilomyces variotii* and *Trichoderma lignorum*). The most technically adventurous was the process developed by ICI in the UK, which started production in 1980. This process used the methylotrophic bacterium, *M. methylotrophus*, to produce a feed protein for chickens, pigs and veal calves, called Pruteen. Production ceased in 1987 for economic reasons, due to the rise in price of methanol, which constituted 59% of the production costs, and a fall in the price of competing soya meal. Nevertheless, this process is worthy of examination due to the advances made in fermentation design and technology during its development. This was, apart from certain systems for wastewater treatment, the world's largest continuous aerobic bioprocess system. It consisted of a 3000 m^3 pressure-cycle airlift fermenter with inner loop and a working fluid volume of 1.5×10^6 L, capable of producing up to 50 000 tonnes of Pruteen per annum (Fig. 14.3). The fermenter weighs in excess of 600 tonnes, is over 60 m high, with a 5 atm pressure difference from the top to the bottom and cost US$80 million in 1979.

Filter-sterilized compressed air was used for both oxygenation and agitation, and all streams into the fermenter were sterilized. The fermentation was performed at pH 6.5–6.9 and 34–37°C with entirely inorganic commercial-grade nutrients. It was operated as a methanol-limited chemostat, the methanol being supplied through numerous distribution points within the fermenter. Bacterial cells were recovered by a novel separation technique, involving initial concentration from 3% (w/w) to 12% (w/w) by flocculation, which was promoted by acid and heat shock. This was followed by centrifugal dewatering, with recycle of water, and air drying. The dried unprocessed product contained 16% nucleic acids and over 70% crude protein.

Strain development of *M. methylotrophus* led to several improvements in its composition and fermentation performance. Protein content of the product was increased by 5%. The cell concentration achieved during fermentation rose from 4 g/L to 30 g/L and the μ_{max} in-

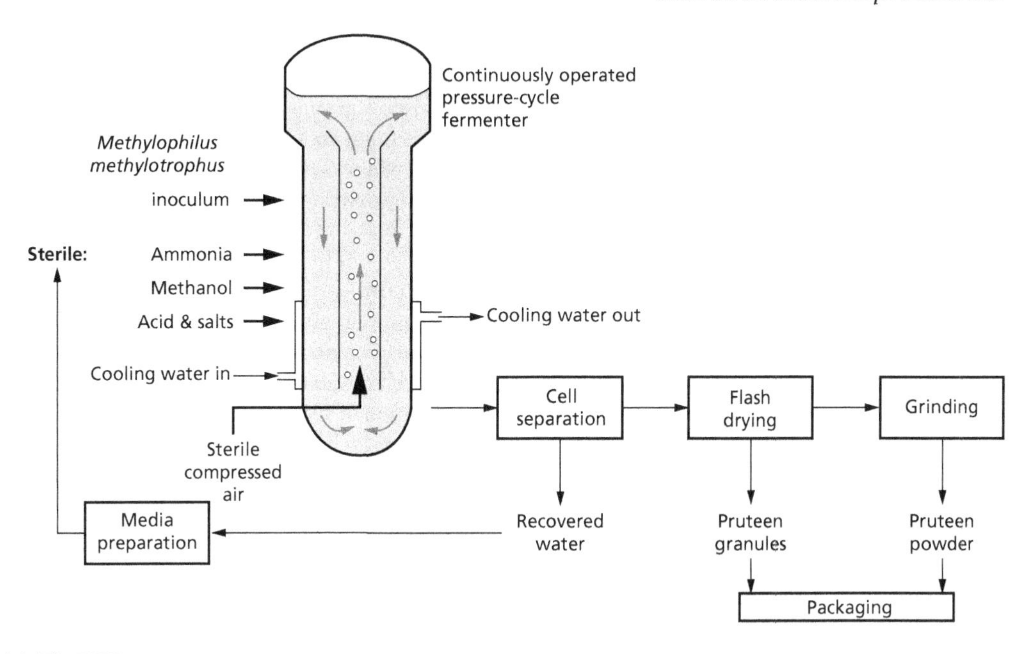

Fig. 14.3 The ICI Pruteen process.

Wild-type

Glutamine synthase (GS)

$$\text{Glutamate} + NH_4^+ + \text{ATP} \rightleftharpoons \text{Glutamine} + \text{ADP} + P_i$$

Glutamine synthetase (GOGAT)

$$\text{Glutamine} + \text{Oxoglutarate} + \text{NAD(P)H} + H^+ \rightleftharpoons 2\ \text{Glutamate} + \text{NAD(P)}^+ + H_2O$$

GOGAT$^-$ mutant

Containing plasmid-borne glutamate dehydrogenase (GDH) from *E. coli*

GDH

$$\text{Oxoglutarate} + \text{NAD(P)H} + H^+ + NH_4^+ \rightleftharpoons \text{Glutamate} + \text{NAD(P)}^+ + H_2O$$

Note: no ATP required

Fig. 14.4 Nitrogen assimilation in wild-type *Methylophilus methylotrophus* and conservation of ATP following introduction of glutamate dehydrogenase into a glutamate synthetase (GOGAT) minus mutant.

creased from 0.38 to 0.5/h. Nitrogen assimilation in the wild type involved glutamine synthetase (GS) and glutamate synthetase (GOGAT) (Fig. 14.4). This assimilation route is wasteful, as a molecule of ATP is required for each ammonium ion assimilated, whereas many other bacteria, including *Escherichia coli*, have a more efficient, energy-conserving route. They contain glutamate dehydrogenase (GDH) which does not require ATP. In order to improve energy efficiency and increase the biomass yield of the process, a mutant of *M. methylotrophus* was produced that lacked GOGAT, into which a plasmid was then introduced, containing the GDH gene from *E. coli*. This increased the biomass yield of the transformant by approximately 3%, a significant improvement, but not as high as theoretically predicted.

QUORN PRODUCTION

The company RHM started research into the production of a fungal mycelium (mycoprotein) for human consumption in 1964. Eventually, following extensive screening studies, a process was developed using *Fusarium venenatum* (formerly *F. graminearum*) ATCC 20334. The process uses only food grade materials and is strictly aseptic. The product, Quorn, with reduced RNA content, was approved for use as a human food protein by the UK Ministry of Agriculture and Food in

1985, over 20 years after the start of the project. The costs of research and gaining product approval have been estimated to be in excess of US$40 million.

Following a joint venture with ICI, and its later buy-out, 1000 tonnes of Quorn are now produced each year in the 40 m^3 airlift fermenter that was formerly used as the Pruteen pilot fermenter (p. 224). This fermenter is operated continuously at 30°C and pH 6.0. Food grade glucose syrup, derived from maize, potato or wheat starch, is used as the carbon source, with supplements of biotin and mineral salts. Ammonia is used to control the pH and act as the nitrogen source. Oxygen is supplied as sterile compressed air and must be controlled within strict limits. If the oxygen level falls too low, ensuing anaerobic metabolism results in the formation of byproducts that give unacceptable flavour and aroma. However, very high dissolved oxygen levels result in reduced productivity. During the fermentation the biomass doubles every 4–5 h and achieves concentrations of 15–20 g/L.

The filamentous structure of the harvested organism is a critically important factor related to eating quality. Mycoprotein is different from other novel bacterial and yeast proteins because of its microfilamentous structure, which can be partially aligned to resemble the microfibrils of meat. This enables it to be processed and flavoured to form meat-substitute foods that have a meat-like texture. The sparsely branched hyphae of *F. venenatum* have an average length of 0.4 mm with a cross-sectional diameter of 10 μm. However, like many other filamentous fungi when grown in prolonged continuous culture, it can become supplanted by highly branched forms, referred to as colonial variants. These colonial variants can arise spontaneously after steady-state growth of 100–1200 h. When this occurs, the fermentation is terminated as the texture of the final product loses its fibrous quality.

The fungal biomass generated contains 10% RNA, which is too high for human consumption (see p. 220). RNA levels are subsequently reduced by a thermal shock at 64°C for 30 min, which renders the organism non-viable and activates the organism's RNases. This results in the breakdown of RNA into nucleotides that diffuse out of the cells into the medium. Thus, RNA concentration is reduced to an acceptable level of less than 2% (w/w). Following RNA reduction, the mycelium is continuously harvested by vacuum filtration. The filter cake formed is a mat of interwoven fungal hyphae, which can be frozen as sheets, formed into various shapes, granulated or powdered.

Mushrooms

Certain mushrooms and other fruiting bodies of filamentous fungi are edible and provide a good source of protein, whereas others have narcotic effects and some are highly toxic. A wide range have been traditionally used for food, but relatively few are grown commercially. In fact, of the hundreds of species that are edible, only about 10 are produced in any quantity. Mushroom production involves a controlled non-axenic solid-substrate fermentation. It is currently the only economically viable product from lignocellulose fermentation. Exploitation of such fruiting fungi for the generation of edible biomass has several advantages:

1 they represent examples of the most efficient conversion of plant wastes into edible food;
2 unlike many other single cell proteins, they are directly edible and many are considered to be food delicacies because of their characteristic texture and flavour;
3 harvesting of fruiting bodies is the easiest possible method of separating edible biomass from the substrate in a solid-state fermentation (see Chapter 6, Solid-substrate fermentation); and
4 compared with animal sources of protein, many have a far superior protein conversion efficiency per unit of land and per unit of time.

Agaricus bisporus

In Europe and the USA *Agaricus bisporus* (button mushroom) accounts for over 90% of total mushroom production value. Agarics are decomposers of cellulosic materials and are naturally found in meadows and woodlands, where they degrade plant debris. They are grown commercially in temperate regions using a substrate of composted straw. A crop is produced within 6 weeks, whereas other mushrooms may take several months or even years to fruit. A closely related species, *Agaricus bitorquis*, is also grown in some areas. It is less prone to certain viruses and the bacterial blotch disease of mushrooms, caused by *Pseudomonas tolaasii*. The *Agaricus* production regime involves the following stages.

1 Inoculum preparation: growth of spawn (inoculum) on sterilized cereal grains.
2 Solid-substrate preparation: composting of straw, manure and fertilizers at 60–70°C for 2 weeks (see Chapter 15, Composting).
3 Substrate 'sterilization', so-called 'peak heating' of compost for 5–7 days.

4 Spawn inoculation into 'sterilized' compost and mycelial growth, referred to as a 'run' at 25°C for 2–3 weeks.
5 Application of a casing (covering) layer of peat and chalk over the substrate.
6 Fruiting body production, fructification, in about four flushes (successive crops) over a period of 4–6 weeks.

Speciality mushrooms

Total mushroom production worldwide has increased in the last 35 years from about 350 000 tonnes in 1965 to now over 500 000 tonnes. The majority of this increase has occurred during the last 15 years and a major shift has occurred in the range of genera cultivated on a commercial scale. *A. bisporus* formerly accounted for almost 70% of the world's supply, but by 1994 it represented less than 50% of world production, as more of the exotic mushrooms became popular.

China is the major producer of speciality mushrooms, producing over half of all edible mushrooms, and by far the widest variety. Levels of production in Europe and North America are quite low, but use of speciality mushrooms is on the increase and the demand is expected to grow. The main speciality genera cultivated are *Lentinula* (Shiitake), *Flammulina* (Enokitake), *Pleurotus* (Oyster mushroom), *Hypsizygus* (Bunashimeji), *Hericium*, *Morchella*, *Volvariella* (Paddy Straw mushroom) and *Grifola* (Maitake). Now there is a demand for the development of improved technology to cultivate these speciality species more efficiently, as traditional practices are not very productive. Some very valuable fungi are obtained only from wild sources and have proved very difficult to cultivate. A particularly valuable example is the black truffle, which is the fruiting body (ascocarp) of *Tuber melanosporum*. These are mycorrhizal ascomycetous fungi that are found in oak woodlands, but unlike mushrooms, truffles are subterranean ovoid structures (approximately 2–8 cm across). Other *Tuber* species are also used, but are far less valuable.

LENTINULA EDODES (SHIITAKE)

Cultivation of the Shiitake mushroom, *Lentinula edodes*, began in China almost a thousand years ago and was then introduced into Japan. They are becoming increasingly popular in the west, and are now grown in Europe and the USA. Worldwide production is approaching 200 000 tonnes/annum. These mushrooms may be used fresh or dried, and apart from culinary use, several medicinal properties have been attributed to Shiitake. Components detected include antihistamines, antitumour and antiviral agents, anticholesterol substances and compounds that inhibit platelet agglutination.

A problem with traditional methods of cultivation on natural logs is the time required before fruiting, which may be several years. In Japan, logs of the shii tree have been used, thus the derivation of the name Shiitake, but most production is now on species of oak. Logs of about 7–15 cm diameter are cut into lengths of about 1 m and drilled with holes spaced at one hole per 500 cm^2. The holes are inoculated with wood piece spawn or sawdust spawn and then usually covered with hot wax to prevent excessive drying. Spawn run, the development of fungal mycelium within the log, takes 6–9 months, after which the logs are transferred to a cooler and more moist 'raising' yard. This change in conditions provides an optimum environment for the growth and development of the mushrooms. The first crop is normally produced in the following year.

Modern production on synthetic logs is much quicker, taking about 4 months. The synthetic logs are prepared from sawdust, straw and corn cobs, along with supplements of wheat bran, rice bran, millet, rye and corn. Water is added to raise the moisture content to around 60% (w/w). The mixture is placed into bags and autoclaved for 2 h at 121°C. After cooling they are inoculated with Shiitake spawn. The inoculum is allowed to develop mycelium for 20–25 days and then the covering bags are removed. After about 4 weeks, exposed substrate blocks begin to form fruiting body primordia about 2 mm under the surface. Stimulation of primordium maturation is promoted by soaking these synthetic logs in water at 12°C for 3–4 h. The first crop or flush of mushrooms is ready to harvest about 10 days after soaking.

Further reading

Papers and reviews

Angelov, A. I., Karadjov, G. I. & Roshkova, Z. G. (1996) Strain selection of baker's yeast with improved technological properties. *Food Research International* 29, 235–239.

Batt, C. A. & Sinskey, A. J. (1984) Use of biotechnology in the production of single cell protein. *Food Technology* 38, 108–111.

Mizuno, T. (1995) Shiitake, *Lentinus edodes*: functional properties for medicinal and food purposes. *Food Review International* 11, 111–128.

Royse, D. J. (1995) Specialty mushrooms: cultivation on synthetic substrate in the USA and Japan. *Interdiscipline Science Reviews* 20, 1–10.

Solomons, G. L. (1983) Single cell protein. *CRC Critical Reviews in Biotechnology* 1, 21–58.

Trinci, P. J. (1991) Quorn mycoprotein. *Mycologist* 5, 106–109.

Books

Chang, S. T. & Hayes, W. A. (eds) (1978) *The Biology and Cultivation of Edible Mushrooms*. Academic Press, New York.

Chang, S. T. & Miles, P. G. (1989) *Edible Mushrooms and their Cultivation*. CRC Press, Boca Raton, FL.

Crueger, W. & Crueger, A. (1989) *Biotechnology: A Textbook of Industrial Microbiology*, 2nd edition. Sinauer, Sunderland, MA.

Demain, A. L., Davies, J. E. & Atlas, R. M. (1999) *Manual of Industrial Microbiology and Biotechnology*. American Society for Microbiology, Washington, DC.

Large, P. J. & Bamforth, C. W. (1988) *Methylotrophy and Biotechnology*. Longman Scientific and Technical, London.

Vonshak, A. (ed.) (1997) Spirulina platensis (Arthrospira): *Physiology, Cell Biology and Biotechnology*. Taylor & Francis, London.

Wainwright, M. (1992) *An Introduction to Fungal Biotechnology*. Wiley, Chichester.

15 Environmental biotechnology

Environmental biotechnology involves the application of biological agents (organisms or their components), along with chemical, physical and engineering processes to maintain, protect or restore the environment. Probably the most important role is in the treatment of solid and liquid wastes from domestic, municipal, agricultural and industrial sources, in order to reduce their potential environmental impact. A major aspect of this is the degradation or elimination of xenobiotic compounds within industrial waste streams and the necessary bioremediation of contaminated land or bodies of water, should these compounds become inadvertently released into the environment. Encompassed within this field is also the application of microbial-based 'clean technology' in mineral biomining and the desulphurization of fuels, along with composting of solid organic wastes and ensiling of plant biomass. This technology also reduces our reliance on synthetic chemical pesticides by allowing the implementation of biological control measures using bioinsecticides, biofungicides, etc.

Waste-water and effluent treatment

Historically the majority of wastes generated by humankind were disposed of directly into the environment. Due to the ever-increasing population this resulted in severe pollution of both land and aquatic environments (lakes, rivers, estuaries and coastal waters), which affected human health. By the 19th century the problem had become so serious that measures had to be taken to treat these wastes. Initially, they were treated to prevent epidemics of water-borne diseases such as typhoid and cholera. Domestic and commercial wastes can now be treated to remove suspended solids and dissolved organic compounds, and to oxidize ammonia (nitrification; see Chapter 3). This renders them chemically and biologically harmless, so that they can be safely discharged directly into the environment. Such treatments prevent transmission of diseases, contamination of land and the direct or indirect pollution of potential potable water supplies. For certain wastes, particularly those derived from the food industry, added-value processes are often involved; marketable materials can be recovered (e.g. fats and proteins), or the waste is converted to valuable products, including fuels, biomass and enzymes.

Before establishing the most suitable method for treating and disposing of any given pollutant, analytical methods are required to assess the polluting strength of the waste. Usually, these tests include the determination of the **biological oxygen demand (BOD)**, **chemical oxygen demand (COD)**, **total suspended solids (TSS)** and **total solids (TS)**, Table 15.1. Other tests may also be performed to determine the levels of specific components such as nitrogen, phosphorus, heavy metals, insecticides and chlorinated compounds.

BOD and COD tests were developed to provide information regarding the amount of oxygen needed to either biologically or chemically oxidize the polluting material present in any given waste-water. The BOD test estimates the amount of oxygen required by aerobic microorganisms to oxidize biodegradable material in polluted waste-waters over a fixed period of time (normally 5 days), at constant temperature (20°C) in the dark. A waste sample, or a diluted sample, is saturated with oxygen and seeded with an inoculum containing a diverse range of microbes. Its oxygen concentration is measured before and after a 5 day incubation period, and the results are expressed as milligrams of oxygen per litre of waste. One major drawback of this test is the long incubation period required. To overcome this problem the more rapid COD test was developed. This determines the amount of oxygen required to chemically oxidize any oxidizable material present in a waste-water. Each test involves adding a known volume of sample, or diluted sample, to a mixture of oxygen-rich potassium dichromate and concentrated sulphuric acid. This is refluxed for 2–4 h and the residual concentration of dichromate is determined by titration with ferrous sulphate or ferrous ammonium sulphate. The concen-

Table 15.1 Terms used in waste-water treatment

Biological oxygen demand (BOD, mg oxygen/L)
Chemical oxygen demand (COD, mg oxygen/L)
Hydraulic retention time (HRT, h)
Mixed liquor suspended solids (MLSS, g/L)
Organic loading rate (OLR, kg BOD/m^3 reactor volume/day)
Sludge loading rate (SLR, kg BOD/kg MLSS/day)
Solids loading rate (S_0LR, kg solids/m^3 reactor volume)
Surface loading rate (S_fLR, m^3 waste-water/m^2 surface area of tank/day)
Total solids (TS, mg solids/L)
Total suspended solids (TSS, mg solids/L)

tration of oxidizable compounds is proportional to the potassium dichromate utilized and the results are expressed in milligrams of oxygen per litre.

Alternate methods for rapidly assessing the polluting strength of waste-waters have been introduced. These include assays for total carbon (TC), total organic carbon (TOC) and total oxygen demand (TOD).

Treatment methods

Until a little over 100 years ago, polluted waste-waters were discharged directly into the environment. In developed countries, there are now strict limits on the quality of waste-waters that are allowed to be directly discharged into water courses and sewage systems. Water authorities or associated organizations are responsible for controlling these 'consent' limits. Within the UK, for example, the consent limit for direct disposal into water courses is normally set at the 20 : 30 standard; that is, 20 mg/L BOD and 30 mg/L TSS. Above these limits suitable treatment must be performed prior to discharge. Numerous methods have been developed to treat these wastes safely prior to disposal. As a result, domestic and industrial waste-water treatment has become one of the major applications of biotechnology in modern times.

Generally, industrial waste-waters have a greater polluting strength than those from domestic sources (sewage). In the case of domestic sewage, apart from small rural locations, the wastes are transported from households via the sewer system to sewage works where treatment occurs. The water authorities charge the householder a fixed charge for this service. However, in the case of industrial waste-waters there are three main options for their disposal:

1 direct disposal to the sewer system for treatment in the municipal treatment facilities;

2 partial treatment of the waste and subsequent disposal into the sewer system; and

3 in-house treatment to reduce the polluting load to a satisfactory level to allow direct disposal into the environment or reuse.

If industial wastes are dispersed into the sewer system, the water authorities charge the producer according to the flow, concentration and composition of the waste-water. Therefore, the more concentrated the polluting load of the waste-water and the greater the flow, the higher are the charges.

When deciding on the most appropriate method of treating any given waste-water, parameters such as the flow (m^3/day), BOD (mg/L), TSS (mg/L), TS (mg/L), temperature, pH, total nitrogen, phosphate and other specific pollutants must be taken into consideration, along with variations that occur in these parameters over a given period of time. The major factors that influence the final choice of treatment plant are the land availability, overall construction and operating costs, performance requirements (e.g. percentage BOD/COD removal) and the human resources needed to run the plant. Options for the treatment of waste-waters include:

1 biological treatments, involving aerobic and/or anaerobic processes;

2 chemical treatments, such as coagulation, flocculation, precipitation and electrochemical processes; and

3 physical treatments, often in the form of screening, sedimentation or incineration.

The method of treatment chosen is normally a function of the waste-water characteristics and more than one method is usually employed. For example, in sewage treatment, all three categories can be utilized at some stage (Fig. 15.1). Generally, any biological waste-water treatment plant treating either domestic sewage or industrial waste-waters can be divided into three main stages:

1 primary treatment (preliminary treatments and primary sedimentation);

2 secondary treatment (aerobic and/or anaerobic biological treatment and secondary sedimentation); and

3 tertiary treatment, which includes processes that, where necessary, remove any remaining inorganic nutrients, especially phosphate and nitrate. These inorganic nutrients have often been discharged into rivers and other water bodies where they can cause excessive higher plant and algal growth. Some algae can also produce toxins that poison fish and livestock.

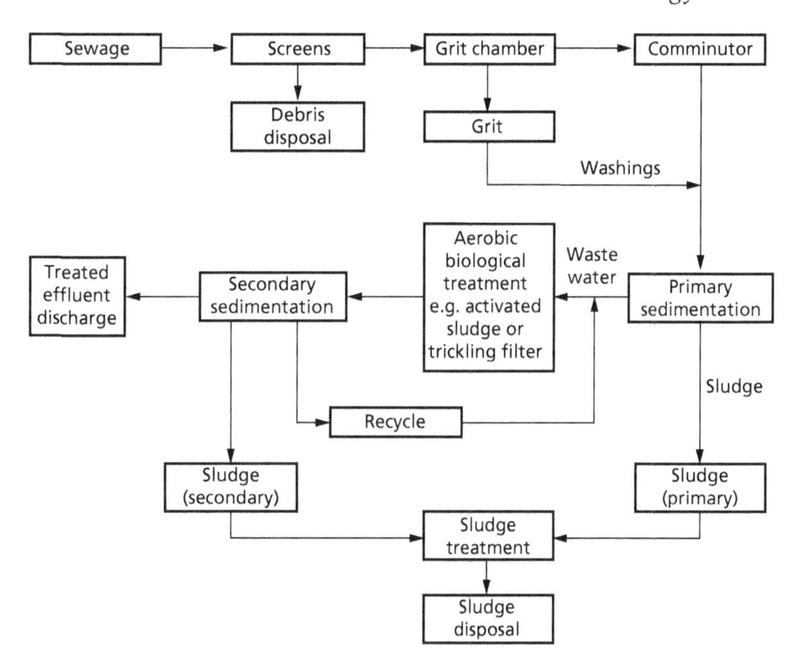

Fig. 15.1 An outline of conventional sewage treatment.

However, not all components of these stages, particularly tertiary treatments, are necessarily employed at every plant, the selection being waste-water dependent.

PRIMARY TREATMENT

Preliminary steps involve the removal of larger floating debris, along with a high percentage of suspended material. Strategies adopted depend on the nature of the specific waste-water and the subsequent treatment processes. The first step normally involves screens that are designed to remove large objects such as rags, papers, etc. Both coarse (6–24 mm apertures) and fine (2–6 mm apertures) microscreens are placed in the inflow channel to the treatment works. The average velocity of liquid in the channels is normally 0.7 m/s and must be maintained above 0.5 m/s to prevent settlement. Solid materials collected on the screens are removed manually or automatically and are either incinerated or buried in landfill sites. Any large particles not removed by screens are reduced to below 0.3 mm in diameter using comminutors (large-scale mechanical blenders). The waste-waters may also pass through grit chambers, which are designed to remove more than 95% of particles greater than 0.2 mm diameter. These are constant velocity chambers that are trapezoidal or V-shaped in cross-section. A constant velocity of 0.3 m/s allows the grit to settle, but other solids remain in suspension. Settled grit is removed from the bottom of the channel and is either washed for use as building material or is disposed of in landfill.

Primary sedimentation

This stage further reduces the suspended solid concentration of the waste-waters while simultaneously reducing the overall BOD. Both organic and inorganic settleable solids are removed by placing the effluent in a continuously fed tank under quiescent conditions. During sedimentation of domestic sewage, depending on the settling characteristics of the waste and operational conditions, 50–70% of the suspended solids are removed with a concurrent reduction in BOD of 30–40%. Sedimentation tanks are normally circular (radial flow) or rectangular tanks (horizontal flow) equipped with mechanical sludge scraping devices to remove the settled sludge.

The main design criteria for primary sedimentation are the **surface loading rate (S_fLR)**, **solids loading rate (S_0LR)** and the operating **hydraulic retention time (HRT)**, which in other fermentation processes may be referred to as the mean residence time. Depending on the individual characteristics of the waste, the S_fLR (i.e. the volume (m^3) of waste-water added to each m^2 surface area of the tank per day) varies within the range 30–45 m^3/m^2/day. Waste-waters with good settling properties operate at higher surface loading rates than those with poor settling characteristics. The same applies to the solids loading rate, which is associated with the quantity of TSS added to each m^2 surface area of the tank per day, and their values are in the range of 2–35 kg $TSS/m^2/day$.

The operating HRT is normally only 1–6 h, as longer HRTs can result in odour formation due to the formation of anaerobic metabolites. For domestic sewage, these processes remove 60% TSS and 30% BOD, but with industrial waste-waters BOD removal varies somewhat depending on the BOD to TSS ratio. Resulting primary sludge normally contains 2–6% (w/v) total solids.

Primary sedimentation removes a significant proportion of the suspended solids to decrease the risk of potential problems in the subsequent stages, particularly blockages, generation of dead zones and damage to pumps. It also reduces the overall BOD loading to the biological stages. Resulting primary sludges must be disposed of safely, normally by incineration or to landfill (following dewatering); alternatively, they may subjected to further biological and physical treatments (see p. 237).

SECONDARY TREATMENTS

Biological treatment utilizes the activities of a mixed microbial population to remove polluting components of waste-waters or reduce their concentration prior to disposal. Non-microbial treatment systems such as biological reed beds exist, but will not be considered here.

In waste-water treatment the microorganisms are usually in the form of aggregates or supported (attached) biofilms and do not occur as individual suspended cells, apart from in some anaerobic reactors. Microbial treatment processes can be divided into two categories, either **aerobic** or **anaerobic**. These may be further separated into suspended homogeneous systems, which include the activated sludge process and anaerobic stirred tank reactor, or attached film processes, e.g. aerobic and anaerobic trickle filters. Efficient operation of any biological treatment plant requires an ability to biodegrade all pollutants presented and tolerate fluctuations in concentration, composition and flow.

Aerobic biological treatment

The basic principle of aerobic treatment is that the waste-water is brought into contact with a mixed microbial population of aerobic organisms and oxygen. Soluble, suspended and colloidal biodegradable materials that contribute to the BOD are then metabolized:

$$\text{aerobic microbes} + \text{BOD} + O_2 \rightarrow \underset{\text{(biomass)}}{\text{new cells}} + CO_2 + \text{residual BOD} + H_2O$$

During the process, part of the biodegraded material is converted into CO_2 (mineralization) and a proportion becomes new biomass (assimilation). Under 'starvation' conditions, some of the microbial biomass (intracellular storage compounds) may also be metabolized; this is referred to as **endogenous respiration**.

A major problem associated with aerobic treatment is the disposal of excess biomass produced during the degradation of the pollutants. Approximately 30–70% of the biodegraded carbon is transformed into new cells, and the remainder is converted to CO_2, the specific values being process dependent. Although the efficiency of the systems relies on the production of new active cells, this simultaneously produces a new form of pollution, the excess waste biomass, which must be safely disposed of. Therefore, aerobic treatment can be classed as only 30–70% efficient, depending upon the specific process. The **deep-shaft process**, developed by ICI, is one attempt to reduce this problem. It uses much of the technology developed for the Pruteen process (see Chapter 14), again using a highly aerobic pressure cycle fermenter, but configured to minimize biomass production. The fermenter, with a depth of 100 m, is sunk in the ground to reduce noise, odour and land usage, and has significantly lower biomass yields. Nevertheless, the most commonly used aerobic processes are still the conventional **activated sludge processes** and **trickle filters** described below.

Homogeneous activated sludge processes. The activated sludge process was originally developed in 1914 by Arnold and Locket. The basic principles of the process are that the waste-water is brought into contact with a mixed microbial population, in the form of a flocculated suspension, within a continuously aerated and agitated tank. These processes are routinely used to treat domestic sewage and industrial waste-waters that have usually undergone primary treatment (Fig. 15.2). Typically, primary treated sewage entering the system contains 150–200 mg/L TSS, 150–200 mg/L BOD, 20–40 mg/L ammoniacal nitrogen and 6–10 mg/L phosphorus. However, these values vary depending on the location and the nature of the wastes deposited into the system.

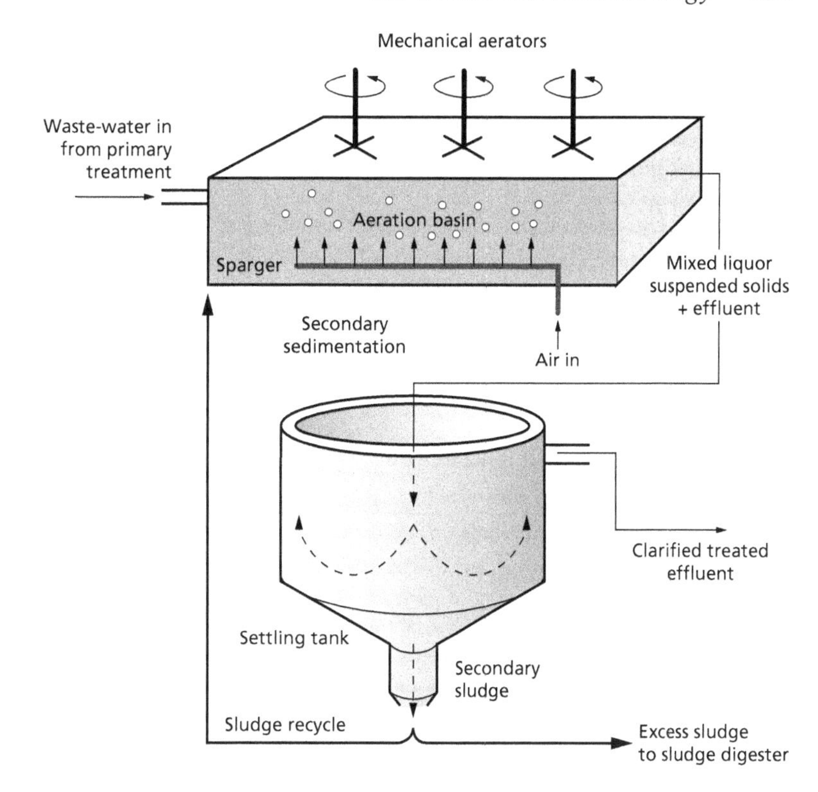

Fig. 15.2 An activated sludge plant.

This is a two-stage process, involving **biological treatment** and **secondary settlement**. Biological treatment is performed in an aerated basin containing a diverse range of flocculated microorganisms, the **mixed liquor suspended solids (MLSS)**, that biodegrade the polluting material present. As the microorganisms grow in the aeration basin they clump (flocculate) together to form stable flocs of activated sludge. The formation of stable flocs of 2–3 mm diameter is essential for the efficient operation of the plant with respect to BOD removal and rapid settlement in the secondary sedimentation stage.

The microorganisms present include a range of bacteria, e.g. carbon oxidizers, filamentous carbon oxidizers, nitrifiers, denitrifiers, etc., along with fungi, protozoans, rotifers, nematodes and algae. Despite being widely used, the microbiology and community structure of activated sluge processes is not well characterized. However, bacteria such as species of *Acinetobacter* and *Zoogloea ramigera* are considered to play a key role in floc formation by the synthesis and secretion of polysaccharide gels. Protozoa act as bacterial scavengers, ensuring low turbidity in the final treated effluent. Some 200 protozoan species have been isolated, but the most important are the ciliated forms, e.g. *Vorticella opercularia*.

Overall, activated sludge must contain a microflora capable of producing all enzyme systems required for the biodegradation of both soluble and insoluble pollutants. These microorganisms should form flocs with good absorbing properties that are stable and settle rapidly.

Secondary settlement occurs after one hydraulic retention time, when the treated effluent from the aeration basin passes into a secondary settlement tank (Fig. 15.2). This is similar in design to primary sedimentation, i.e. S_fLR, 15–30 m^3/m^2/day; S_oLR, 50–100 kg/m^2/day; HRT, 2–4 h. Here the flocculated microorganisms rapidly settle to form a secondary sludge, normally containing 1–3% (w/v) total solids, and a clarified supernatant. Often the supernatant is suitable for final disposal, but where necessary it can be subjected to a tertiary treatment to remove inorganic nutrients.

Depending on the design sludge loading rate (SLR), a proportion of the settled flocculated MLSS (secondary sludge) is returned to the aeration basin to maintain the required operating MLSS (microbial biomass) concentration. This allows a high concentration of biomass to be maintained in the aeration basin independent of the growth rate of the microorganisms, thus preventing microbial wash-out. These systems are comparable to stirred tank bioreactors with biomass recycle, as mentioned in Chapter 6.

Numerous designs and configurations of activated sludge plants exist, but they vary in only four key aspects: sludge loading rate (see below), MLSS concentration (kg/m^3), configuration and method of oxygen supply. The main parameters that have to be taken into account when designing activated sludge systems are:

hydraulic loading rate (HLR, kg BOD/m^3/day)

$$= \frac{\text{raw BOD (kg/m}^3) \times \text{flow rate (m}^3\text{/day)}}{\text{aeration tank volume (m}^3)}$$

hydraulic retention time (HRT, h)

$$= \frac{\text{reactor volume (m}^3)}{\text{flow (m}^3\text{/h)}}$$

sludge loading rate (SLR, kg BOD/kg MLSS/day)

$$= \frac{\text{raw BOD (kg/m}^3) \times \text{flow rate (m}^3\text{/day)}}{\text{MLSS (kg/m}^3) \times \text{aeration tank volume (m}^3)}$$

The operating SLR affects the level of treatment achieved and is basically the food to biomass ratio, which is the mass (kg) of BOD provided per kilogram of biomass (MLSS) per day. Therefore, as a rule and up to a limit, the more food (BOD) that is added to each kilogram of MLSS, the faster the microorganisms grow (see Chapter 2, Microbial growth). However, for maximum purification (percentage BOD removal) the food to biomass ratio should be low. This maintains the cells in a partially starved state, thereby 'encouraged' to actively metabolize any biodegradable pollutants present, to produce a low residual BOD (i.e. the substrate is limiting). Conversely, with increasing food to biomass ratios, the food availability increases, allowing higher microbial growth rates and greater biomass yields. Also, as substrate is no longer limiting its residual concentration in the final effluent increases.

Modes of operation of activated sludge plants. There are three main modes of operation for activated sludge plants: conventional, extended aeration and high rate treatment. The major difference is the operating SLR (see Table 15.2 for typical data for the three modes of operation). However, percentage BOD removal, HRT, biomass yield and sludge age (residence time) vary depending on the nature of the waste being treated. This is particularly true for industrial waste-waters whose concentration and composition vary considerably. As a rule, the more concentrated the influent BOD, the longer the required operating HRT to achieve the necessary degree of treatment and the lower the sludge age.

Conventional processing is used for complete treatment of waste-waters such as domestic sewage. Here the lower the SLR operated, the greater the level of purification obtained. Over 95% removal of BOD can often be achieved at the lower end of the SLR range (0.25 kg BOD/m^3/day), falling to 85% removal at the higher end of the range (0.5 kg BOD/m^3/day).

Extended aeration operates at a lower SLR than conventional plants and achieves approximately the same degree of purification, but the operating HRT is signifi-

Table 15.2 Typical data from activated sludge plants treating domestic sewage

Type	SLR*	MLSS (g/L)	Sludge age (days)	% BOD removal	HRT (h)	Biomass yield ($Y_{x/s}$†)	Effluent quality
Conventional	0.25–0.5	1–3.5	3–8	85–95	6–12	0.5	Fully nitrified at the low end of the SLR range and non-nitrified at high end; high quality 20 : 30 standard‡
Extended aeration	0.05–0.2	3.5–5	60–90	85–95	24	0.2–0.3	High quality fully nitrified
High rate	0.5–5	1–5	1–3	60–80	2	0.5–0.7	Non-nitrified

*SLR, kg BOD/kg MLSS/day.
†Biomass yield, kg biomass/kg BOD removed.
‡20 mg/L BOD and 30 mg/L TSS.

cantly longer. First impressions suggest that this system has no advantages over conventional treatment, particularly as this system is often more costly to construct and operate. Higher costs are due to increased HRTs that require a reactor with a greater volume and, consequently, more energy for aeration. Nevertheless, the main advantage of this system is the significantly reduced biomass yield (0.2–0.3 kg biomass per kilogram of BOD removed). This is approximately 50% of that found in conventional plants and substantially reduces the costs of its disposal. The reduced biomass yield is a function of the lower SLR, which maintains the cells in starvation conditions, so that a proportion of cells respire endogenously.

High rate treatment is mostly used for the partial processing of strong industrial waste-waters and is designed to remove only 60–80% of BOD. Consequently, the treated waste-waters normally remain too polluted for direct disposal into the environment. The high food to biomass ratio (kg BOD:kg MLSS) favours the faster-growing microorganisms and results in an increased biomass yield (0.5–0.7 kg biomass per kilogram of BOD). However, the high SLR produces short HRTs.

Generally, for any given waste-water, the lower the SLR, the larger the aeration basin volume required. This results in longer operating HRTs, reduced biomass yields and improved percentage BOD removal. The opposite occurs as the SLR increases. A smaller aeration basin volume is required, which gives reduced HRTs, increased biomass yields and lower BOD removal rates. Any specific SLR value chosen is therefore a function of the degree of treatment required, land availability, running costs and the cost of disposal of the excess sludge generated.

Dissolved oxygen in activated sludge plants. The operating dissolved oxygen (DO) level is a function of the value chosen in the design requirements. If the objective is full nitrification, a dissolved oxygen concentration of at least 2 mg/L is necessary. However, where only carbon oxidation and denitrification is required, a lower DO will suffice. The DO concentration required within the aeration basin can be determined by either mathematically modelling the system to predict oxygen demand at different times of the day, or by use of feedback mechanisms incorporating oxygen electrodes. Oxygen requirements may be supplied by mechanical aerators installed in the aeration basin, bubble diffusers (spargers), or a combination of the two (Fig. 15.2).

Trickle filters. The basic principle of aerobic trickle filters is that a microbial population is allowed to develop as a biofilm on an inert support material within a biological reactor (Fig. 15.3). Polluted waste-water is continuously sprayed over the surface of the support material and percolates (trickles) through the filter bed, where it is biodegraded by the microbial population. Aeration is achieved by exploiting the difference in temperature between the inside and outside of the reactor, resulting in a counter current of air. High microbial activity within the reactor causes a rise in temperature, and the warm air rises and allows fresh air to enter at the bottom of the reactor.

As treatment proceeds, the biofilm grows and increases in depth until a critical thickness is achieved, at which point oxygen becomes limiting at the surface of the support material. This results in the biomass falling off, called sloughing, after which the biofilm starts to redevelop.

Microbial populations vary considerably depending on the position within the filter. At the top, a range of microorganisms develop, including bacteria, fungi, protozoans and algae; along with macroorganisms, especially insects and their larvae. Below the surface, carbon-oxidizing microorganisms predominate, whereas nitrifiers are mostly found at the bottom of the filter. Overall, a highly complex food chain is created within these filters.

The three most important features of the packing material (inert support material) are as follows.

1 The specific surface area to volume ratio for biological attachment: the larger the surface area, the greater the biomass concentration per unit volume of the reactor and therefore the faster the rate of biodegradation.

2 The voidage volume: a high void space is required to prevent clogging and short-circuiting of the waste-water as it passes through the filter bed. Also, a high voidage volume aids oxygen transfer.

3 The density of the support material: the more dense the support material, the stronger the construction has to be to support and contain the total weight.

Trickle filters can be designed to operate under two modes of operation. Low rate filters almost invariably have stone or some other dense mineral medium with a low surface area and high density, whereas high rate filters use plastic media with high voidage and high surface areas. As these systems are non-homogeneous and have complex ecology, it is impossible to quantify the total biomass concentration attached to the inert support material. When designing such filters it is not prac-

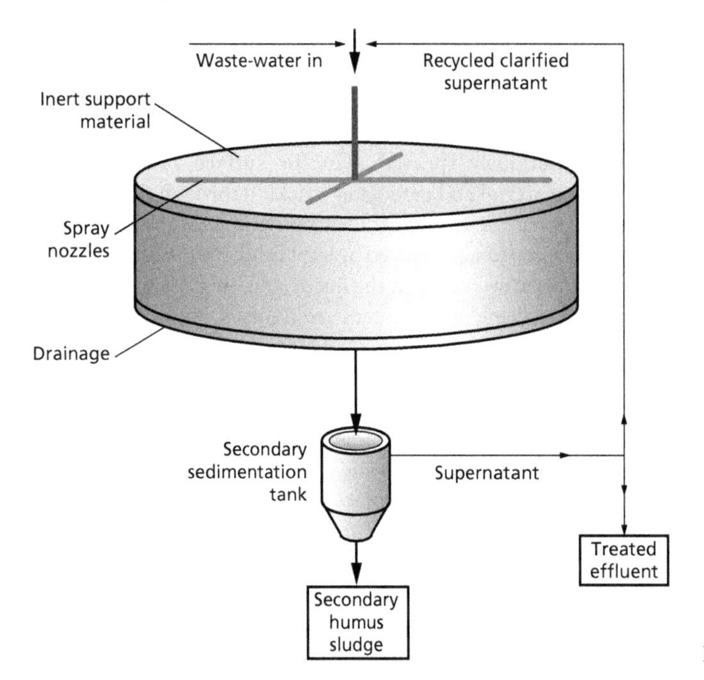

Fig. 15.3 An aerobic trickle filter.

tical to use the sludge loading rate (kilogram of BOD per kilogram of MLSS per day) as in the activated sludge process, because the biomass concentration is unknown. Therefore, it is normal to use the **organic loading rate** (**OLR**, kilogram of BOD/m^3/day), which is the mass (kg) of BOD added to each m^3 working volume of the reactor per day. This does not take into account the microbial biomass concentration within the bioreactor.

Low rate filters as used in sewage works are usually designed to produce effluents of high quality. They employ mineral support material (e.g. slag and granite), which develop a mature biological film within 4–24 weeks. These mineral low rate filters are normally circular or rectangular. Their depth is often restricted to 1.5–2.5 m, due to the dense nature of the support material and the associated construction costs. Most circular filters do not have diameters greater than 40 m and rectangular filters are less than 75 m long and 45 m wide. Mineral support media usually have surface areas in the region of 80–110 m^2/m^3 and a voidage of 45–55%. This relatively low voidage can result in filter blockages, known as ponding. When operating such filters it is important that the biomass should not be allowed to dry out as it affects their overall efficiency. Therefore, recycling of clarified supernatant from secondary sedimentation is often required (Fig. 15.3).

Low rate filters operate with an OLR of 0.06–0.12 kg BOD/m^3/day, a wetting rate of 0.5–4.0 m^3/m^2/day, and an HRT in the region of 20–60 min. They remove 90–95% of the BOD, resulting in high-quality effluent.

High rate trickle filters are often used for treating concentrated industrial waste-waters, acting as a 'roughing' process rather than a complete treatment, comparable with the high rate activated sludge process. They remove 50–80% of the BOD and their OLR is about 10-fold higher than with low rate filters. To overcome problems associated with low rate filters (dense support material, low surface area for attachment, potential ponding problems, and limited depth), plastic support materials have been developed. These plastic materials are chemically stable, but are gradually degraded by light. Their low density reduces associated civil engineering costs and permits filter depths of 6–9 m, which minimize land requirements. In operation, a high voidage volume, normally greater then 95%, reduces ponding. The very large surface area for microbial attachment, normally in the range 100–300 m^2/m^3, also results in high biomass concentrations. This allows greater OLRs to be used, while maintaining good levels of BOD removal. However, the large surface area of the support material necessitates higher wetting rates, in order to keep the biomass moist. Recycling of clarified supernatant from the secondary sedimentation tank satisfies this requirement.

Overall, high rate filters are defined as those where the operating OLRs are in excess of 0.6 kg BOD/m^3/day, but this may reach as high as 10 kg BOD/m^3/day. However, the higher the operating OLR, the lower the degree of purification attained. For example, depending on the nature of the waste, at an OLR of 1 kg BOD/m^3/day, BOD removal efficiencies of 80–90% can be expected, falling to approximately 50% with OLRs of 3–6 kg BOD/m^3/day.

Anaerobic waste-water treatment

Anaerobic treatment is often performed on sludges (see p. 238) and high-strength industrial waste-waters (see Table 15.3 for a comparison with aerobic treatments). As the name implies, it utilizes the activity of both facultative and obligate anaerobic microorganisms, in the absence of oxygen. They biodegrade the polluting waste-water and/or sludges to generate methane, carbon dioxide and biomass. The microbiology of these anaerobic digestion processes is complex. Their efficient and stable operation requires a microbial population containing at least three different interacting microbial groups. The three trophic groups broadly involved in the sequential biodegradation of the polluting material are as follows.

1 **Fermentative/hydrolytic bacteria (group 1)**: a group of facultative and obligate bacteria able to secrete extracellular enzymes that hydrolyse complex polymers, such as proteins, lipids and polysaccharides. Their products include a range of volatile fatty acids (VFAs; mainly acetic, butyric and propionic acids), along with CO_2, hydrogen, methanol and trace compounds.

Table 15.3 Comparison between aerobic and anaerobic waste-water treatment

Aerobic	Anaerobic
Approximately 50% of biodegraded carbon is converted into new cells	5–10% of biodegraded carbon is converted into new cells
Approximately 50% of biodegraded carbon is converted into CO_2	90–95% biodegraded carbon is is converted into methane and CO_2
Fast growing (minutes to hours)	Slow growing (days)
No energy production	Energy produced in the form of methane

2 **Acetogenic bacteria (group 2)**, including species of *Syntrophomonas* and *Syntrophobacter*. They metabolize the end-products from the group 1 microorganisms, primarily forming acetic acid, CO_2 and hydrogen.

3 **Methanogenic bacteria (group 3)**: a selection of strict obligate anaerobes associated with methane production, classified into two types. The **acetotrophs** such as *Methanosarcina* break down acetic acid into methane and CO_2, whereas the **hydrogenotrophs** reduce CO_2 with the simultaneous oxidation of hydrogen to form methane and water. For example, *Methanobacterium thermoautotrophicum* synthesizes methane from CO_2 and hydrogen in seven steps. This bacterium has several ways of catalysing these reactions, depending upon the specific environmental conditions.

The composition of biogas produced is dependent on the substrates biodegraded. However, as a general rule, biogas composition is normally 70% methane and 30% CO_2. In the past, and still in some plants, it is burnt and wasted, whereas many modern plants use it to generate heat or electricity (see Chapter 10). This is useful as, unlike aerobic treatment, optimum efficiency of anaerobic reactors demands an input of heat, because mesophilic and thermophilic reactors normally operate at 30–37°C and 40–45°C, respectively.

Several reactor designs exist for anaerobic treatment. They range from simple conventional mixed sludge reactors, utilized primarily to treat solid waste materials such as primary and secondary sludges, to the more efficient contact digester, anaerobic filter or anaerobic upflow sludge digester. A major problem associated with these anaerobic treatments is that their efficient and stable operation requires a balanced interaction between all three groups of microorganisms mentioned above. In such mixed culture anaerobic ecosystems there is a high degree of 'cross-feeding', involving inter-species hydrogen transfer that benefits the whole system. Changes to factors that influence the activity of any one of these groups can result in system failure. In particular, the rate of acid production must equal the rate of acid utilization. However, if the rate of acid production by the group 1 and 2 microorganisms exceeds the rate of acid consumption by the slower-growing methanogenic group, system failure eventually occurs due to acid accumulation.

Sludge treatment and disposal

During primary sedimentation and biological treatment stages vast quantities of sludges can be generated. These primary and secondary sludges are highly polluting and, depending on the final method of disposal, require further treatment. The characteristics of these two sludges vary considerably.

Primary sludges are non-biological, containing 2–6% (w/v) TS and are easily thickened, whereas biological secondary sludges contain 0.5–1.0% (w/v) TS and are difficult to thicken.

The properties of sludges generated during primary sedimentation and aerobic biological treatment depend on the raw waste-water characteristics and the biological process used. However, it is important to realize that each sludge has its own individual characteristics and that no two sludges are the same. Also, the composition of sludges often changes with time.

Major considerations are the thickening characteristics of a sludge, described by the **sludge volume index (SVI)**, defined as the volume occupied by 1 g of sludge after settling for 30 min in a 1 L Imoff cone. A sludge with an SVI of below 100 ml/g is considered to have good settling qualities and sludges with an SVI greater than 200 ml/g have poor settling characteristics. As primary sludges usually have good settling properties and secondary sludges exhibit poor settling characteristics, the two are often mixed to produce an overall sludge with average settling qualities.

Depending on the size of the treatment plant some or all of the following sludge treatment stages can be implemented prior to final disposal: thickening, digestion or stabilization, conditioning and dewatering.

Sludge thickening. The aim of sludge thickening is to significantly reduce the volume of sludge before further treatment while retaining the solids content.

Sludge thickening leads to reductions in:

1 the size of stabilization vessel required;
2 quantities of stabilizing chemicals necessary;
3 heat requirements for stabilization;
4 land requirements for subsequent treatment stages; and
5 transport or pumping costs.

Routine methods available for sludge thickening include the use of gravity settlement. This is very similar in design to primary and secondary sedimentation, as the sludges are placed in a tank where the solids are allowed to settle under quiescent conditions. To prevent hindered settlement, the sludges are gently agitated by rotating picket fences to allow the liquid to pass upwards and the solids to move downwards. Dissolved air flotation may also be employed, which involves supersaturating the sludge with air under a pressure of 2–4 atm. When it is transferred into a tank at atmospheric pressure the gases in solution are released. As they rise, solid particles become attached at the gas–liquid interface and float to the surface where they can be collected.

The final choice of thickening method depends on the nature of the sludges to be treated. Nevertheless, data in Table 15.4 clearly show the effect of sludge thickening. Its significance is further emphasized if we consider the case of a secondary sludge with a 0.8% TS (w/v). If, after thickening, the concentration is increased to 4% (w/v), this results in an 80% reduction in the volume of sludge requiring subsequent treatment. Further increase in concentration to 16% (w/v) achieves a 95% reduction in sludge volume.

Sludge stabilization. The main aims of sludge stabilization are to reduce the solids content of the sludges, to destroy pathogens when present, and to eliminate or reduce the potential for odour formation and putrification. Several methods are available, including:

1 biological processes, utilizing either aerobic or anaerobic microorganisms;
2 chemical oxidation, e.g. using chlorine;
3 chemical stabilization, e.g. by dosing with lime; and
4 physical treatment, e.g. heating.

Biological sludge stabilization utilizes the activity of either aerobic or anaerobic microorganisms to biodegrade the constituent organic matter. In the case of anaerobic treatment, the microbiology is the same as that described earlier for anaerobic waste-water treatment. The anaerobic bacteria metabolize the primary and secondary sludges into potentially useful methane, plus carbon dioxide and new cells. The normal reactor configuration is a stirred tank, heated to 35–37°C, operating at an HRT in excess of 20 days. During the process, approximately 70% of the total solids and between 30 and 50% of the volatile solids are removed. It is

Table 15.4 Typical plant data for sludges

	Primary sludges % TS (w/v)	Secondary sludges % TS (w/v)
Raw sludges	1–6	0.5–1.0
Following gravity settlement	6–9	3–4
Following air flotation	–	3–6

estimated that around 1 m^3 of biogas is produced per kilogram of volatile solids removed.

Aerobic stabilization is suitable for secondary sludges where microbial cells are the major components, because it relies on their ability to respire endogenously in the absence of 'food' and results in the oxidation of 75–80% of the cell biomass. However, a major disadvantage of treating primary sludges by aerobic methods is that they contain a high proportion of biodegradable components, which will be oxidized by the microorganisms. Consequently, the microbial biomass increases until the external nutrients become depleted; only then will they respire endogenously. A further disadvantage is the long operating HRT (15 days), during which the microorganisms must be supplied with oxygen. Most aerobic plants operate at ambient temperatures, but plants operating at 30–50°C have been used, resulting in shorter operating HRTs of 3–4 days.

Sludge dewatering. Prior to ultimate disposal, sludges are dewatered to further increase the solids content to 50% (w/v). The main difference between thickening and dewatering is that after thickening the sludge still behaves as a liquid, but after dewatering it should behave as a solid. Dewatering can be accomplished using sand beds, filter presses, centrifugation or lagoons, and its beneficial effects include:

1 lower transport costs through a reduction in sludge volume;
2 the product is suitable for disposal onto land, as it is odourless with reduced potential for environmental pollution;
3 increased fuel value, as it becomes easier to incinerate and has a higher calorific value; and
4 reduced leachate problems when disposed to landfill.

Disposal of sludges and other solid wastes. Methods routinely used for the disposal of final sludges and other solid wastes are as follows.

1 Landfilling, which is also used for other agricultural, industrial and urban wastes. However, there are potential pollution problems as materials can leach into adjacent water courses when unsuitable or ill-prepared sites are used. Also, it is becoming increasingly expensive due to a lack of suitable landfill sites. Nevertheless, landfilling has potential as a means of methane production. This may be more actively pursued in the future provided that problems associated with the establishment of suitable microbial populations can be overcome, possibly by inoculation with appropriate methanogens.
2 Incineration is routinely used for solids and well-dewatered sludges with solids contents in excess of 30% (w/v). For sludges, the system operates with limited energy input due to their high calorific value, leading to self combustion. However, there may be problems with the disposal of residue ash, as it often contains high concentrations of heavy metals.
3 Biologically stabilized dewatered sludge may be used as a low-cost fertilizer and soil conditioner on agricultural land, often incorporated with composted solid organic agricultural and household wastes (see below). This mode of disposal is becoming increasingly popular, but the regulations regarding disposal are very stringent, particularly regarding nitrogen, phosphorus and heavy metal concentrations, and pathogen content. These factors obviously influence the period between application and subsequent use of the land. The soil characteristics, topography of the land and potential risk of polluting water courses are additional considerations, along with any possible nuisance to the community.

Composting

Composting of solid organic wastes has been carried out for thousands of years as part of agricultural practices for the processing of green plant wastes to provide a useful soil conditioning material. Compost is also used as substrate for the commercial production of mushrooms, e.g. *Agaricus bisporus* (see Chapter 14). In some countries, it provides a valuable means of municipal solid waste management, by conversion into a useful product, as opposed to incineration or disposal in landfill.

Compost production involves simple technology; large-scale commercial operations use aerated piles, tunnel systems or rotating drums. The process essentially entails the controlled oxidation of organic matter to produce a stable and humified product.

$$\text{Organic matter} + O_2 \rightarrow \text{stabilized product} + CO_2 + H_2O$$

Initially, the substrate should contain a carbon to nitrogen ratio of 30 : 1, which may be achieved by appropriate mixing of plant material and other organic wastes with nitrogen-rich manures. These aerobic solid-substrate fermentations utilize a complex range of indigenous microbial activities and involve a succession of various bacteria, including actinomycetes, and fungi. The microbial consortium that develops is influenced by the substrate composition and the temperature.

Initially, mesophilic bacteria and fungi predominate, and their metabolic activities generate heat. The rise in temperature ultimately suppresses their growth and they become superseded by thermophilic bacteria (e.g. *Bacillus stearothermophilus*, *Thermus* species and actinomycetes) and fungi (e.g. *Rhizomucor pusillus*). The temperature continues to rise to the 'peak heating' phase (70–80°C), which encourages the proliferation of other thermophiles. From this point the temperature slowly declines. At 40–60°C additional thermophilic fungi become involved, especially species of *Aspergillus*, *Chaetomium* and *Humicola*, whose cellulases and hemicellulases are responsible for a substantial amount of plant cell wall degradation. Finally, some 20–30 days after initiation, the process undergoes a second mesophilic phase, as the material becomes recolonized by a range of mesophilic microorganisms.

In the preparation of compost for mushroom cultivation (e.g. *Agaricus bisporus*), the degradation of cellulose must be prevented, because it later serves as the carbon source for the mushrooms. Consequently, the process is essentially halted following peak heating. High levels of bacteria are also helpful in mushroom compost, as they serve as good sources of organic nitrogen for the basidiomycetes.

To prevent the development of anaerobic conditions, the compost must be regularly turned, or aerated, and the moisture level maintained at around 55% (w/v). As temperatures can rise to 70–80°C, this has the beneficial effect of killing weed seeds and pathogenic organisms. These processes take 9–12 months to complete; the volume of composted material becomes reduced by 50%, and the result is a humified and sanitized product. Apart from their uses in agriculture and horticulture, composts provide a rich microflora and microfauna that is useful for aiding bioremediation processes (see p. 241). Nowadays commercial inoculants may be used to control compost production, rather than merely relying upon indigenous microorganisms.

Ensiling

Silage is important as a winter feed for cattle that is produced by controlled anaerobic fermentation of plant material. The plant biomass used varies from country to country and includes grasses, legumes and plant wastes. Ensiling aims to preserve the material with a minimum loss of nutritional value. This is achieved by ensuring rapid development of both anaerobic conditions and low pH, thereby inhibiting undesirable spoilage organisms, particularly Enterobacteriaceae and clostridia. The fermentation is traditionally performed by indigenous lactic acid bacteria whose numbers rise to 10^9/g within 2–4 days, after which their numbers slowly decline. During this period they ferment water-soluble carbohydrates to produce lactic acid (preferably to levels of 6–8% of the silage dry matter), along with some acetic acid and ethanol. The final pH depends upon the plant materials used.

Numerous silage additives are available that may aid the process, including non-protein nitrogen sources (ammonia, urea, etc.), acids (acetic, propionic, formic, etc.), microbial enzymes (especially cellulases) and microbial inoculants. Commercial inoculants containing one or more species of homofermentative lactic acid bacteria, e.g. *Pediococcus*, *Streptococcus* and some *Lactobacillus* species, may now be used to speed the fermentation and improve silage quality. ICI, for example, markets several inoculants such as Ecosyl, a freeze-dried preparation of *Lactobacillus plantarum*.

Biodegradation of xenobiotics

Resistance to biodegradation is seen in some natural organic compounds such as lignocellulose, but is more apparent with xenobiotics. These are man-made compounds with structures that microorganisms have never been exposed to. Consequently, many of these are recalcitrant, remaining unchanged in the environment. Ever-increasing industrial activity is continually producing new synthetic compounds and those xenobiotic compounds that are potentially dangerous pollutants include some detergents, pesticides, halogenated aliphatics, aromatics, nitrosamines, nitroaromatics, chloroaromatics, polychlorinated biphenyls, phthalate esters and polycyclic aromatic hydrocarbons. Many of these enter the environment from industrial wastewaters. Properties that increase their resistance to biodegradation are polymerization, particularly branched polymers, the presence of stable bonds not subject to hydrolysis or other cleavage, and heterocyclic, aromatic and polycyclic components. Also, the presence of chlorine, nitro and sulphonate groups usually increases recalcitrance.

Normally, biodegradation of materials by microorganisms involves:

1 initial proximity, allowing adsorption or physical access to the substrate;

2 secretion of extracellular enzymes to degrade the substrates, especially polymers;

3 uptake, via transport systems, of soluble substrates or hydrolysis products of polymers; and
4 intracellular metabolism, often inducible and plasmid encoded.

In the case of recalcitrant compounds, one or more of these steps is missing, or the concentrations of the pollutant may be directly toxic or it is bioconverted to toxic products. In some instances, the environmental conditions may be unsuitable and/or other nutrients or essential cometabolites may not be available. Complete mineralization of some compounds may be unnecessary. Partial degradation or even formation of non-toxic complexes may be sufficient to render them harmless.

The development of methods to treat industrial waste-waters and sites contaminated with xenobiotics has been made possible by extensive research. Many of these biodegradation processes do not rely on single organisms, in most cases consortia of microorganisms are used, many of which have been isolated from the environment. Genetic modification of these biodegraders, to introduce novel properties, can lead to improved rates of biodegradation. This has been aided by the elucidation of degradative metabolic pathways and transport mechanisms. In addition, improvements have been achieved in treating waste-waters by using immobilized degrading or inactivating systems incorporating microbial cells and/or their enzymes.

Bioremediation

With the onset of the Industrial Revolution, an ever-increasing proportion of the Earth's surface became contaminated with natural and xenobiotic toxic chemicals. Examples include polluted aquifers and other water bodies, mining areas, and industrial sites contaminated with petrochemical products, pesticides, heavy metals, radionuclides and many of the xenobiotics mentioned above.

The basic principle of bioremediation involves utilizing the activity of microorganisms naturally present in the soil and water, or selected organisms inoculated into the environment, to biodegrade or detoxify contaminating compounds *in situ*. In the majority of cases a consortium of microorganisms will be involved in the biodegradation of the contaminant, rather than a single species. To optimize the process, promotion of the growth of indigenous microorganisms is necessary. It can be achieved by the addition of key nutrients such as nitrogen and phosphorus, which are normally present in growth-limiting concentrations. This enables the natural microbial flora to develop and metabolize the contaminant. Alternatively, known biodegraders of the contaminant that have been identified, isolated and their activities optimized can be used as an inoculant. For example, a recent addition to the growing list of microorganisms able to sequester or reduce metals is *Geobacter metallireducens*. This bacterium can remove uranium, a radioactive waste, from drainage waters in mining operations and from contaminated groundwaters. However, the most radiation-resistant bacterium known is *Deinococcus radiodurans*; this organism is also being developed to help clean up soil and water contaminated by solvents, heavy metals and radioactive waste. A genetically engineered strain of *D. radiodurans* has been produced which can detoxify mercury (genes derived from *Escherichia coli*) and degrade toluene (genes derived from *Pseudomonas putida*) in radioactive environments.

The use of genetically engineered microorganisms has been investigated in other areas of bioremediation. However, legislative problems, public concern regarding the release of genetically modified organisms into the environment and possible genetic instability have restricted their application. Nevertheless, due to comparatively low cost and generally positive environmental impact, bioremediation offers an attractive alternative or adjunct to conventional clean-up technologies. Its implementation has been successful at many sites, particularly those contaminated with petroleum products. However, it is not always the technology of choice, as accurate predictions of degradation rate are rarely possible.

Biomining (mineral leaching)

Conventional mineral mining involves extracting crushed mineral ores which have been dug from the earth, using high temperatures (smelting) or toxic chemicals that have often caused environmental pollution. More recently, the mining industry has been turning to more efficient and 'environmentally friendly' methods for extracting minerals, utilizing microorganisms that leach metals from the ores. This is not a new method, for it appears that the Romans recovered bacterially leached copper from mines around the Mediterranean over 2000 years ago and the first large-scale leaching operation was developed in the 18th century at the Rio Tinto mines in Spain. However, it is only in the last 50 years that the role of microorganisms in the conversion of metals to a soluble form has been recognized and the full potential of microbial extraction has begun to be exploited.

Nowadays bacteria are deliberately employed to recover copper and other metals from low-grade ores, slags, sludges and wastes. Such methods will be increasingly important as sources of high-grade ores become depleted. Microbial leaching methods exhibit improved recovery rates, reduced operating costs, require less energy and present far fewer pollution problems. Currently, 25% of all copper extracted is produced through bioprocessing, which has an annual value in excess of US$1 billion. These methods are also being used in the economic extraction of uranium, nickel, mercury and gold, often from very low grade ores that were formerly considered to be worthless. The extraction of many other metals is also being examined. This alternative approach to mining, ore processing and waste-water treatment is being aided by further developments in molecular biology, microbial physiology and process engineering.

Many metal ores are highly insoluble metal sulphides or are found in association with iron sulphide (FeS_2, pyrite). Consequently, acidophilic sulphur-oxidizing bacteria, particularly *Thiobacillus ferrooxidans*, play a major role in metal solubilization (leaching), although several thermophilic bacteria and mixed culture systems are also under investigation. Two aspects of its metabolism are especially relevant to a role in metal leaching. First, *T. ferrooxidans* is a chemoautotroph, obtaining its energy and reducing power via the oxidation of both ferrous (Fe^{2+}) ions and reduced forms of sulphur, e.g. sulphide (S^{2-}) (see Chapter 3), with carbon dioxide serving as its inorganic carbon source. Second, it is an acidophile, active in the pH range 1.5–3.5 with an optimum pH of 2.3. *T. ferrooxidans* is also mesophilic, preferring the temperature range 30–35°C, and is naturally present in certain sulphur-containing materials. Its action on ferrous sulphide releases sulphuric acid and an oxidizing solution of ferric ions, which can leach out metals from crude ore. An example is the solubilization of copper from chalcopyrite, $CuFeS_2$ (Fig. 15.4).

Commercial microbial leaching and oxidation of minerals is carried out on several different scales with various degrees of control. These involve dumps, heaps, vats and *in situ* mining. For dump extraction, low-grade copper ores, for example, usually containing less than 0.4% (w/w) copper, are deposited in natural valleys or basins with impermeable floors to ensure that there is no contamination of groundwater by the leachate (Fig. 15.5). Dumps may have built-in aeration facilities and a system for spraying the top surface with acidified water (pH 1.5–3.0). This percolates through the dump carrying dissolved O_2 and CO_2, whereas nitrogen and phosphate nutrients for growth are normally supplied by the ore. These conditions provide a favourable environment for the growth of leaching bacteria. At pH 1.5–5.0 *Thiobacillus* species soon proliferate throughout the

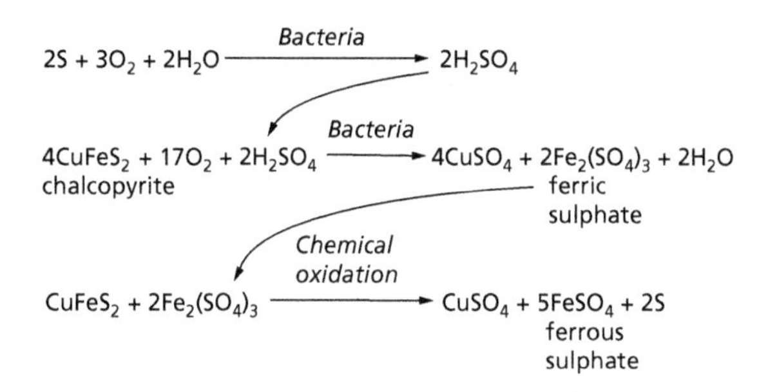

Fig. 15.4 Leaching of copper from chalcopyrite.

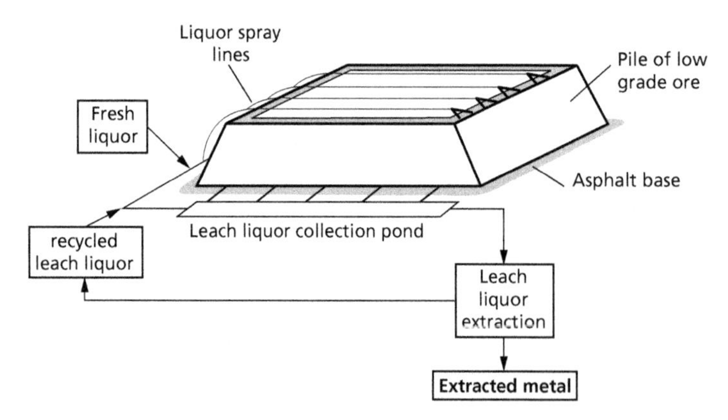

Fig. 15.5 Microbial extraction of metals from low-grade ores.

pile. The role of other bacteria, notably *Leptospirillium ferrooxidans*, has not been fully elucidated. Also, heterotrophs may be involved, but their role is likely to be more important above pH 5.0, along with the activities of *Gallionella* species.

Resulting ferric ions (Fe^{3+}) and sulphuric acid solubilize the copper, and leached metal-enriched solutions are collected in catch streams near the base of the pile. These solutions are concentrated by precipitation, addition of organic solvents, ion-exchange or microbial recovery methods. The metal-free acid solution is then recycled to the top of the pile. Engineering problems associated with large-scale dump leaching include:

1 supplying sufficient oxygen, which is the growth-limiting factor;

2 ensuring even percolation through the dump without channelling; and

3 prevention of overheating in the dump, as the temperatures can exceed 80°C.

Heap leaching is carried out on higher-quality ores. It involves much smaller operations than dumps, maximum size 10^8 kg, and the process takes months rather than years. Crushed ore is piled to 2–3 m and the top is flooded with acidified water containing bacterial inoculum. The leach solution percolates through the heap and is continually recycled. Aeration, via incorporated air lines, is more efficient than in dumps and the whole process is more easily controlled.

The use of batch or continuous stirred tank reactors offers the possibility of highly efficient leaching under controlled conditions. However, control of *in situ* leaching is much more difficult to achieve. This is carried out underground, often in disused mines where up to 30% of metals, e.g. copper or uranium, may remain in walls and supporting structures. Mines are flooded and a metal-rich, acidic leachate is pumped out some 3–4 months later. Applicability of *in situ* leaching is obviously dependent on geological factors: a porous rock is essential to allow efficient percolation of the leach solution. The process has advantages in terms of energy costs and environmental impact, and may be used to support conventional mining operations. Copper is currently recovered from mines in Arizona and Canada by *in situ* leaching. It may be possible to recover copper from disused mines in Cornwall, UK, using this technology.

Bioprocessing of ores releases a great deal of heat that may impede or kill the bacteria currently used. Therefore, to increase the efficiency of biomining, microbial strains better suited to large-scale operations are being sought. Prime candidates are thermophiles that would thrive in this oxidative environment, e.g. sulphur-metabolizing archaeans such as *Sulfolobus*. These organisms are effective metal sulphide leachers that are extremely thermophilic with the ability to operate over a wide pH range of 1–6. Also, attempts are being made to find or genetically engineer bacterial strains that are resistant to heavy metals. Mercury, cadmium and arsenic are particularly toxic to the microorganisms currently used and slow the bioprocessing. However, some microorganisms possess enzymes that protect against heavy metals or have systems for their removal. If the genes that afford this protection against heavy metals can be identified, suitable resistant biomining strains could be engineered.

Alkali leaching of metals can be used when the ores contain a high proportion of calcite, limestone or other acid-consuming carbonates, and is suitable for extracting certain uranium, molybdenum, radium and selenium ores. These processes primarily employ algae to solubilize the metals, particularly species of *Spirogyra*, *Oscillatoria* and *Chara*.

Microbial desulphurization of coal

Coal has a high sulphur content which generates environmentally polluting SO_2 when burned. The SO_2 may be removed using flue gas desulphurization equipment following coal combustion. However, the installation of this equipment in developing countries is often prohibitively expensive and continuing SO_2 emissions are causing major environmental problems. Therefore, low-cost technologies have been sought for stripping the sulphur from coal and other fuels prior to combustion. One such method is microbial desulphurization.

The sulphur in coal is in two forms: organic sulphur that is bonded with the carbon atoms and inorganic mineral sulphur, mostly as pyrite (FeS_2). Microbial desulphurization targets the pyrite and the best known microorganism that dissolves pyrite is *T. ferrooxidans* (see Chapter 3 and Biomining above). This leaching method can be applied where large amounts of land are available for coal processing and storage, but is not suitable if rapid desulphurization is required. Thermoacidophilic archaeans such as *Sulfolobus acidocaldarius* can also be utilized for this purpose. Their higher temperature of operation improves the rate of pyritic sulphur oxidation.

Increased use of pulverized coal and the coal-based liquid fuel that combines coal with water (coal water fuel, CWF), has led to an alternate microbial desulphurization process, the 'flotation separation method'. This involves preparation of a suspension of pulverized coal in water, which is held in a separation tank and sparged with air. Particles of coal adhere to air bubbles and float, whereas any hydrophilic ash particles sink. Pyrite particles have the same surface properties as coal and would normally float. However, bacteria added to the suspension selectively adhere to the pyrite particles, making them become hydrophilic and they sink. Consequently, these processes remove both ash components and about 70% of the pyrite.

Bioinsecticides

The worldwide annual costs of the chemical control of both insect pests in agriculture and insect vectors of animal and human diseases have been estimated at over US$350 million. In addition to the high cost, the use of conventional chemical pesticides can be problematic, due to their lack of specificity, persistence in the environment and accumulation to damaging levels in higher animals, particularly birds. Control of insects by use of microbial pathogens (bacteria, fungi, protozoa and viruses) can have a number of potential advantages over synthetic chemical insecticides. They are specific, relatively cheap to produce, many have a narrow activity spectrum and they present no residue problems. The development of resistance is also less likely. However, many of these pathogens cannot be easily cultured on a sufficiently large scale for practical application. To date, the most successful bioinsecticides are common spore-forming, Gram-positive, rod-shaped soil bacteria in the genus *Bacillus*. These bacteria produce crystalline protein endotoxins during sporulation. The insecticidal endotoxins are not contact poisons and must be eaten by target insects to have an effect. Some insecticidal products, composed of a single *Bacillus* species or subspecies, may be active against an entire order of insects, whereas others may be effective against a few species or even just a single species.

Since the 1960s, the most widely used microbial insecticides are preparations of *Bacillus thuringiensis*. Over 3000 tonnes are produced annually by fermentation processes in the USA, Europe, Russia and China, via aerobic submerged fermentations designed for high percentage sporulation and δ-endotoxin accumulation. They operate at 30°C and pH 7.2–7.6 using low-cost complex media, normally containing starch, corn steep liquor, casein and yeast extract. Certain strains also produce a β-endotoxin during the growth phase, which is a non-specific nucleoside toxin. Sporulation, with associated δ-endotoxin formation, occurs after 25–30 h and achieves concentrations of over 10^9 spores/ml.

Most commercial *B. thuringiensis* products contain both protein toxin and spores, whereas others consist of just the toxin component. All preparations are applied at levels of 4–6 g/ha for the control of over 40 different insects that cause agricultural and forestry diseases, and some carriers of human disease. *B. thuringiensis* endotoxins are specific to a small subset of insects and quickly break down to non-toxic compounds when exposed to ultraviolet light and other environmental factors. These characteristics have earned the products a reputation for being environmentally friendly.

The polypeptide δ-endotoxin is in the form of a crystal or crystal complex which is produced alongside a spore during sporulation. This parasporal body may constitute up to 30% of cell mass. Genes encoding these insecticidal crystal proteins (ICPs) are located on large plasmids. ICPs from different strains vary in size and crystal shape, and even slight variations of one or two amino acid residues can result in dramatic effects on their insecticidal activity.

ICPs require both solubilization and activation before they become biologically active toxins. For most, solubilization occurs in the highly alkaline environment of the insect midgut and activation is through proteolysis by insect gut enzymes. This activity converts the 130 000 Da protoxin to the active 55 000–65 000 Da form. Their safety, with regard to humans and animals, other than insects, is in part due to the highly acidic conditions found in human and animal guts, which are unfavourable environments for these toxins. The low pH solubilizes and denatures the ICPs, rendering them susceptible to hydrolysis by gut proteases.

In the insect gut, it is thought that the active toxin recognizes specific receptors on the surface of insect midgut epithelial cells. A hexagonal pore complex, composed of six active toxin units, forms through the cell membrane. This results in the loss of potassium ions and small molecules, which affects the insect's ability to control osmoregulation. Poisoned insects may die quickly, or stop feeding and die within 2–3 days from the effects of septicemia (blood poisoning).

Three structurally distinct domains of the toxins have been shown by crystallographic studies. Domain I consists of seven α-helices and is thought to be associated with membrane interactions, insertion of the toxin into the insect's midgut epithelium and pore formation.

Domain II is in the form of a triangular column of three β-sheets and may be involved in receptor binding. Domain III is composed of antiparallel β-strands, and along with domain II, is implicated in insect specificity and stability. Insect resistance to specific ICPs may be due to altered receptor binding specificities.

B. thuringiensis does not reproduce and persist in the environment in sufficient quantities to provide continuing control of target pests. The bacteria may multiply in the infected host, but few spores or crystalline toxins are released when a poisoned insect dies. Therefore, *B. thuringiensis* products must be applied much like conventional chemical insecticides.

Until the early 1980s, commercial *B. thuringiensis* products were effective only against the caterpillars of butterflies and moths (Lepidoptera). However, additional isolates that kill other types of insect pests have been identified and developed. The nature of the endotoxin differs among *B. thuringiensis* subspecies and the characteristics of these specific endotoxins determine which insects will be poisoned. Commercially available *B. thuringiensis* formulations now fall into three broad categories.

1 Formulations that kill caterpillars. The best known and most widely used are formulated from *B. thuringiensis* var. *kurstaki* isolates that are toxic only to larvae of butterflies and moths. They are used to control many common leaf-feeding caterpillars.

2 Formulations that kill mosquito, black fly and fungus gnat larvae, prepared from *B. thuringiensis* var. *israelensis*. However, they do not kill larval stages of 'higher' flies, e.g. house flies. These preparations have been particularly useful in controlling the vectors of malaria and river blindness, mosquito and blackfly, respectively.

3 Formulations prepared from *B. thuringiensis* var. *san diego* and *B. thuringiensis* var. *enebrionis*, which are toxic to certain beetles within the order Coleoptera.

Advances in genetic engineering have afforded improved prospects for developing new *B. thuringiensis* insecticides and an ability to deliver the toxins to the target insects in a variety of ways. Many strains of *B. thuringiensis* possess multiple ICP genes, thus producing either multiple insecticidal crystals or mixed crystals containing several different but related ICPs, each one having a unique spectrum of insecticidal activity. It is now possible to identify those that are particularly active against various target insects and to use genetic techniques to construct strains carrying several ICPs selected for optimized activity towards the desired insect targets. This also has the benefit of controlling the development of insect resistance.

Genes for the production of ICPs can also be incorporated into other bacteria that are part of the plant's natural microflora. When inoculated onto the crop plant these modified bacteria grow and multiply, providing continued protection against insect pests. However, concerns about the release of live genetically engineered organisms into the environment has led to an alternate approach to toxin delivery. *Pseudomonas fluorescens*, engineered to express *B. thuringiensis* toxin, are cultivated and then chemically killed to fix the toxin within the cells. This microencapsulates and stabilizes the toxin, reducing degradation when applied to plant leaves and improving the storage life of the product. Alternatively, genes coding the production of ICPs have been inserted directly into the chromosomes of certain crop plants, e.g. maize and canola (rapeseed), again providing season-long control. However, under these conditions development of insect resistance may become a potential problem, because the insect populations are continuously exposed to the toxin.

Toxin complexes from the Gram-negative bacteria *Photorhabdus luminescens* and *Xenorhabdus nematophilus* are rather different from those of *Bacillus* species and exhibit potential as future insecticidal agents, should resistance to established insecticidal toxins become widespread.

Other bioinsecticides include *Bacillus popillae* preparations, which are primarily used to control Japanese beetles, a major pest in the USA. These bacteria persist in the environment and provide long-lasting control. However, unlike *B. thuringiensis*, *B. popillae* is not easily cultivated in artificial media and spores are usually commercially produced in living insect larvae. Also, their mode of action does not appear to directly involve toxins. Ingested spores germinate within the gut and go on to invade the larvae.

Microorganisms are used to control many other agricultural pests and diseases, including fungal pathogens. For example, commercially available preparations of the fungi *Pythium oligandrum* and *Ampelomyces quisqualis* are used to control several plant-pathogenic fungi such as *Botrytis cinerea*. Spores of *Coniothyrium minitans* are also prepared for the biocontrol of the fungal plant pathogen *Sclerotinia sclerotiorum*.

Further reading

Papers and reviews

Brim, H., McFarlan, S. C., Frederickson, J. K., Minton, K. W., Zhai, M., Wackett, L. P. & Daly, M. J. (2000) Engineering *Deinococcus radiodurans* for metal remediation in radioactive mixed waste environments. *Nature Biotechnology* **18**, 85–90.

Chakrabarty, A. M. (1996) Microbial degradation of toxic chemicals: Evolutionary insights and practical considerations. *ASM News* **62**, 130–136.

Chen, W. & Mulchandani, A. (1998) The use of live biocatalysts for pesticide detoxification. *Trends in Biotechnology* **16**, 71–76.

Eggen, T. & Majcherczyk, A. (1998) Removal of polycyclic aromatic hydrocarbons (PAH) in contaminated soil by white rot fungus *Pleurotus ostreatus*. *International Biodeterioration and Biodegradation* **41**, 111–117.

Fang, H. H. (1991) Treatment of wastewater from a whey processing plant using activated sludge and anaerobic processes. *Journal of Dairy Science* **74**, 2015–2019.

Groenhof, A. C. (1998) Composting: Renaissance of an age-old technology. *Biologist* **45**, 164–167.

Knowles, B. H. (1994) Mechanisms of action of *Bacillus thuringiensis* insecticidal delta-toxins. In *Advances in Insect Physiology* (ed. P. D. Evans), Vol. 24, pp. 275–308. Academic Press, London.

Kroyer, G. T. (1993) Bioconversion of food processing wastes: recent research and new developments. An overview. *Microbiology Europe* Sept/Oct, 30–34.

Lee, M. D., Odom, J. M. & Buchanan, R. J., Jr (1998) New perspectives on microbial dehalogenation of chlorinated solvents: Insights from the field. *Annual Review of Microbiology* **52**, 423–452.

Lettinga, G. (1995) Anaerobic digestion and wastewater treatment systems. *Antonie van Leeuwenhoek: Journal of Microbiology and Serology* **67**, 2–28.

McFarland, B. L. (1999) Biodesulphurization. *Current Opinion in Microbiology* **2**, 257–264.

Rawlins, D. E. & Silver, S. (1995) Mining with microbes. *Bio Technology* **13**, 773–778.

Reineke, W. (1998) Development of hybrid strains for the mineralization of chloroaromatics by patchwork assembly. *Annual Review of Microbiology* **52**, 287–331.

Spain, J. C. (1995) Biodegradation of nitroaromatic compounds. *Annual Review of Microbiology* **49**, 523–555.

Stams, A. J. M. & Oude Elferink, S. J. W. H. (1997) Understanding and advancing wastewater treatment. *Current Opinion in Biotechnology* **8**, 328–334.

Steffan, R. J., Timmis, K. N. & Unterman, R. (1994) Designing microorganisms for the treatment of toxic wastes. *Annual Review of Microbiology* **48**, 525–557.

Verstraete, W., De Beer, D., Pena, M., Lettinga, G & Lens, P. (1996) Anaerobic bioprocessing of organic wastes. *World Journal of Microbial Biotechnology* **12**, 221–238.

White, C., Sayer, J. A. & Gadd, G. M. (1997) Microbial solubilization and immobilization of toxic metals: key biogeochemical processes for treatment of contamination. *FEMS Microbiological Reviews* **20**, 503–516.

Books

Alexander, M. (1994) *Biodegradation and Bioremediation*. Academic Press, New York.

Bousher, A., Chandra, M. & Edyvean, R. (1995) *Biodeterioration and Biodegradation* 9. Institution of Chemical Engineers, Rugby.

Bull, A. T., Marrs, B. & Kurane, R. (eds) (1998) *Biotechnology for Clean Industrial Products and Processes: Towards Industrial Sustainability*. OECD Publications, Paris.

Characklis, W. G. & Marshall, K. C. (1990) *Biofilms*. Wiley, New York.

Cheremisinoff, N. P. (1996) *Biotechnology for Waste and Wastewater Treatment*. Noyes Publications, Westwood, NJ.

Demain, A. L., Davies, J. E. & Atlas, R. M. (1999) *Manual of Industrial Microbiology and Biotechnology*. American Society for Microbiology, Washington DC.

Glazer, A. N. & Nikaido, H. (1995) *Microbial Biotechnology*. W. H. Freeman, New York.

Hardman, D., McEldowny, S. & Waite, S. (1993) *Pollution: Ecology and Biotreatment*. Longman Scientific and Technical, Harlow.

Kim, L. (ed.) (1993) *Advanced Engineered Pesticides*. Marcel Dekker, New York.

Kirkwood, R. C. & Longley, A. J. (eds) (1995) *Clean Technology and the Environment*. Blackie Academic and Professional, London.

Mitchell, R. (ed.) (1992) *Environmental Microbiology*. Wiley–Liss, New York.

Moo-Young, M., Anderson, W. A. & Chakrabarty, A. M. (1996) *Environmental Biotechnology: Principles and Applications*. Kluwer Academic, Boston.

Scragg, A. (1999) *Environmental Biotechnology*. Longman Scientific and Technical, Harlow.

Wainwright, M. (1999) An Introduction to Environmental Biotechnology. Kluwer Academic Publishers, Boston.

16 Microbial biodeterioration of materials and its control

The term biodeterioration has been in use for only about 25–30 years. It may be defined as any undesirable change occurring in a natural or processed material of economic importance, brought about by the activities of living organisms whether plants, animals or microorganisms. Here we will consider only microbial biodeterioration, from which few materials are immune. Natural materials subject to biodeterioration include animal products (bone, fur, leather and wool), plant products (wood, cotton and other fibres), stored unprocessed foodstuffs (grain, potatoes, and other fruits and vegetables) and stone. Most refined/processed products are also susceptible, including building materials (brick, concrete and mortar), cellulosic materials (chipboard, paper and card) and petrochemical products (fuels and lubricants); along with glass, metals, paints, plastics, rubber, pharmaceuticals, cosmetics and toiletries, and miscellaneous products such as microchips. Biodeterioration of processed foods is not examined here, as it is normally considered within food microbiology and food hygiene.

Biodeterioration essentially involves negative aspects of microbial activities and is often confused with biodegradation. Biodegradation is a term which, as far as humankind is concerned, is related to positive/useful activities of microorganisms, utilizing their 'breakdown' abilities, in particular, to transform wastes into useful end-products or more acceptable/less hazardous forms. The organisms involved and their activities are often the same as those associated with biodeterioration, but it is the location of the events that is different. For example, fungi decaying building timbers constitute biodeterioration, whereas similar activities on fallen timber or wood-wastes may be considered to be useful biodegradation. A second example is bacterial biodeterioration of in-use cutting oils and lubricants, compared with microbial activities on oil spillages, which is considered as biodegradation and is important in bioremediation.

In many instances, the biodeterioration of solid materials follows the formation of a surface biofilm, which may consist of a heterogeneous mixture of various microorganisms. The biofilm is initiated through the adhesion of microorganisms to a surface, usually aided by their secretion of extracellular polysaccharides (glycocalyx). Once established, the glycocalyx provides a protective physical barrier and also helps protect against chemical biocides.

Damage inflicted by microorganisms may be due to one or more of the following factors.

1 Mechanical damage/physical disruption, e.g. fungal hyphae growing through wall plaster.

2 Chemical/biochemical processes, where the organism utilizes the material as a carbon and/or energy source, or damages it through the release of hydrolytic or corrosive agents, such as acids and enzymes.

3 Soiling/fouling, involving simply the presence of microorganisms or their deposition of byproducts that are not physically destructive, but that nevertheless devalue or toxify the material, e.g. wood-staining fungi, fungal toxins contaminating grain, nuts, etc.

Biodeteriogenesis, rather like disease, involves stages of 'infection' or contamination, incubation and manifestation (development of observable symptoms). Economic aspects of biodeterioration that must be considered include the cost of preventing an attack (use of biocides or preservatives; see Chapter 2), replacement of materials and remedial treatment or restoration.

Biodeterioration of stored plant food materials

Loss of plant food material prior to harvest is essentially covered by the field of plant pathology, and as mentioned above, microbial problems associated with processed food materials are handled within the realms of food microbiology and food hygiene. However, post-harvest decay and spoilage of stored unprocessed plant material is usually considered within biodeterioration. The microbial damage here ranges from complete loss to effects that may necessitate the downgrading of a product, e.g. from human food to animal feed.

Processed foods can be preserved by various treatments: drying, smoking, addition of salts or sugars, pickling, heat pasteurization or sterilization, freezing, use of chemical preservatives, etc. (see Chapter 2). However, for stored unprocessed foods, losses are mostly limited by controlling the immediate storage environment by chilling or using inert gases and, more recently, by irradiation. Post-harvest biodeterioration by microorganisms can be especially problematical in tropical and subtropical regions. This is largely due to the higher temperatures and moisture levels, and often the storage facilities are poorer.

Stored unprocessed plant food materials are essentially divided into three categories.

1 Durable products, e.g. grain, oilseeds and legumes.
2 Semi-perishable products, e.g. root vegetables and apples.
3 Perishable products, e.g. soft fruits and salad vegetables.

Losses of durable crops in storage may be up to 10% in temperate regions, rising to 20% in tropical regions. For perishable items, the losses in tropical regions may be up to 50%, or even higher for certain products.

Two major world crops worthy of particular consideration are potatoes and cereals, which are semiperishable and durable, respectively.

The Irish or European potato (*Solanum tuberosum* L.)

This is a vital universal food plant and is the most important of all vegetable crops. It is cultivated worldwide, but is a primary crop of temperate zones. All potato varieties are clones, propagated vegetatively by 'seed tubers' or 'seed pieces' (portions of a tuber) and consequently, are vulnerable to pathogens. The potato tuber is essentially a stout underground stem with an apical end and a stem end. Around the tuber are spirally arranged buds ('eyes') that point towards the apical end. Covering the tuber is a protective periderm (skin/peel) surrounding the flesh, which is composed of an outer cortex, medulla and inner pith regions.

Main-crop potato tubers are harvested in late autumn and must be stored for several months in order to supply the year round consumer and processing markets. Suitable storage must take into account the high moisture content, and the fact that the potato is living, thus requiring oxygen for respiration, which leads to the generation of water vapour and CO_2.

The majority of potential bacterial and fungal storage pathogens cannot penetrate the protective intact periderm that covers each tuber. However, at lifting in the field, tubers are prone to periderm damage, which leaves them open to microbial attack. Potatoes destined to be the next season's 'seed' can be treated with fungicides (e.g. thiabendazole) and other pesticide, as they are not to be consumed. However, the food crop cannot be treated in this way. Usually, they are held at 12–14°C and 85% humidity for 2 weeks to allow curing (suberization), leading to wound healing of damaged periderm. Subsequently, potatoes are maintained at 5–6°C in order to prevent sprouting. Ventilation with forced air at 95% humidity prevents both heating and tuber shrinkage. Tubers cannot be held at a lower temperature, as this induces 'low-temperature sweetening' (starch conversion to sugars), which reduces cooking quality, e.g. blackening on frying due to caramelization of sugars.

Losses of tubers in storage are primarily due to bacterial and fungal pathogens. Bacterial soft rots are caused by *Erwinia* species, ring rot is induced by *Corynebacterium* species and development of common scab is the result of infection in the field by *Streptomyces scabies*. The fungal problems include *Phytophthora erythroseptica* (pink rot), *Phytophthora infestans* (late blight) and *Phoma* species (gangrene). Probably the most important storage pathogen in recent years is dry rot caused by *Fusarium* species, e.g. *Fusarium solani* var. *coeruleum*, which is found wherever potatoes are grown. This organism affects only stored tubers and cannot invade undamaged periderm, lenticels or suberized surfaces, and must enter via wounds from soil on the tuber surface. The incidence of dry rot is directly related to periderm damage sustained prior to storage. Initially, lesions develop as brown/black flecks on the tuber surface, later forming large hollow cavities within, and the tuber surface appears wrinkled with numerous white tufts of mycelium. Secondary infection by soft rot bacteria may also occur. Overall, losses are the result of varietal susceptibility of tubers to the pathogen, the extent of wounding, the time at which wounding occurs and the level of microbial inoculum in the soil.

Cereals

True cereals include barley (*Hordeum vulgare*), maize

(*Zea mays*), oats (*Avena sativa*), rye (*Secale cereale*), wheat (*Triticum aestivum*) and rice (*Oryza sativa*). Other crops often included in this category, but which are not true cereals, include millet (*Setaria italica*), sorghum (*Sorghum vulgare*) and buckwheat (*Fagopyrum sagittatum*). Botanically, true cereal grains are fruits, not seeds, with a fused pericarp and testa forming a leathery husk (see Fig. 12.1). Overall, they are probably the most important group of food plants. Cereals have a major advantage over crops such as potato, because they naturally have a low moisture content, which on further drying allows long periods of storage without deterioration. They too are living, but even drying to 12% moisture does not affect their viability, which is important for the seed stock.

Provided that storage conditions are suitable, losses during storage rarely exceed 5%, but may occasionally reach as high as 30% in some areas of Africa and India. Losses of quality and quantity are predominantly due to fungi; other microorganisms normally cause relatively minor problems. The nature of the microbial damage includes:

1 a decrease in viability, which is important for the seed stock;
2 discoloration, particularly of the embryo, due to invasion by fungal mycelium;
3 biochemical changes, such as the production of fatty acids, giving rancid odour and flavour;
4 loss of mass; and
5 production of mycotoxins.

The storage fungi involved in cereal biodeterioration are mostly xerotolerant species of *Aspergillus*, *Fusarium* and *Penicillium*. They mainly develop from dormant spores on the outside of the grain, or from dormant mycelium lying under the surrounding pericarp. Factors influencing fungal growth include moisture content of the grain, temperature, length of storage time, level of fungal contamination, quantity of foreign debris (broken seeds, plant fragments, soil, etc.), and the activities of insects and mites.

For each species of fungus there is a minimum moisture content below which it cannot grow. The most xerotolerant are *Aspergillus restrictus* and *A. halophilicus*. Cereals are normally stored at 4–7°C following drying to around 12% (w/w) moisture, which allows for 'moisture pick-up' during storage. At 14.0–14.5% (w/w) moisture, fungal growth is slow, but above this level even small incremental elevations in moisture content can result in pronounced increases in fungal growth. Activities of insects and mites can promote moisture increase through their respiration, and they also distribute fungal spores and mycelium.

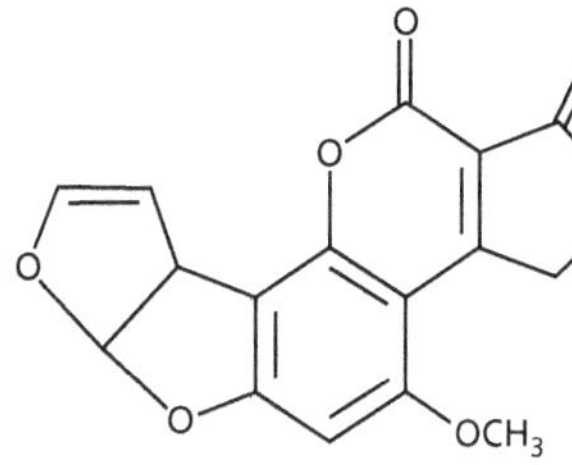

Fig. 16.1 Aflatoxin B_1.

Mycotoxin levels tend to be higher in organically grown cereals than in those where fungicides have been used, and in developing countries where storage conditions are less rigorously controlled. Any mycotoxins produced are not destroyed by cooking or processing and their concentrations in stored cereals are directly related to levels of fungal growth. Consequently, contaminated cereals should be destroyed, not down-graded for animal feed. However, chemical degradation treatment of mycotoxins has been attempted in contaminated peanut meal destined for animal feed.

Mycotoxins that may be generated include several aflatoxins (bis-furocoumarin metabolites) from *Aspergillus* species such as *A. flavus*, e.g. aflatoxin B_1 (Fig. 16.1). Many of these compounds have LD_{50} values of less than 50 μg/kg. They cause liver damage and are considered to be carcinogenic. *Fusarium* toxins, such as T-2, F-2 and zearalenone, are also highly toxic, causing alimentary toxic aleukia in humans and oestrogenic syndrome in pigs. The penicillium toxins include the liver-damaging rubrotoxin from *Penicillium rubrum*, and penicillic acid, which produces haemorrhagic syndrome in poultry.

Non food animal products

These products include animal hides, leather, wool and animal glues, all of which are predominantly proteinaceous materials and subject to attack from microorganisms secreting appropriate proteases.

Leather

Freshly stripped hides contain fatty and proteinaceous debris that is rapidly colonized by lipolytic and proteolytic microorganisms from the immediate environment, which is particularly conducive to microbial contamination. If precautions are not rapidly implemented the hides may be damaged through surface etching that reduces their value. To avoid this, fresh hides are normally sprayed with a suitable biocide and cooled prior to storage. However, biocides, although killing the microorganisms, may not necessarily inactivate their hydrolytic enzymes.

Leather production from the hides involves soaking in water, a stage particularly susceptible to attack by *Bacillus* species, especially *B. subtilis*, *B. megaterium* and *B. pumilis*. They secrete several extracellular enzymes that may remain active long after the death of their producer organisms. The hides then undergo liming, deliming and tanning to impregnate them with a variety of chemicals (e.g. chrome-tanning), and are finally dried. Provided the humidity is greater than 80%, even finished leather can be damaged by microorganisms, which may reduce tensile strength, etch and discolour. As the pH of finished leather is fairly acidic, fungi such as *Mucor*, *Rhizopus* and *Aspergillus* species are primary biodeteriogens, whereas bacteria are usually secondary colonizers. Leather shoes are susceptible due to the damp condition occurring both inside and outside, especially in tropical environments.

Wool, fur and feathers

All of these materials are primarily composed of the cystine-rich protein keratin, and the main agents of their biodeterioration are keratinophilic fungi, such as *Trichophyton* species, and certain bacteria, particularly *Streptomyces* species. These are soil-borne microorganisms and animal dermatophytes. The damage inflicted includes pigmentation, odour production and loss of tensile strength.

Worldwide losses of wool alone amount to tens of millions of dollars per annum. However, wool is less prone to attack once cleaned and degreased to remove lanolin. Biodeterioration following manufacture into clothing and furnishings is now less problematical, due to the incorporation of biocides during processing, which are often quaternary ammonium compounds or bronopol (2-bromo-2-nitropropane-1,3-diol).

Stone and related building materials

Stone buildings, monuments and natural rocks, along with the related processed building materials (concrete, brick and mortar) are all prone to microbial attack. Algae and cyanobacteria, especially species of *Pleurococcus* and *Oscillatoria*, and lichens are general colonizers of surface detritus associated with these materials. Fungi, including *Botrytis*, *Penicillium* and *Trichoderma* species, may also be involved, particularly on interior surfaces. This microbial growth causes soiling and simple changes to the appearance of the material, which can be especially troublesome in tropical and subtropical conditions. It may necessitate frequent painting, or cleaning with biocidal washes of bleach, phenolics or organo-tin compounds.

The physical presence of organisms may also cause excessive expansion and contraction associated with wetting and drying of colonies. Entrapment of water within the colonies and the cracks that they create can lead to enhanced frost damage. In addition, hyphal penetration into the surface layers of these materials can result in flaking and widening of cracks. This may be promoted by their excretion of corrosive metabolites. For example, several organic acids solubilize calcium carbonate, and oxalic and citric acid solubilize silicates. Bacterial damage to concrete and stone containing oxidizable sulphur compounds results from *Desulfovibrio desulfuricans* producing H_2S, which is then oxidized to sulphuric acid by *Thiobacillus thiooxidans*. Their activities also damage steel rods within reinforced concrete. Nitrifying bacteria, *Nitrobacter* and *Nitrosomonas* species, may also cause damage by solubilizing calcium-based rock, as their oxidation of ammonia to nitrate leads to the formation of a relatively soluble salt, calcium nitrate.

Cellulosic materials

Plant cell biomass derived materials include wood, card, paper and plant fibre textiles; along with carboxymethyl cellulose, methylhydroxypropyl cellulose and microcrystalline cellulose, used to thicken cosmetics, paints, etc. (see Chapter 10, Bioconversion of lignocellulose). Susceptibility of these materials to microbial attack depends on the environmental conditions, the presence of biocidal agents (chlorine compounds, phenolics, organosulphides, etc.) and their physical form. For example, natural wood is generally more resistant to attack than processed materials such as chipboard and paper.

Cellulosic materials are primarily susceptible to fungal attack, although some bacteria, notably *Cellulomonas* and *Cellvibrio* species, may be involved under very damp conditions. Many fungi produce an array of extracellular hydrolytic enzymes, notably cellulase, a complex of several enzymes including exo-β-1,4-

glucanase, endo-β-1,4-glucanase and β-glucosidase, and hemicellulases, e.g. xylanase and xylosidase. These may persist even if the producing organism is killed.

Damage to wood ranges from loss of quality to major reductions in strength. Staining caused by fungi such as *Chlorociboria* and *Ceratocystis* species may result in down-grading with a consequent loss of value, but the strength of the wood remains unaffected. However, many other microbial activities have major effects on material strength. These include fungal dry rot and wet rot of building materials. Dry rot is caused by the basidiomycete, *Serpula lacrymans*, that requires moisture contents of 25–40%. It is the material that the fungus has decayed and moved on from that is dry. *S. lacrymans* attacks softwoods in particular and is very difficult to eradicate from a building even when using powerful biocides and extensively stripping and removal of apparently uninfected surrounding material. Wet rots require moisture contents in excess of 50% and are caused by several different basidiomycetes, e.g. *Coniophora puteana*. They do not spread beyond the damp areas and are far easier to treat. Further growth of these wet-rot microorganisms is prevented if the cause of dampness in the material is eradicated.

Fuels and lubricants

Fuels and lubricants are primarily produced from petrochemicals and are attacked only if water is present. Consequently, lubricants such as cutting oils are more prone to biodeterioration. They consist of an emulsion of oil finely dispersed in water, in the presence of an emulsifying agent. When other fuels and lubricants become contaminated with water, it forms an aqueous layer below the hydrophobic hydrocarbon. Those hydrocarbons with shorter chain lengths, e.g. petrol/gasoline, are normally less susceptible to biodeterioration than longer hydrocarbons, e.g. diesel oil and kerosene, as are products containing phenolics.

Biodeterioration may occur in storage or in use. Microbial growth can result in the formation of mycelial mats and sludge that may cause blockage of fuel lines and filters, the production of corrosive metabolites and reduced lubrication or degraded fuel properties. A surprising number of microorganisms can utilize these materials. *Aspergillus* species and several other genera have been isolated from kerosene, but the organism often found to be primarily responsible for biodeterioration is the fungus *Hormoconis* (formerly *Cladosporium*) *resinae*. This is a soil microorganism that can degrade a wide range of compounds, from linear and branched alkanes to aromatic rings. It can be a particular problem in aviation fuel, and in both fuels and lubricants for marine diesel engines. Biodeterioration of these materials may be controlled by incorporating biocides such as organoboron compounds and isothiazolones.

Metals

Microorganisms are known to initiate and enhance the corrosion of metals through the formation of biofilms. There are three main routes for microbial corrosion: concentration cells, release of corrosive metabolic products and removal of cathodic hydrogen by sulphate-reducing bacteria.

Microbial concentration cells arise from an oxygen gradient that develops as a microbial colony, in contact with the metal, utilizes the available oxygen. The borders of the colony have access to oxygen and become cathodic, whereas the oxygen-limited centre becomes anodic. This leads to metal ion formation, which usually produce insoluble hydroxides. Iron corrosion may involve bacteria such as *Gallionella*, a chemolithotroph that obtains energy through the oxidation of ferrous ions to ferric ions and forms insoluble ferric hydroxide deposits. Microbial metabolic products, especially inorganic and organic acids, can also be involved in metal corrosion. For example, sulphur oxidizers produce highly corrosive sulphuric acid and *H. resinae*, a fuel degrader (see above), produces a wide range of organic acids that have been implicated in the corrosion of metal fuel tanks. Sulphate-reducing bacteria convert sulphate in anaerobic reduced environments to sulphide. If these organisms are in contact with iron, notably steel sewage pipes, iron sulphide is generated, which is corrosive to mild steel.

Plastics

Plastics are a diverse group of polymeric materials that include polyethylenes, polystyrene, polyvinyl chloride (PVC) and polyesters. They are generally durable, low-cost polymers that have replaced many traditional materials for clothing, furniture, packaging, building, etc. They contain the main polymer, copolymers, and a range of other additives such as fillers, plasticizers and antioxidants. More recently, several plastics have been designed to be biodegradable, in order to reduce environmental problems (see Chapter 10, Polyhydroxyalkanoates). Hence, for these, there must be a balance between stability during their 'working life' and degradability when no longer required.

Many plastics are largely resistant to microbial attack, but their additives are usually more susceptible. Initially, microorganisms often metabolize these additives and form a surface biofilm. For example, plasticized PVC becomes stained by the surface growth of *Streptomyces rubireticuli* and polyamides (e.g. nylon) are discoloured by the action of *Penicillium* species. Regenerated celluloses, including cellulose acetate, cellulose nitrate, cellophane and rayon, are relatively susceptible to microbial attack. Of the polyesters, polycaprolactone and polybutylene adipate are more readily degraded by bacteria and fungi, whereas polyethylene terephthalate (PET), used extensively for bottles, is resistant.

Biodeterioration of cosmetics and pharmaceuticals

Pharmaceuticals are products that are used internally and externally (topically) for therapeutic purposes. They may be divided into items that are:

1 **non-sterile**: solids (tablets, capsules and powders), liquids (suspensions and syrups), creams and lotions; and

2 **sterile**: injectables (parenterals), both single dose and multidose drugs, intravenous infusions, etc., along with products for use in and around the eye area, including drops, lotions, ointment, washes and contact lens cleaning solutions.

Cosmetics and toiletries are used externally for cleansing and decorative purposes, but some products are invariably swallowed and contact mucous membranes. They are non-sterile products in the form of liquids, creams, lotions, semisolids, solids, powders and sticks.

Many non-sterile pharmaceuticals and cosmetics are very similar as they are composed of the same basic formulations, apart from the active ingredient, which is often a minor component in terms of mass. This is particularly true for creams, ointments and lotions. These items are also manufactured, packaged and stored in similar ways. A further common feature is the restriction in the range of preservatives that can be incorporated. Most of these products have very complex formulations, which exacerbates problems associated with preservation. They often contain a large proportion of water, along with animal, plant or mineral oils, natural gums, thickening agents, suspending agents, carbohydrates, aroma and flavouring agents. In addition, the cosmetic preparations may contain protein hydrolysates, milk, beer, egg, plant extracts, etc. Overall, these product formulations provide ideal conditions for microbial growth, as not only are there copious quantities of potential nutrients, but the pH is usually close to neutral.

The biodeterioration losses within the pharmaceutical and cosmetics industries are difficult to determine. Manufacturers are obviously very sensitive about this topic, as public loss of confidence can be disastrous and very difficult to re-establish. However, the total worldwide value of products from these industries is in the order of hundreds of billions of dollars per annum. If one merely estimates annual losses at only 1%, the economic implications are highly significant.

These products may be rendered unsaleable if one of four events occurs.

1 The presence of low levels of acutely pathogenic microorganisms or higher levels of opportunistic pathogens.

2 Contamination with toxic microbial metabolites that can persist even when the producer microorganisms are dead or removed.

3 The occurrence of detectable physical and/or chemical changes.

4 Loss of function of the active ingredient(s).

The presence of pathogens or potential pathogens

Health hazards associated with these products are dependent upon the type of microorganisms present, their concentration, the route of administration and the health of the user/patient.

Obviously, the presence of only a few cells of some highly pathogenic microorganisms would be sufficient to cause disease. Even relatively low numbers of some opportunistic pathogens can be dangerous, particularly if the product is used by an individual who is ill and whose immune system is already compromised. This is particularly important when using products on hospital patients, the old and the very young. For example, the use of creams or ointments contaminated with low levels of pseudomonads on healthy individuals is unlikely to be harmful. However, if applied to patients with skin damage through dermatoses or burns, the consequences could be fatal. Even antiseptic and disinfectant solutions are not immune. Dilute solutions of quaternary ammonium compounds that may be used as disinfectants for catheters, etc., have been found to contain high levels of pseudomonads. These microorganisms, especially *Pseudomonas aeruginosa*, are also dangerous if they contaminate eye products, such as eye drops and washes, mascara, etc.

The presence of microbial toxins

Low levels of mycotoxins, particularly in products that are to be ingested, pose a major problem, especially as they remain after the producer fungus has been killed or removed. Endotoxins are also problematical; some are often referred to as pyrogens as they indirectly induce fever. They are mainly fragments of microbial lipopolysaccharides liberated from mostly Gram-negative bacteria, although some may originate from fungi and other microorganisms. It is vitally important to eliminate them from injectable products. Consequently, ultrapure water, free from these materials, is used in the manufacture of intravenous drugs and infusions.

Chemical and physicochemical changes

The rate of any change depends upon the chemical structure of the ingredients, the physicochemical characteristics of the product and the inoculum of biodeteriogen that initiates attack. However, any organisms found when measurable changes become detectable may not necessarily be those responsible for initiating the attack: they may already be dead. In fact, several organisms may have been involved in sequential attack from initiation to the point when changes become detectable. Often a limited initial attack is followed by the actions of more destructive groups.

Observable effects of microbial biodeterioration include odours resulting from the production of organic acids, fatty acids, amines, ammonia and hydrogen sulphide. Formation of ammonia or acids may ultimately alter the pH, which can markedly affect physical properties, including viscosity and colour. Spoiled creams and lotions may also develop lumps and slime, or gas bubbles may be generated. These products composed of oil–water emulsions can become unstable, ultimately 'cracking' to form separate oil and water phases.

Oils and fats are particularly susceptible to microbial attack when dispersed in aqueous formulations, being degraded to glycerol and fatty acids by lipases. The fatty acids may then be broken down via β-oxidation to form odorous ketones. Even semisolids with high fat/oil content are not immune, as fungi may grow in surface films of moisture.

Many thickening and suspending agents are prone to depolymerization by specific extracellular enzymes excreted by microorganisms, e.g. amylases and cellulases. Also, humectants may support microbial growth, notably glycerol and sorbitol, which are used extensively in toothpaste and many other products.

Loss of function of the active ingredient

The active ingredient is often present at relatively low concentrations, particularly in pharmaceuticals, but if it is attacked the product is rendered useless. These active ingredient may be therapeutic or antimicrobial agents, and in cosmetics and toiletries they are mostly detergents, colouring agents, etc. Examples of attack on active ingredients include breakdown of antibiotics in capsules, tablets and in solution. Penicillins and cephalosporins are obviously prone to attack by organisms that produce β-lactamases, whereas chloramphenicol is inactivated by the action of chloramphenicol acetyltransferase that is secreted by certain microorganisms. In addition, aspirin and heroin are both broken down by the bacterium *Acinetobacter lwoffii*. Atropine, an alkaloid used in eye preparations, is degraded by *Pseudomonas* and *Corynebacterium* species, and hydrocortisone is attacked by the fungus *Cladosporium herbarum*.

In many cosmetics and toiletries, the active ingredient is a detergent. Shampoos, for example, often contain the anionic detergent, sodium dodecyl sulphate (SDS), which is prone to attack by alkyl sulphatases from bacteria, especially *Pseudomonas*, *Citrobacter* and *Aerobacter* species. Breakdown results in reduced lathering properties and may generate unpleasant odours, particularly hydrogen sulphide resulting from the reduction of sulphate. For detergents, there must be a compromise between resistance to attack in use, and biodegradability once the material enters the sewage system and environment.

Preservatives, disinfectants and antiseptic compounds may be subject to attack at in-use concentrations and when further diluted. Many of these compounds are more resistant if nitrated or halogenated.

Factors influencing microbial spoilage

Microbial growth is determined by the nutrient status of the product formulation, its pH, oxygen concentration, water activity, temperature and the efficacy of the preservative system employed. The size and composition of the microbial inoculum is obviously of major importance. Sources include the raw materials, and contamination during stages of manufacture, packaging and storage, and while in use. The life of the product from manufacture to sale, and the duration and treatment received while in use are crucially important. Some of these products can spend several months or longer in a warm humid bathroom. During this time they may be subject to repeated challenge by microorganisms. For example, a jar of hand cream is challenged each time the user takes a sample.

Moisture content of a product or, more specifically, water activity (A_w) influences proliferation of microorganisms. Water activity can be reduced by the presence of solutes such as salts or polyethylene glycol, but it is difficult to prevent moisture films from developing on the surface of products. Consequently, surface biodeterioration can occur on what may essentially be a resistant product.

Redox potential, the oxidation–reduction balance, depends upon both oxygen content and the nature of the ingredients. Most biodeterioration of cosmetics and pharmaceuticals is due to aerobic and facultative microorganisms. Anaerobic spoilage, unlike in the food industry, is relatively rare. Removal of oxygen would therefore prevent most spoilage. This may be achieved during manufacturing stages through the use of carbon dioxide or nitrogen blankets for bulk storage, and final products can also be packaged within an anoxic atmosphere.

The pH of a product obviously influences the range of microorganisms that will grow. However, most of these products are at pH 5–8, allowing the growth of a very wide range of microorganisms. Product pH also determines the effectiveness of the preservatives. Many of those allowed become less effective as the pH is raised and are fairly ineffective above pH 7.0. Some product components may assist microbial survival. Proteins, suspending agents, kaolin, magnesium trisilicate, aluminium hydroxide, etc., and low concentrations of some surfactants, allow microbial adsorption, which may aid in avoiding the action of preservative agents.

Package design also influences the susceptibility of a product to biodeterioration. Some plastics are permeable to oxygen and moisture, which may support microbial activities. Cap liners are often made of cork or board that readily absorbs moisture, and can be ideal sites for the growth of microorganisms. From here they can go on to invade the product. Therefore, cap liners must be impregnated with preservative to avoid this problem.

Assessment of microbial contamination (microbiological quality control)

Microbial quality control is conducted to:

1 monitor microbial contamination of raw materials;

2 monitor and confirm the efficacy of operations such as sterilization;

3 control the danger from pathogenic microorganisms by confirming their absence; and

4 verify the expected storage life and provide an estimate of perishability.

For products that must be sterile, the detection of viable organisms is crucially important, as the presence of any viable organism is evidence of process failure. In some products the sterilization treatment may be severe, e.g. heat treatments. Consequently, if any organisms do survive, they will be damaged and their survival may not be confirmed until they have had a long period of recovery. Hence, a product tested shortly after a sterilization process may apparently be free from viable organisms, but it is only after days or weeks that survivors recover sufficiently to become detectable. Therefore, the timing of tests is vitally important in determining the efficacy of sterilization operations.

Assessment of viable microorganisms in non-sterile products presents several problems that include ensuring homogeneity of samples, suitable sample preparation prior to testing and, where necessary, neutralization of added preservatives. Assessment may involve total microbial counts and/or estimation of groups, e.g. coliforms, or the detection of specific microorganisms. Additional evidence of microbial activity within a product may come from changes occurring to the pH, viscosity, emulsion stability, etc. However, by the time that changes of this nature occur, biodeterioration is normally well under way.

As previously mentioned, spoilage may be through the presence of dangerous metabolic products of microorganisms that have since died, often killed during processing. Sensitive analytical tests are available for the detection of very low levels of mycotoxins. The presence of endotoxins (pyrogens) can be determined by febrile response in animals such as rabbits, or now more readily via the Limulus test (amoebocyte lysate assay). It utilizes amoebocytes from the blood of the horseshoe crab (*Limulus polyphemus*), which clot in the presence of endotoxins. This test is very sensitive and can detect as little as 10^{-12} g/ml. The detection of microbial enzymes is often important, as they too can persist after cell death. Many hydrolytic enzymes can pose major problems to product storage life, e.g. cellulases.

Other tests that may be performed on the products are 'challenge tests'. These essentially examine the effectiveness of the preservative system used to prevent growth of microorganisms. They may utilize known product biodeteriogens, or a group of standard organisms is frequently used in the testing of cosmetics and pharmaceuticals, which normally includes *Pseudomonas aeruginosa* (Gram-negative), *Staphylococcus aureus* (Gram-positive), *Candida albicans* (yeast) and *Aspergillus niger* (filamentous fungus). The tests may involve a single challenge with the organism(s). Alternatively, repeated sequential challenges may be given in order to simulate in-use treatment.

Factors affecting the performance of preservatives

The aim of incorporating preservatives into many of these products is to kill or inhibit the growth of microorganisms that enter during the repeated use of the product, as relatively few products are of the single dose type. As little preservative as possible should be used and wherever feasible the least toxic one that will do the job. Underdosing can lead to the development of resistance by microorganisms; overdosing with preservative agents is uneconomical, has greater toxicity and can lead to greater environmental pollution. Hence, the need for extensive testing and evaluation of preservative systems in new product formulations prior to manufacture.

Preservatives should not be expected to sterilize formulations that are heavily contaminated as a result of using poor raw materials or poor standards of manufacture. Ideally, preservatives should be:

1 broad spectrum, effective against all microorganisms likely to enter the product;
2 free from toxicity, irritancy and allergenicity to the user;
3 stable, able to withstand production procedures and the 'life' of the product;
4 compatible with all other formulation ingredients; and
5 free from odour and flavour, and should not affect the chemical or physical properties of the product.

A very limited number of preservatives are approved for use in cosmetics and pharmaceuticals (Table 16.1). The actual choice of preservative will depend on its activity spectrum, solubility, stability, volatility, toxicity, irritancy, colour, odour, taste, the pH of the formulation and the cost. As previously mentioned, the pH is very important, as preservatives approved for use often have a fairly narrow pH range over which they are effective.

Most preservatives interact with other formulation ingredients as well as microbial cells. In many cases the actual quantity of preservative that is 'available' for antimicrobial action may be only a small portion of the total preservative present in the product (Fig. 16.2). Nevertheless, the remaining 'unavailable' preservative can still contribute to the possible toxicity or irritancy to the user.

Many cosmetic and pharmaceutical products are multiphase systems. These are difficult to preserve, particularly where there are oil–water emulsions and surfactant micelles. In emulsified products 'available' preservative may be considerably reduced because many undergo partition, predominantly into the oil phase, or are solubilized into surfactant micelles. Once within the oil phase or micelle, the preservative is not immediately available to act on contaminating microorganisms. The extent of the distribution is dependent upon the partition coefficient, and many preservatives are more soluble in vegetable oils than mineral oils (Table 16.2). Further loss of 'available' preservative may occur through weak bonding with non-micellar surfactants and hydrocarbon polymers, and adsorption to particulate material.

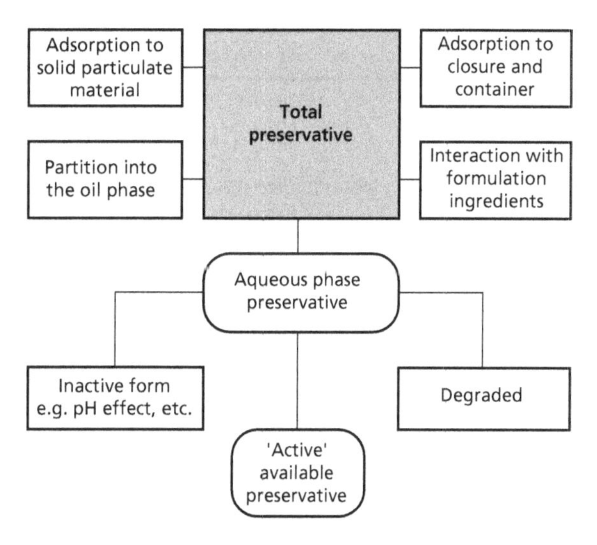

Fig. 16.2 Factors influencing the availability of a preservative within a product.

Table 16.1 Preservatives for cosmetics and pharmaceuticals

Alcohols	Widely used as disinfectants and antiseptics, but will not kill endospores; e.g. ethanol (50–70%, v/v) and isopropanol (50–70%, v/v) denature proteins and solubilize lipids
Formaldehyde	Highly reactive, reacts with NH_2, SH and COOH groups and kills endospores at high concentration. Possible carcinogen. Formerly used quite extensively to preserve complex systems. Formaldehyde donors, e.g. Germall 115-imidazolidinyl urea, are now preferred for use in cosmetics
Hypochlorite	Forms hypochlorous acid (HClO), a strong oxidizing agent, i.e. $HClO \rightarrow HCl + O$ (nascent oxygen). Used as a general disinfectant
Silver nitrate ($AgNO_3$)	General antiseptic and used in pharmaceutical products for the eyes
Phenyl mercuric nitrate/ mercuric chloride and related compounds	Used for injectables at 0.002% (w/v), e.g. Thiomersal
Quaternary ammonium compounds, e.g. cetrimide and benzalkonium chloride	Are cationic surfactants that disrupt cell membranes. Used as skin antiseptics and disinfectants, also as a preservative for ophthalmic preparations, eye drops and contact lens cleaning solutions at up to 0.01% (w/v). Activity reduced by organic matter, more effective against Gram-positive bacteria than Gram-negative bacteria, yeasts or moulds. They have a rapid effect and their activity is enhanced by EDTA, benzyl alcohol and other compounds. Benzalkonium chloride acts synergistically with phenyl mercuric nitrate
Phenolic compounds	Denature proteins and disrupt cell membranes. Used as antiseptics at low concentrations, and as disinfectants at higher concentrations. Phenol (0.5%, w/v) and ortho-cresol (0.3%, w/v) are used as preservatives in injectables, creams and ointments. Chlorocresol (0.1%, w/v) is more effective than phenol and ortho-cresol, and is used in eye drops and injectables, creams and ointments. Chloroxylenol is used in many cosmetics, its activity against Gram-negative bacteria is enhanced by adding EDTA
Bronopol (2-bromo-2-nitropropan-1,3-diol)	A broad-spectrum preservative, active against Gram-positive and Gram-negative bacteria, and fungi over a wide pH range of 3–8. Affects membranes and dehydrogenase enzymes, used at 0.01–0.02% (w/v). Preservative of cosmetics and pharmaceuticals, and many other materials
Organic acids and derivative	Includes propionic acid and propionates, sorbic acid and sorbates (0.2%, w/v), benzoic acid and benzoates (0.1%, w/v for products at pH 5.0 or less) and parabens, especially for food, pharmaceuticals and cosmetics
Nisin	Effective against Gram-positive bacteria, particularly spore formers. Active at acid pH, causing membrane disruption; possible role in toothpaste, skin creams and lotions

Packaging can also influence the effectiveness of the preservative. Some preservatives adsorb to plastics and even glass, e.g. quaternary ammonium compounds adsorb to both of these. Also, their adsorption to rubber, cork or card cap inserts can reduce preservative action. However, some formulation components aid preservation, performing as activators or collaborators. Useful examples include cosolvents, e.g. ethanol, isopropanol and other alcohols, and ethylene glycol. Also, ethylenediamine tetraacetic acid (EDTA) is used in certain cosmetic and pharmaceutical preparations and is well known to improve the effectiveness of several antimicrobial agents, particularly against Gram-negative bacteria.

Table 16.2 The partition of preservatives into liquid paraffin and vegetable oil

	Approximate partition coefficient (K^c_w) in	
Preservative	Liquid paraffin	Vegetable oil
Phenol	0.1	5
Chlorocresol	1.5	120
Chloroxylenol	6.5	600
Propyl paraben	3.0	300

Oil–water partition coefficient* $K^c_w = C_o/C_w$ = concentration in oil/concentration in water.
* At equilibrium.

Often there are great benefits to be gained from the use of a combination of preservatives, including:
1 increase in the spectrum of inhibition;
2 reduction in irritancy and toxicity, as lower levels of each preservative are used;
3 prevention of the development of resistant microorganisms;
4 synergistic preservative effects;
5 prolongation of preservative action; and
6 compensation for physicochemical limitations.

Biodeterioration testing

The testing of materials for resistance to biodeterioration and the examination of the efficacy of biocidal treatments are obviously of major importance in determining the 'life expectancy' of materials in use. Various techniques have been devised, including environmental simulation tests, in order to estimate how the material performs in use.

Soil burial tests are useful techniques for solid building, furnishing and clothing materials. Here the item is buried in soil for a certain period and thus subjected to potential attack from a very wide range of naturally occurring soil microorganisms. After this treatment the material must be tested for loss of properties. These may include changes in colour, tensile strength or elasticity, loss of mass, etc. Alternatively, the material may be specifically challenged with a known biodeteriogen or group of bioteriogens. This is usually performed under very favourable conditions for the microorganisms, with optimum environmental conditions of humidity, temperature, etc., and with a good supply of nutrients. Again, after a specified period of exposure to the microbial activities, the material is assessed for loss or retention of its properties.

Further reading

Papers and reviews

Ascaso, C., Wierzchos, J. & Castello, R. (1998) Study of the biogenic weathering of calcareous litharenite stones caused by lichen and endolithic microorganisms. *International Biodeterioration and Biodegradation* **42**, 29–38.

Blake, R. C., Norton, W. N. & Howard, G. T. (1998) Adherence and growth of a *Bacillus* species on an insoluble polyester polyurethane. *International Biodeterioration and Biodegradation* **42**, 63–73.

Cain, R. B. (1994) Biodegradation of detergents. *Current Opinion in Microbiology* **5**, 266–274.

Davis, J. L., Nica, D., Shields, K. & Roberts, D. J. (1998) Analysis of concrete from corroded sewer pipe. *International Biodeterioration and Biodegradation* **42**, 75–84.

Gaylarde, C. C. & Morton, L. H. G. (1999) Deteriogenic biofilms on buildings and their control: a review. *Biofouling* **14**, 59–74.

Gu, J., Ford, T. E., Berke, N. S. & Mitchell, R. (1998) Biodeterioration of concrete by the fungus *Fusarium*. *International Biodeterioration and Biodegradation* **41**, 101–109.

Margesin, R. & Schinner, F. (1998) Biodegradation of the anionic surfactant sodium dodecyl sulphate at low temperatures. *International Biodeterioration and Biodegradation* **41**, 139–143.

Morton, L. H. G. & Surman, S. B. (1992) The role of biofilms in biodeterioration. In: *Proceedings of the International Symposium on Surface Properties of Biomaterials*, Manchester, UK, May 1992 (eds R. West & G. Batts). Butterworth-Heinemann, Oxford.

Morton, L. H. G., Greenway, D. L. A., Gaylarde, C. C. & Surman, S. B. (1998) Consideration of some implications of the resistance of biofilms to biocides. *International Biodeterioration and Biodegradation* **41**, 247–259.

Schink, B., Janssen, P. H. & Frings, J. (1992) Microbial degradation of natural and new synthetic polymers. *FEMS Microbiological Reviews* **9**, 311–316.

Books

Allsopp, D. & Seal, K. J. (1986) *Introduction to Biodeterioration*. Edward Arnold, London.

Bousher, A., Chandra, M. & Edyvean, R. (1995) *Biodeterioration and Biodegradation 9*. Institution of Chemical Engineers, Rugby.

Characklis, W. G. & Marshall, K. C. (1990) *Biofilms*. Wiley, New York.

Dexter, S.C. (1986) *Biologically Induced Corrosion*. National Association of Corrosion Engineers, Houston, TX.

Heitz, E., Flemming, H.-C. & Sand, W. (1996) *Microbially Influenced Corrosion of Materials*. Springer-Verlag, London.

Hopton, J. W. & Hill, E. C. (eds) (1987) *Industrial Microbiological Testing*. Blackwell Scientific Publications, Oxford.

Houghton, D. R., Smith, R. N. & Eggins, H. O. W. (eds) (1988) *Biodeterioration 7* (International Biodeterioration Symposium). Elsevier Applied Science, London.

Rose, A. H. (ed.) (1981) *Microbial Biodeterioration*. Academic Press, London.

Rossmoore, H. W. (ed.) (1991) *Biodeterioration and Biodegradation 8: Proceedings of the 8th International Biodeterioration and Biodegradation Symposium*, Windsor, Ontario, Canada, 26–31 August 1990. Elsevier Applied Science, London.

17 Animal and plant cell culture

Animal cell culture

Animal cell culture is predominantly used in the production of a wide range of health-care products, including therapeutic and analytical proteins, monoclonal antibodies and viral vaccines (Table 17.1), and for *in vitro* toxicity testing. Culture of animal cells entails the dispersion of tissues into a cell suspension by mechanical means and/or use of enzymes such as proteases, e.g. trypsin and collagenase. The separated cells may then be grown as a primary culture. Some cells are anchorage-dependent and must be cultured as a monolayer on solid support materials. These materials normally have large surface to volume ratios and can be perfused with, or suspended in, liquid nutrient medium. Anchorage-independent cells are grown as a suspension directly in the liquid culture medium. Cultivated primary cells may be repeatedly subcultured to form a cell line or strain. Most primary cell lines have a finite life, and will eventually fail to divide and die. This phenomenon has been attributed to apoptosis or programmed cell death. However, cells can be transformed to become continuous cell lines by infection with oncogenic viruses, treatment with carcinogenic chemicals or by propagating cells from tumours. Commonly used robust cell lines include those derived from baby hamster kidney (BHK), Chinese hamster ovary (CHO) and human fibroblasts or epithelial cells.

Animal cells are typically eukaryotic. At 10–100 μm in diameter, they are rather large compared with most microorganisms. In culture, animal cells have relatively slow growth rates, with doubling times of 5–10 h. Their lack of a cell wall also makes them more sensitive to shear forces and bubble rupture. Environmental conditions required for optimum growth vary quite widely. The optimum temperature and pH for mammalian cells are 37°C and a pH of about 7.3, whereas insect cells usually grow best at 28°C and pH 6.2. Fish cells normally tolerate a wide temperature range of 25–35°C and a pH of 7.0–7.5. Culture media must be well buffered and isotonic, although in some instances stress-inducing hyperosmotic conditions may enhance productivity. Animal cells, particularly those from mammals, usually demand very complex media (see Chapter 5), but genetic modification of cells is being used to reduce their reliance on various animal proteins, especially those derived from serum.

Animal cells that are anchorage-independent can be cultured freely suspended in modified stirred tanks or airlift systems of capacities up to 10 000 and 2000 L, respectively. Anchorage-dependent cells must be immobilized or entrapped, using stationary monolayer cultures, hollow-fibre systems, ceramic cartridges, microcapsules and microbeads (see below, Monoclonal antibodies). Fed-batch suspension and perfusion culture predominate as modes of operation, particularly for the large-scale production of recombinant therapeutic proteins and monoclonal antibodies. Older viral vaccine production processes were mostly simple batch operations, but these too are moving to fed-batch suspension or continuous perfusion systems.

A key factor that often limits animal cell cultures is the accumulation of inhibitory compounds that include lactic acid, ammonium ions, carbon dioxide and methylgloxal, derived from triose phosphate. Oxygen requirements vary somewhat depending on the culture conditions and fermentation system. For suspension culture in conventionally stirred and aerated fermenters, high turbulence must be created in order to maintain gas transfer rates. However, the undesirable mechanical stresses generated, particularly shear forces, can be reduced if stirring is performed using sheets of nylon rather than metal paddles and if aeration is via bubble-free methods. In bubble-free systems, the air is passed through hollow fibres formed from hydrophobic membranes that are submerged in the fermenter.

Monoclonal antibodies

Antibodies are extremely specific, as each one binds to a

particular antigen (Fig. 17.1). This trait enables them to be used for many purposes, including disease diagnosis and therapy, as the basis of many bioassays for the detection of drugs, viral and bacterial constituents, and in the purification of enzymes and other proteins.

Conventional methods for antibody production involve 'raising' them in live animals, e.g. horse, rabbit, etc. An antigen preparation is injected into the animal and then, after antibodies have been formed, they are purified from collected blood serum (antiserum). However, there are problems with this method. It yields antisera that contain undesired substances, produces a very small amount of usable antibody and is **polyclonal**, a product of many different cells. As each antigen may have several different epitopes (antigenic determinants), polyclonal preparations are heterogeneous mixtures of antibody molecules, containing components specific for each epitope of the antigen. Far superior **monoclonal** antibodies (MAbs) have been available for over 20 years. They are called monoclonal because they originate, following appropriate screening and selection, from only one cell clone. Consequently, they are specific for a single epitope. Monoclonal antibodies have numerous uses as analytical or therapeutic agents. MAbs are potentially more effective than conventional drugs in fighting disease, as many drugs attack not only the foreign substance, but also the body's own cells, sometimes producing undesirable side-effects and allergic reactions, whereas MAbs attack only the target molecule and consequently have no or fewer side-effects.

Table 17.1 Examples of animal cell products

Enzymes Asparaginase, collagenase, hyaluronidase, tyrosine hydroxylase and urokinase
Growth factors Epidermal growth factors, nerve growth factors
Hormones Calcitonin, follicle-stimulating hormone, relaxin, thyroxine
Immunobiological regulators, such as interferon (anticancer glycoprotein). Lymphokines (hormonal proteins regulating immune responses of human body); interleukins (anticancer agents)
Monoclonal antibodies Produced by hybridoma cells and used for diagnostic assay systems, therapeutics, biological separations, e.g. purification of interferon by affinity chromatography
Insecticides, e.g. baculoviruses
Tissue plasminogen activator (prevents blood clotting)
Viruses, for viral vaccines for human (influenza, measles, mumps, poliomyelitis, rabies, rubella, etc.) and animal (foot-and-mouth disease, canine distemper, swine fever, etc.) use. Retroviruses or adenoviruses are also produced for use in gene therapy

MAbs are obtained from antibody producing cells, B-lymphocytes, isolated from mammalian spleen, which have been fused (hybridized) with suitable 'immortal' cells, usually myeloma cells (malignant plasma cells) that grow incessantly in culture. The resulting hybrid cell, hybridoma, synthesizes large amounts of antibody and has the ability to grow continually.

MAbs are prepared using either various forms of direct *in vitro* culture of hybridoma cells or by a process known as ascites production (see below). Both routes allow the manufacture of large quantities of pure MAbs. Production of MAbs may begin with the immunization of a mouse or hamster with the desired antigen (Fig. 17.2). Where necessary, immunization protocols can be adopted that promote the generation of antibody-forming cells with specificity for rare epitopes or poor immunogens. Within 3–4 weeks sufficient antibody-producing B-lymphocytes will have been generated, which can then be extracted from the spleen. Alternatively, instead of using live animals, *in vitro* immunization may be used. These methods are often more

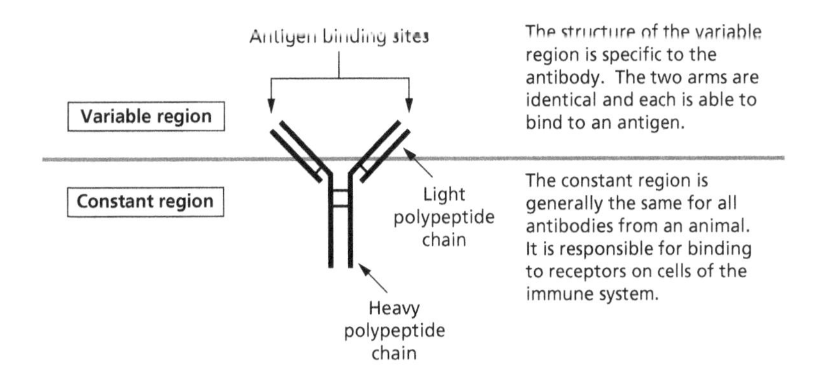

Fig. 17.1 Antibody structure.

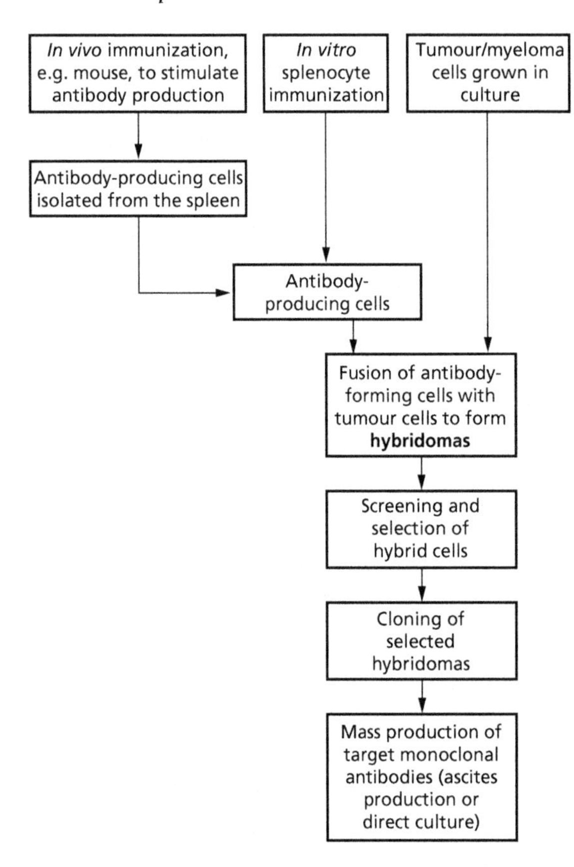

Fig. 17.2 An outline of the production of monoclonal antibodies.

reproducible, reduce time for the production of antibody-producing cells and allow the formation of human antibody-producing cells.

The antibody-producing cells, from whatever source, are then fused with myeloma cells. This can be achieved by incubating them together in the presence of polyethylene glycol, followed by selection of only fused cells. Selected hybrids are cloned and then screened for the desired antibody. Improved rates of fusion, and specific fusion of selected pairs of cells, have been made possible by using electrofusion and laser fusion techniques. These methods reduce the time necessary for selection and screening postfusion.

Cultivation of selected stable hybridomas may be via one of two routes. Ascites production methods involve injecting the hybridoma cells into the peritoneal cavity of mice or other rodents, where they proliferate and secrete antibody into the ascitic fluid. This fluid can contain antibody levels up to 15 mg/ml and about 5 ml of ascitic fluid can be extracted from each animal. The second approach involves direct culture as hybridoma cell suspensions or as immobilized forms. Suspension cultures are usually performed in stirred tank or airlift fermenters and immobilization may take the form of cell anchorage within hollow fibres or other matrices. Antibody yields from direct culture are usually quite low compared with ascites production. However, they have the advantage of cheaper downstream processing and less risk of contamination. Culture in stirred tank reactors is usually conducted in vessels not exceeding 1000 L capacity. Cell densities approach 10^7 cells/ml and MAb yields are about 30 mg/L. Airlift fermenters of 100–1000 L capacity, some of which now operate continuously, produce lower cell densities of 2×10^6 cells/ml, but MAb yields are higher at 40–500 mg/L.

Hollow-fibre immobilization provides a densely packed cell system that simulates *in vivo* capillary networks of tissues. The hollow fibres are often made of cellulose acetate, polysulphone or acrylic polymers. They have lumen of 200 μm and their walls are 50–75 μm thick, with molecular weight cut-offs of $(10–100) \times 10^3$ Da (see Fig. 7.10). Cell densities of 10^6–10^7 cells/ml can be achieved and these systems often operate for several weeks or even months. Hollow-fibre systems also have the advantage that relatively high concentrations of MAbs are collected in the culture fluid and the cells remain separated from the MAbs, making downstream processing easier. The cost of MAb production by these methods is often much lower. However, they are not problem free. Fibres can become blocked, nutrient or product diffusion may be rate-limiting and toxic metabolites can accumulate.

Cell immobilization within the matrix of ceramic cartridges has also been attempted, which can achieve cell densities of over 10^9 cells/ml and may operate continuously for about 2 months. Systems using cells entrapped within microcapsules or microbeads, such as alginate gel beads, are also used for MAb production. Alginate beads are prepared by mixing the cells with sodium alginate at a 'seeding density' of 2×10^6 cells/ml. The beads of gel are then formed by adding the suspension, dropwise, into calcium chloride solution. The seeded beads are then transferred to a bioreactor for MAb production.

Depending upon the method of production, it may be necessary to remove the cells from the medium into which the antibodies have been excreted using centrifugation or filtration. Cell-free medium is then concentrated by ultrafiltration and the antibodies are purified by several chromatographic steps. This usually involves affinity chromatography, both cation and anion exchange chromatography, often followed by a hydrophobic interaction chromatographic step. Finally, gel filtration is performed to produce a pure MAb preparation (see Chapter 7, Protein purification). In some cases, a conjugate is then coupled to the antibody, which may be a toxin for therapeutic use, or, for diagnostic purposes, an enzyme, radiolabel or other detectable conjugate.

It is also possible to produce bi-specific MAbs that simultaneously react with two different antigens. This is achieved by fusion of two separate hybridomas or of a hybridoma with a B-lymphocyte. Alternatively, two distinct antibodies can be joined by chemically cross-linking, or through antibody amalgamation. In the latter, the two antibody 'arms' (Fig. 17.1), each one composed of a light and a heavy chain, are separated and each is reassociated with an 'arm' from a different antibody. In addition, genetic engineering has brought about the development of novel antibody constructs that can be expressed by myeloma cell lines. For example, rodent–human chimaeric antibodies may be produced that are better tolerated when used in chemotherapy. Also, antibodies can be directly produced that have an added novel function, such as the incorporation of an enzyme activity.

Plant cell culture

Plants are major sources of pharmaceuticals, dyes, food colours, food flavours, enzymes, polysaccharides, fragrances, insecticides and herbicides. Many are secondary metabolites that are complex low molecular weight organic compounds (Table 17.2). In some cases, their conventional agricultural production or gathering from the wild has a number of problems and supply cannot be ensured. Production via plant cell culture offers several important advantages, particularly as the control of product supply is independent of the availability of the plant itself. The product can be produced under controlled and optimized conditions, and producer strains are more readily improved. Plant cell cultures also have potential as bioconversion tools.

Table 17.2 Examples of products of plant cell culture

Product	Producer organism
Anthraquinones (pigments)	*Morinda citrifolia*
Capsaicin (flavour)	*Capsicum* species
Digoxin (cardiotonic)	*Digitalis lanata*
Ginsenosides and ginseng biomass (herbal medicine)	*Panax ginseng*
Quinoline alkaloids (flavour and pharmaceutical use)	*Cinchona ledgeriana*
Shikonin (pigment and pharmaceutical use)	*Lithospermum erythrorhizon*
Taxol (cancer therapy)	*Taxus cuspidata*
Vinblastine and vincristine (cancer therapy)	*Catharanthus lanata*

Plant cells are larger and grow more slowly than most microorganisms, being 10–100 μm in diameter with doubling times of 20–100 h. Consequently, batch fermentations often last 2–4 weeks, which may present problems in excluding microbial contaminants. The cells also have a higher water content, which is partly due to the presence of a large central vacuole that can occupy as much as 95% of the intracellular volume. In suspension culture they tend to aggregate and are more sensitive to shear. Therefore, reactors developed for microorganisms are often unsuitable without appropriate modification to the agitation system. Airlift reactors usually provide a more suitable environment, supplying good mixing while subjecting cells to less mechanical stress.

In some instances, only low levels of a product are formed in suspension culture. However, cell cultivation in systems that allow immobilization within a matrix, self-immobilization to a surface, or merely allow cell aggregation, often produces a more suitable environment for product biosynthesis. These conditions resemble those *in vivo*, where plant cells are in close proximity to each other and organized into tissues. However, cultivation by these methods may be slower and the control of the microenvironment is often more difficult to achieve.

In most fermentations, the plant cells are grown heterotrophically, as it is often simpler to perform than autotrophic culture in photobioreactors and allows greater control. Usually, a two-phase culture is operated for the production of secondary metabolites. The first phase uses a medium optimized for growth and biomass generation. Conditions are then changed to support slow growth, or no growth, which favours secondary metabolite production and accumulation. Factors that can be manipulated to improve secondary metabolite formation include light, temperature, aeration and agitation, and most importantly, media composition.

Each undifferentiated plant cell is totipotent, as under specific conditions, when provided with the appropriate combination and quantity of hormones (mostly auxins, gibberellins and cytokinins), it can regenerate a whole plant. Production of specific metabolites, in quantity, is often associated with certain stages of differentiation. Therefore, the provision of the appropriate hormonal balance is usually vital in bringing about the necessary changes to the metabolism of undifferentiated cells in culture. Auxins normally have the greatest effect on secondary metabolism. Usually, the concentration of auxin is lowered or a weaker auxin is used during the production phase. The influence of other groups of hormones, gibberellins and cytokinins, appear to be more variable. Phosphate and nitrogen limitation is often effective in further improving the yield of many secondary products. Also, *in vivo* several secondary metabolites are produced in response to stress, particularly attack by pathogens and osmotic changes. Consequently, an increase in sucrose concentration or the addition of fungal elicitors can be used to trigger the formation of certain secondary 'stress' metabolites by plant cell cultures. For example, the polysaccharide elicitor called Pms, isolated from *Phytophthora megasperma* var. *soja*, stimulates isoflavanoid production in soya bean cell cultures.

Many plant products, particularly secondary metabolites, are not excreted into the fermentation medium. They are either stored intracellularly, often within a vacuole, or become incorporated into the cell wall. Therefore, cells must be harvested and processed to extract the target product. However, attempts are being made to extract products *in situ* during the fermentation. These extractive fermentations use absorbants, solvents or mechanisms that temporarily destabilize the cell membranes. Such integrated methods may prevent product degradation and enhance its production.

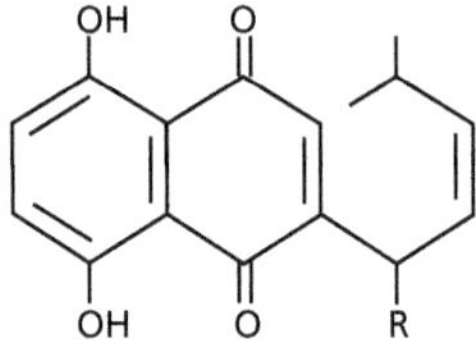

Compound	R
Deoxyshikonin	–H
Shikonin	–OH
Acetylshikonin	$-OCOCH_3$
Isobutylshikonin	$-OCOCH(CH_3)_2$
β,β-Dimethylacrylshikonin	$-OCOCH{=}C(CH_3)_2$
Isovalerylshikonin	$-OCOCH_2CH(CH_3)_2$
α-Methyl-n-butylshikonin	$-OCOCH(CH_3)CH_2CH_3$
β-Hydroxyisovalerylshikonin	$-OCOCH_2C(OH)(CH_3)_2$

Fig. 17.3 The structure of shikonin and some derivatives.

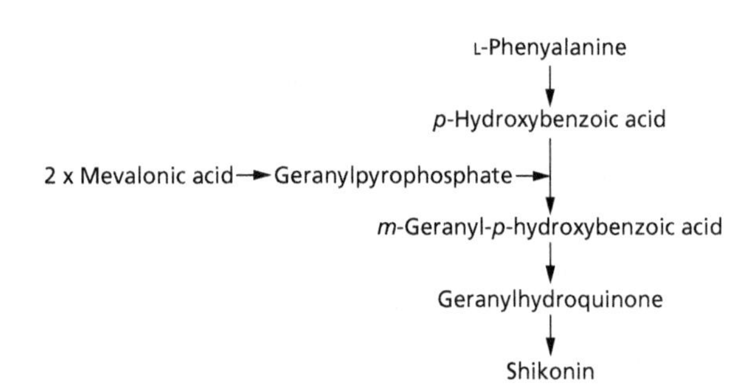

Fig. 17.4 The biosynthesis of shikonin.

Shikonin production

Several attempts have been made to commercialize the production of plant metabolites from plant cell culture in bioreactors, but few have achieved commercial success. One successful operation is the production of shikonin, which is manufactured on an industrial scale in Japan. The product is a low molecular weight organic compound, a red naphthoquinone (Fig. 17.3). This pigment is found in the roots of *Lithospermum erythrorhizon*, a plant that grows wild in Japan, Korea and China. It has therapeutic applications as an anti-inflammatory and antibacterial agent, and promotes the healing of wounds and burns. Shikonin is also used as a dye, and for colouring food and cosmetics.

The development of plant cell culture as the method of manufacture of shikonin resulted from the exhaustion of the wild plant supply. A market was present, but little or no product was available. Agricultural production of the plant is difficult, takes 3–7 years and produces less product, only 1–2% on a dry weight basis. In contrast, cell cultures take 3 weeks and generate shikonin at levels of 12–14%.

Shikonin is synthesized from *p*-hydroxybenzoic acid and geranylpyrophosphate, derived from L-phenylalanine and two molecules of mevalonic acid, respectively (Fig. 17.4). Once produced, the pigment accumulates as granules outside the cytoplasmic membrane within the cell wall. Shikonin biosynthesis is affected by both physical and chemical factors, including temperature, growth rate, light, carbon and nitrogen sources, hormones and endogenous levels of rosmarinic acid. On solid Linsmaier–Skoog agar medium, red shikonin-containing callus is produced, but when transferred to liquid medium containing the same components, except agar, no pigment is formed. Transfer from solid to liquid media can sometimes cause changes in the metabolic activities of certain cells, but in this case it was found that the addition of a little agar to liquid medium resulted in pigment formation. The active components eliciting pigment synthesis were found to be oligogalacturonides (agaropectins, acidic polysaccharides). This phenomenon appears to involve copper ions. *In vitro*, shikonin production is also promoted by methyl jasmonate and the addition of fungal elicitors, such as extracts of *Aspergillus niger*.

Industrial production uses selected cell lines that produce high levels of shikonin. The process is separated into two distinct phases. Both are performed in stirred tank reactors at 25°C, with a dissolved oxygen concentration of 6.0–6.4 mg/L. The first phase is carried out in a 200-L capacity fermenter using a medium designed to generate 'white' biomass and is completed within about 7–9 days. These cells are then transferred to a larger fermenter of 750 L capacity. Medium for the second phase was developed to obviate the need for added agar to elicit shikonin formation. It contains higher levels of borate, calcium, copper, iron, sulphate and sucrose than are present during the first phase. However, only low levels of phosphate are provided, and addition of vitamins, manganese and iodine is unnecessary. Ammonium ions are also omitted as they have been found to inhibit shikonin production. The second stage takes 12–14 days for conversion into red pigment-containing cells. The cell yield approaches 20 g/L with a shikonin yield of about 2.5 g/L.

Attempts are being made to extract the product *in situ* during the fermentation, which can improve the overall yield by up to six-fold. However, in the current process, cells are harvested by filtration at the end of fermentation. Shikonin and its derivatives are then extracted from the cells using hexane and component shikonin derivatives are hydrolysed with 2% (w/v) potassium hydroxide. This is followed by crystallization of the pure shikonin. The fermentation-derived product contains fewer shikonin derivatives than the whole plant extract and is a purer product.

Chemical synthesis of shikonin has been achieved, but is uneconomic at present, as is any recombinant DNA system. Nevertheless, they both pose future threats to the fermentation process.

Table 17.3 Potentially valuable algal products

Antibiotics	Aponin (antialgal) from *Gomphosphaeria aponica* Chlorellin (antibacterial) from *Chlorella* species Gallotannin (antiviral) from *Spirogyra* species Malyngolide (antifungal) from *Lyngbya majuscula*
Metabolites	D-amino acids Glycerol Glycollate
Pigments	β-carotene Phycocyanin (blue food colourant) from *Spirulina platensis* Phycoerythrin
Toxins	Anatoxin from *Anabaena flos-aquae* Aplysiatoxin from *Nostoc muscorum* Microcystin from *Microcystis aeruginosa*

Microalgae

In some parts of the world, certain microalgae have been cultured in open ponds as sources of food and desirable metabolites. However, they remain a largely untapped source of valuable products, particularly unsaturated fatty acids, antibiotics, toxins, pigments, polysaccharides and enzymes (Table 17.3). Their pigments are valuable as colourants for food and cosmetics. Pigments such as phycobilins can also be covalently linked to antibodies, biotin, avidin or lectins as fluorescent markers for diagnostic purposes. In addition, some microalgae can not only bioaccumulate pesticides, but are capable of their biotransformation. In fact, microalgae have considerable potential as biocatalysts for performing a wide range of biotransformations, particularly following their immobilization. This may be achieved by active entrapment in alginate or polyacrylamide gels and adsorption to polyurethane blocks or glass fibre mats, which is particularly useful for filamentous forms. Immobilization provides physical separation of the biocatalyst from the product stream, extended use and reuse, and allows high cell densities to be maintained.

Alternatives to animal cell and plant cell culture

Animal cell culture is expensive to perform due to the requirement for costly complex growth media and slow growth rates, which along with relatively low cell culture densities, generally results in poor product yield. Consequently, the cloning of genes needed for the production of animal cell metabolites into a suitable microorganism is probably a more efficient route of production. Yeasts are potential hosts as they are readily grown in culture and have established vector systems. Other attractive features are their strong promoters, ability to secrete proteins and performance of many post-translational modifications. They produce few extracellular proteins of their own, therefore engineered proteins are relatively easily purified from spent medium. Proteins already made using yeasts include vaccines (hepatitis B and malaria), diagnostics for hepatitis C and HIV, and therapeutics (insulin) (see Chapter 11). In addition, strains of *Escherichia coli*, *Saccharomyces cerevisiae* and *Pichia pastoris*, engineered to produce large quantities of human or humanized MAbs, overcome many of the problems associated with conventional MAb production. Recombinant MAbs, produced by these microorganisms, are likely to play an increasing role as therapeutic agents and for diagnostic purposes.

Volumetric productivities of conventional plant cell cultures are also generally lower than for microorganisms. In addition, there are often difficulties in selecting true single-cell clones, and rapid screening is much more problematic than for microorganisms, which readily grow on solid media. A more productive approach in the future may be to employ 'hairy root' cultures. Dicotyledonous plants can be transformed by the soil bacterium *Agrobacterium rhizogenes* to form hairy root tissue that is capable of unlimited growth in culture, is stable and does not require hormones. Hairy root cultures have potential as sources of many natural plant products and heterologous products, provided that suitable large-scale, low capital cost, bioreactors can be developed. Alternatively, chemical synthesis may be a viable alternate route of production for certain compounds. Also, as a result of advances in genetic engineering technology, some plant products may now be more economically produced using genetically modified bacteria or yeasts. However, this is not necessarily straightforward, as many plant products are synthesized via complex, often poorly characterized, pathways.

Further reading

Papers and reviews

Brodelius, P. & Pedersen, H. (1993) Increasing secondary metabolite production in plant-cell culture by redirecting transport. *Trends in Biotechnology* **11**, 30–36.

Chang, H.-N. & Sim, S.-J. (1995) Extractive plant cell culture. *Current Opinion in Biotechnology* **6**, 209–212.

Hu, W.-S. & Aunins, J. G. (1997) Large-scale mammalian cell culture. *Current Opinion in Biotechnology* **8**, 148–153.

Keen, M. J. & Rapson, N. T. (1995) Development of a serum-free culture medium for large scale production of recombinant protein from a Chinese hamster ovary line. *Cytotechnology* **17**, 153–163.

Koths, K. (1995) Recombinant proteins for medical use: the attractions and challenges. *Current Opinion in Biotechnology* **6**, 681–687.

Semple, K. T., Cain, R. B. & Shmidt, S. (1999) Biodegradation of aromatic compounds by microalgae. *FEMS Microbiology Letters* **170**, 291–300.

Shanks, J. V. & Morgan, J. (1999) Plant 'hairy root' culture. *Current Opinion in Biotechnology* **10**, 151–155.

Shimizu, Y. (1996) Microalgal metabolites: a new perspective. *Annual Review of Microbiology* **50**, 431–465.

Trill, J. J., Shatzman, A. & Ganguly, S. (1995). Production of monoclonal antibodies in COS and CHO cells. *Current Opinion in Biotechnology* **6**, 553–560.

Books

Becker, E. W. (1994) *Microalgae: Biotechnology and Microbiology*. Cambridge University Press, Cambridge.

Dixon, R. A. & Gonzales, R. A. (eds) (1994) *Plant Cell Culture: a Practical Approach*, 2nd edition. IRL Press, Oxford.

Fu, T.-J., Singh, G. & Curtis, W. R. (eds) (1999) *Plant Cell Culture for the Production of Food Ingredients*. Kluwer Academic/Plenum, New York.

Ritter, M. A. & Ladyman, H. M. (1995) *Monoclonal Antibodies. Production, Engineering and Clinical Application*. Cambridge University Press, Cambridge.

Spier, R. E. & Griffith, J. B. (eds) (1990) *Animal Cell Biotechnology*, Vol. 4. Academic Press, London.

Vlak, J. M., de Gooijer, C. D., Tramper, J. & Miltenburger, H. G. (1996) *Insect Cell Cultures. Fundamental and Applied Aspects*. Kluwer Academic, Dordrecht.

Vonshak, A. (ed.) (1997) Spirulina platensis (Arthrospira): *Physiology, Cell-biology and Biotechnology*. Taylor & Francis, London.

Index

Page numbers in **bold** refer to pages on which tables appear; page numbers in *italics* refer to illustrations.